THE LANGUAGE OF ACHILLES
AND OTHER PAPERS

The late Adam Parry

The Language of
Achilles
AND OTHER PAPERS

ADAM M. PARRY

CLARENDON PRESS · OXFORD

1989

Oxford University Press, Walton Street, Oxford OX2 6DP

Oxford New York Toronto
Delhi Bombay Calcutta Madras Karachi
Petaling Jaya Singapore Hong Kong Tokyo
Nairobi Dar es Salaam Cape Town
Melbourne Auckland

and associated companies in
Berlin Ibadan

Oxford is a trade mark of Oxford University Press

Published in the United States
by Oxford University Press, New York

This collection © Oxford University Press 1989

British Library Cataloguing in Publication Data

Parry, Adam
The language of Achilles and other papers
1. Classical literatures, —Critical studies
I. Title
880.09
ISBN 0–19–814892–5

Library of Congress Cataloging in Publication Data
Parry, Adam.
[Essays. Selections]
The language of Achilles and other papers/Adam M. Parry.
Includes index.
1. Classical philology. I. Title.
PA27.P257 1989   880'.9'001—dc19   88–36833
ISBN 0–19–814892–5

Set by Latimer Trend & Company Limited, Plymouth

Printed and bound in
Great Britain by Biddles Ltd,
Guildford and King's Lynn

Foreword

PROFESSOR ERIC HAVELOCK, who was best qualified to do it, wrote an excellent obituary notice of Adam Milman Parry (1928–71) and his wife Anne Reinberg Amory Parry (*Yale Classical Studies* 24 (1975, pp. ix–xv); but I cannot help prefacing what I say about Adam's work with a few personal reminiscences. His character somehow invites the less formal American rather than the severer English mode of treatment.

Adam Parry was not only a scholar of rare gifts, but one of the most delightful and accomplished human beings I have known. Handsome, elegant, and witty, he had a most unusual charm, by no means diminished by a streak of bohemian wildness. He once told me that the only three things he could have been were a classical scholar, a lawyer, and a beach-comber: he had a mixture of the qualities needed for all three professions. Soon after my first arrival in New Haven as Visiting Professor in 1964, I was invited to dinner by Adam, of whom I had heard much from Maurice Bowra, and his first wife Barbara. By way of preparation for campus life an experienced friend had lent me Alison Lurie's first novel, *Love and Friendship*, which I had much enjoyed. I asked the Parrys if they too had read it, and learned that they had. Then I asked if it were not perhaps a *roman à clef*, and Adam gravely told me that he thought it was. Had I known the Parrys a very little longer, I would have realized that they had been the model for a couple in the book: these people's children managed to burn down a college house, and when their parents were given another, they burned down that one also; this episode, I was told, was founded upon fact.

Life with Adam was full of excitement. Not long after that, I was driving with him in his car, and Adam cut in in front of another driver in proper Massachusetts fashion. A moment afterwards the car was stopped by a traffic light, and the other driver jumped out of his car and advanced towards us. He was a roughneck wearing a check shirt and a baseball cap, and carried a pistol, which he pointed at Adam. Just then the light changed, and Adam coolly drove off; luckily the man was too surprised to fire.

Adam took a strong left-wing line in politics, thus displaying a kind of aristocratic rebellion against the section of American society which had fewest intellectual interests. One good result of this was that during the rebellious sixties, when Adam was chairman of the Yale classics department, the students found that they had no one to rebel against.

Anne Parry as a young girl had been with her parents high up in the Andes, where her father was building a railway, and rations were dropped daily from the air. Wandering down the line of huts, the young Anne saw what use each woman made of the identical rations served to all; in consequence, she became a superb cook. As a hostess, she was a perfectionist, sometimes spending two days in preparing for a dinner-party. Guests at these parties, some of whom had come up from New York, were carefully selected; if one of them turned out not to be an entertaining talker, she would make up for this by being a beautiful young woman. The conversation was exceedingly amusing; with such a host and hostess, people tended to be at their best.

In the back of one's mind there was always a faint flavour of danger when Adam was about; he had a fatal penchant for taking risks. In the spring of 1971 Eric and Christine Havelock gave a splendid luncheon-party at their house in rural Connecticut, more or less equidistant between Poughkeepsie and New Haven; Adam and Anne were there, and were in their best form. Afterwards we all filed out to watch them ride away on one of Adam's motorcycles; as they vanished, we were all silent, and looked round at one another in anxiety. That summer they were both killed in a motorcycle accident near Colmar.

Adam used to say that if his famous father had lived he himself would never have become a classical scholar. Being the son of the most distinguished American Hellenist of his time was not an unmixed blessing, particularly since Adam came to be discontented not with his father's work but with the way in which other scholars had arrested his investigation at the point it had reached at the time of his sudden death and with their use of his legend to enforce the stultifying dogma. With great courage and intelligence, he succeeded in placing his father's achievement in its right perspective; and during his short life assured himself of an honoured place in the history of American and indeed of international scholarship.

The article 'Landscape in Greek Poetry' (No. 2) already shows Parry writing in a relaxed and elegant manner and using his familiarity with several languages and literatures, as well as with the better sort of modern literary criticism, to throw light upon his subject. He argues that in early Greek poetry 'landscape, as a distinct element, plays no part', descriptions of nature being used 'sparingly and briefly, as a direct metaphor for things human'. Then, taking a hint from the unusually wide significance given to the term *pastoral* by Sir William Empson, he puts forward the view that Greek pastoral 'presents an idyllic nature in which the poet can move by proxy, and in which he is content'. Thinking of the *Phaedrus*, most effectively quoted at the beginning of his article, he suggests that pastoral in this sense originates with Plato.

Adam Parry's thesis '*Logos* and *Ergon* in Thucydides' won him the degree of D.Phil at Harvard in 1957. He intended to expand it into a major study called *The Mind of Thucydides*, and so it remained unpublished till ten years after his death. In the thesis a succinct study of the *logos/ergon* antithesis from the beginning of Greek literature down to the fourth century leads up to an account of the way Thucydides makes use of it. The popular distinction between these notions, as Parry calls it, is simple, and rejects *logos* in favour of an *ergon* which is assumed to be easily knowable; the literary distinction, however, 'presents *logos* and *ergon* as analytically disjunct, but of equal or near-equal importance', and further suggests that '*ergon* is unknowable except through *logos*'. This leads up to an account of the complicated way in which Thucydides uses the antithesis, showing how a statesman like Themistocles or Pericles tries to make his *logoi* express and conduce to *erga* and how a historian like Thucydides in a different way sets out to do the same. Then Parry goes on to examine the antithesis as it appears in the first two books of the History, rounding off his work with a brief summary of the part it plays in the remainder of the work. Avoiding the *naïveté* of critics who have supposed that Thucydides thought that his book would enable readers to predict the future, he remarks that Thucydides believed that it would be of some assistance in showing in a general way how things tend to happen. Pericles, he thinks, praises the Athenians for having, more than any other community, made their *erga* correspond with their *logoi*. The close relationship between Thucydides' antithetical style, making full use of abstract nouns used to characterize concrete events, and his tragic view of history was further developed in a later essay on 'Thucydides' Use of Abstract Language' (No. 11). In a posthumously published article on 'Thucydides' Historical Perspective' (No. 17), Parry powerfully brings out the apparently paradoxical notion that though Pericles' policy finally led to disaster the Athenians, in the view of Thucydides, had been right to follow it. Following in the tracks of F. M. Cornford's *Thucydides Mythistoricus*, Parry argued that the History of Thucydides is essentially a tragic history. It was not necessarily influenced by tragedy, as Cornford had supposed; it was enough that Thucydides was always conscious of Homer as a predecessor. In another Thucydidean article (No. 13), Parry refuted the contention of Sir Denys Page that the description of the Plague in Book Two is written in the technical language of the doctors. In Thucydides' tragic view of events, according to Parry, the Plague has the dramatic function of following the idealism of the Funeral Speech by an unforeseeable disaster that no intelligence could have predicted or controlled.

In his article 'Thucydides as "History" and as "Literature"' (*History and Theory* 22 (1983), 60f. = *The Greeks and Their Legacy* (1988), 60f.), Sir Kenneth Dover writes about the new kind of study of Thucydides, no longer focused upon the problem of when each part of the History was written, which has become fashionable in recent years. Dover does not mention Parry, but since the earliest of the works cited as exemplifying this approach appeared only as late as 1966, Parry may be regarded as a pioneer, and his failure to complete the major work which he had planned as a grievous loss to scholarship.

But the most important work left by Parry is his contribution to Homeric study. The early article on 'The Language of

Achilles' (No. 1) shows him already troubled by the problem of how, if it was true that the Homeric poems came into being without the aid of writing as a result of improvisation facilitated by the existence of numerous standard formulas and orally preserved, they could contain great poetry. If there were certain things which the inherited vocabulary of oral poets did not allow them to say, how could poets whose area of operation was thus restricted produce poetry of high quality? Taking the truth of the accepted orthodoxy for granted, Parry was obliged to conclude that Achilles in his reply to the embassy in Book Nine 'has no language in which to express his disillusionment': but as M. D. Reeve (*CQ*, NS 23 (1973), 193f.) later pointed out, Achilles *does* express his disillusionment. Long before Reeve's article appeared, Parry had seen that the article could have reached a very different conclusion, if its acceptance of the then current dogmas about oral poetry had been abandoned. But Parry's article is, as Reeve said, one of the more important contributions to Homeric studies in the last fifty years, for it raised the question which was bound eventually to lead to the reconsideration of the current theories about oral poetry and the formula.

Parry's literary skill and fine appreciation of Homer are also displayed in his castigation of two bad translations (No. 4) and of one good but not wholly satisfying version (No. 7); it was obvious enough that Robert Graves' Homer was not worthy of him, but at that time it was highly desirable that someone should point out that the undeservedly popular translation of E. V. Rieu rested on Rieu's discovery that Homer was really Trollope.

Three years later Parry again touched on Homeric problems in his article 'Have We Homer's *Iliad*?' (No. 10). Guided by the fine feeling for poetry, a quality not shared by all modern Homeric scholars, which helped him to perceive the essential unity of the poems and the close relation between their parts, Parry wrote, speaking of his father, that 'it is up to us not to stop where he stopped'. His father's work, he says, was at first designed to describe the *tradition* of Greek heroic song: 'he is not to be blamed', he adds, 'for not having stressed, or only in rare moments, the distinctiveness of Homer'. The danger of confusing Homer with another kind of poet, a poet with a style not depending, as Homer's does, upon an oral tradition, should not lead us to confuse Homer with the tradition which he followed. Parry argues that 'the Ionian singer of 725 BC, trained in the use of a formulary technique far more subtle and elaborate than any other we know, would, if he had learned to manipulate the magical σήματα which had come to him from the Phoenicians, not be inclined to change his thoughts of modes of expression at all'; and he recognizes that modern research into oral poetry in no way rules out 'the notion explored by Lesky, Whitman, Wade–Gery, and Bowra, that Homer himself knew the art of writing'.

This conclusion is argued for in detail in Adam Parry's splendid introduction to his father's collected papers (No. 15), which appeared during the year of his death. Starting with an admirable sketch of the history of the Homeric Question, Parry goes on to describe his father's work and its effects, taking advantage of Sterling Dow's discovery in the library at Berkeley of the thesis with which Milman Parry obtained the degree of MA before he went to France. After surveying the influence of his father's work up to the time of writing, Adam Parry remarks that to the lover of literature 'the whole argument may well appear so narrowly technical as to miss somehow the fundamental issue, which is the poetry of Homer, and how Parry's work, and that of his successors, will affect our reading of it'. Next he sets out to answer this question. Milman Parry, he says, 'unlike many, if not most, of his followers, possessed an acute sensitivity to the poetry of Homer': but he was too much concerned with the tradition to pay special attention to the individual poet, so that his followers 'have often cheerfully adopted his limitations along with his constructive arguments', with the result that intelligent appreciation of the poems has been inhibited.

The arguments of the introduction were strengthened by the posthumously published article 'Language and Characterization in Homer' (No. 18), in which Parry shows how mistaken it is to think that certain Homeric phrases, metrically convenient and often recurring as they are, carry no meaning, and goes on to demonstrate Homer's power of characterization, mainly through a perceptive study of the portrayal of Menelaus.

To understand the nature of Adam Parry's achievement one must take account of the intellectual climate of the time at which he wrote. At a time when the leading Hellenists of England and America accepted a view of oral and formulaic poetry that rendered any literary discussion of Homer's work impossible, Adam Parry saw that his father had proved that the Homeric poems belonged to a tradition that had long been oral, but that he had not proved that these poems were themselves oral poems. At a time when American national pride had invested heavily in a theory that counted as the main American achievement in Greek scholarship, it took courage for the son of the man canonized as the author of the theory to argue that the theory had got out of hand.

Adam Parry also made an effective contribution to the understanding of Virgil. The early article, 'The Two Voices in Virgil's *Aeneid*' (No. 8), while avoiding the error of those who credit Virgil with a secret wish to undermine Augustus, contains a sensitive appreciation of the tragic element in the poem. Parry brings out what he was later (p. 266) to call 'the strange amalgam of triumph and sadness, of confidence and nostalgia, of the martial tones of Roman and Augustan achievement, and of the poignant notes of personal loss and renunciation'. Still more notable is the article 'The Idea of Art in Virgil's *Georgics*' (No. 16), in which he sets out to reveal 'the total thrust and meaning of the work, the way in which all the short descriptions and short narratives, natural, mythological, · historical, and philosophical, the expressions of different attitudes, sometimes in apparent contradiction to each other, are orchestrated into a complex but unified vision of the world, which is hard for us to grasp'; the leading idea of the article is that Zeus' condemnation of man to work for his living, first described by Hesiod, is in a sense a good thing, because work necessitates art. The poem, he concludes, 'is ultimately about the life of man in this world, about a kind of art in living which can confront the absurdity and cruelty of both nature and civilization, and yet render our existence satisfactory and beautiful.'

HUGH LLOYD-JONES

Christ Church, Oxford
2 May 1989

Contents

I

The Language of Achilles

I WISH in this paper to explore some of the implications of the formulaic theory of Greek epic verse.[1] In doing so, I shall take the theory itself largely for granted.

Let us first consider a famous passage from the end of Book 8 of the *Iliad*, the one describing the Trojan watch-fires:

Οἱ δὲ μέγα φρονέοντες ἐπὶ πτολέμοιο γεφύρας
ἥατο παννύχιοι, πυρὰ δέ σφισι καίετο πολλά.
ὡς δ' ὅτ' ἐν οὐρανῷ ἄστρα φαεινὴν ἀμφὶ σελήνην
φαίνετ' ἀριπρεπέα, ὅτε τ' ἔπλετο νήνεμος αἰθήρ·
ἔκ τ' ἔφανεν πᾶσαι σκοπιαὶ καὶ πρώονες ἄκροι
καὶ νάπαι· οὐρανόθεν δ' ἄρ' ὑπερράγη ἄσπετος αἰθήρ,
πάντα δὲ εἴδεται ἄστρα, γέγηθε δέ τε φρένα ποιμήν·
τόσσα μεσηγὺ νεῶν ἠδὲ Ξάνθοιο ῥοάων
Τρώων καιόντων πυρὰ φαίνετο Ἰλιόθι πρό.
Χίλι' ἄρ' ἐν πεδίῳ πυρὰ καίετο, πὰρ δὲ ἑκάστῳ
ἥατο πεντήκοντα σέλᾳ πυρὸς αἰθομένοιο.
ἵπποι δὲ κρῖ λευκὸν ἐρεπτόμενοι καὶ ὀλύρας
ἑσταότες παρ' ὄχεσφιν ἐΰθρονον Ἠῶ μίμνον.

(*Il.* 8. 553–65) |

These lines could be shown, by an examination of parallel passages, to be almost entirely made up of formulaic elements.[2] That they are so amazingly beautiful is of course the

---

[1] The theory is the work of Milman Parry, although it had been partly suggested by Heinrich Düntzer (*Homerische Abhandlungen* [Leipzig, 1872] 508 ff.). It is substantially stated in *L'Epithète traditionelle dans Homère* (Paris, 1928), and is elaborated in a number of articles written in America. For an account of Milman Parry's work and a complete bibliography, see A. B. Lord, 'Homer, Parry, and Huso', *AJA* 52 (1948), 34–44 [and below, ch. 15].

[2] Cf. e.g. (with 8. 553) *Il.* 15. 703, 14. 421, 8. 378; (554) 2. 2, 3. 149, 8. 561, 8. 562; (555) 4. 422, 5. 437, 17. 367; (556) 4. 278, 12. 271; (557–8) 16. 299–300; (559) 13. 98, *Od.* 6. 106; (560) *Il.* 6. 4; (563) *Od.* 21. 246, *Il.* 22. 150; (564) 5. 196; (565) 8. 41, 12. 367, *Od.* 18.318

Transactions of the American Philological Association, 87 (1956), 1–7. Reprinted by permission of the American Philological Association.

consequence of Homer's art in arranging these formulae. But I wish to speak now of the quality of their beauty. Here is a straight English translation:

And they with high thoughts upon the bridges of war sat all night long, and they had fires burning in great number. As when in heaven the stars around the splendid moon shine out clear and brilliant, when the upper air is still; and all the lookout places are visible, and the steep promontories, and the mountain dells; and from the heaven downward the infinite air breaks open; and the shepherd is delighted in his heart: so many, between the ships and the streams of Xanthus, were the Trojans burning shining fires before the walls of Ilium. A thousand of them were burning in the plain, and by each one were sitting fifty men, in the light of the blazing fire. And the horses, munching white barley and wheat, stood by the chariots, awaiting the thronèd Dawn.

The feeling of this passage is that the multitude of Trojan watch-fires is something marvellous and brilliant, that fills the heart with gladness. But this description, we remember, comes at the point in the story when the situation of the Achaeans is for the first time obviously perilous; and it is followed by Book 9, where Agamemnon in desperation makes his extravagant and vain offer to Achilles, if he will save the army. The imminent disaster of the Achaeans is embodied in these very fires. Yet Homer pauses in the dramatic trajectory of his narrative to represent not the horror of the fires, but their glory. I suggest that this is due precisely to the formulaic language he employs. There is a single best way to describe a multitude of shining fires; there are established phrases, each with its special and economical purpose, to compose such a description. Homer may arrange these with consummate art; but the nature of his craft does not incline him, or even allow him, to change them, or in any way to present the particular dramatic significance of the fires in this situation.[3] Instead, he presents the constant qualities of all such fires. |

---

[3] A comparison with Alexander Pope's translation makes this strikingly clear. Mr Pope turns the bright and beautiful fires into a nightmare. The formulaic lines 562 and 563, for example, become:

> A thousand piles the dusky horrors gild
> And shoot a shady lustre o'er the field.

The formulaic character of Homer's language means that everything in the world is regularly presented as all men (all men within the poem, that is) commonly perceive it. The style of Homer emphasizes constantly the accepted attitude toward each thing in the world, and this makes for a great unity of experience.

Moral standards and the values of life are essentially agreed on by everyone in the *Iliad*. The morality of the hero is set forth by Sarpedon in Book 12 (310–28). Sarpedon's speech there to Glaucus is divided into two parts. The first expresses the strictly social aspect of the Homeric prince's life: his subjects pay him honour in palpable forms, and he must make himself worthy of this honour by deeds of valour. The second part expresses a more metaphysical aspect: it is the hero's awareness of the imminence of death that leads him to scorn death in action.

The second part of Sarpedon's speech is by far the more famous: perhaps no passage in the *Iliad* is better known.[4] But the first part is equally important for an understanding of the poem. Its assumption is, first, that honour can be fully embodied in the tangible expressions of it (the best seat at the feast, the fullest cups of wine, the finest cuts of meat, and so forth);[5] for everyone agrees on the meaning of these tangible expressions; and second, that there is a perfect correspondence between individual prowess and social honour. For this too is universally agreed on. I need not add that most of Sarpedon's speech, particularly the first part, is made up of | traditional

> Full fifty guards each flaming pile attend,
> Whose umber'd arms, by fits, thick flashes send.

Pope's lines are not to be despised; his turning the description to dramatic account is in full agreement with modern literary standards, and is very appropriate to an art that is unlike Homer's.

[4] This is at least true of the English-speaking world. The handsomely theatrical quotation of the passage by Lord Granville is known to us from Robert Wood's *Essay on the Original Genius of Homer*. (The story is referred to by Matthew Arnold in *On Translating Homer*.) Milman Parry, in *Harvard Alumni Bulletin* 38 (1936), 779–80, criticizes Lord Granville for lack of historical understanding of the passage. The criticism seems unfair: Granville probably understood the passage well when he made an identification between himself and the Homeric hero, an identification not so hard to make, considering how concrete were the rewards, and even how strenuous the obligations, of a successful 18th-century English aristocrat.

[5] *Il.* 12. 311 = 8. 162.

formulae, and that the same thoughts, in the same words, appear in other places in the *Iliad*.[6]

The unity of experience is thus made manifest to us by a common language. Men say the same things about the same things, and so the world to them, from its most concrete to its most metaphysical parts, is one. There is no need, as there is in Plato's day, for a man to 'define his terms'. And accordingly, speech and reality need not be divided into two opposing realms of experience, as we find them divided in the fifth century by the analytic distinction of *logos* and *ergon*;[7] for the formulaic expressions which all men use are felt to be in perfect accordance with reality, to be an adequate representation of it.

Let us examine this last proposition. The epic heroes live a life of action. Speech, counsel and monologue are seen as a form of action. Phoenix tells in Book 9 of the *Iliad* how he was enjoined to bring up Achilles and *teach him all things*, all things, that is, a hero need know, *to be a speaker of words and a doer of deeds* (9. 442–3). Phoenix here makes a practical separation, but no real distinction in kind: the hero must know how to do things—in the accepted manner; and how to talk about things—in the accepted manner. The two are complementary halves of a hero's abilities, and the obverse and reverse of his great purpose: to acquire prestige among his fellows.[8]

Speech is a form of action, and, since the economy of the formulaic style confines speech to accepted patterns which all men assume to be true, there need never be a fundamental distinction between speech and reality; or between thought and reality—for thought | and speech are not distinguished; or

---

[6] Cf. e.g. (with 12. 310) *Il.* 9. 38; (310–11) 8. 161–2; (312) *Od.* 8. 173; (313–14) *Il.* 20. 184–5; (315) 12. 321, 12. 324; (317) 7. 300; (318) 22. 304, *Od.* 1. 241, *Il.* 5. 332; (320) 11. 668; etc.

A propos of *Il.* 12. 313: that Homer's τέμενος, both the word and the institution, was known to historical Mycenaean society, is now demonstrated by the decipherments of Linear B script. See Ventris and Chadwick, 'Evidence for Greek Dialect in the Mycenaean Archives', *JHS* 73 (1953), 84–103, esp. 99, top.

[7] The distinction becomes a permanent feature of the thought of fifth-century writers as diverse as Gorgias, Thucydides, and Sophocles, part of the syntax of their speech. Cf. Gorgias, B3 Diels, s. 84; Thucydides 2. 35 and *passim*; Sophocles, *Philoctetes* 96–9 and 307–13.

[8] A good example of this is the word κυδιάνειρα, used regularly in a formula with πόλεμος. But when Achilles withdraws from action in Book 1, it is said of him: οὔτε ποτ᾽ εἰς ἀγορὴν πωλέσκετο κυδιάνειραν / οὔτε ποτ᾽ ἐς πόλεμον ... (1. 490–1).

between appearance and reality—for the language of society is the way society makes things seem.

If such a distinction did openly exist, we should know where to find it: it would be in the character of Odysseus, the hero who by the end of the fifth century has become the type of the Sophist, the man who substitutes an illusory speech for the realities of life.[9] But in Homer, at least in the *Iliad*, Odysseus is a great and honourable warrior. His being a master of words is simply a manifestation of this. What words he speaks are felt as a clear reflection of reality, because, like those of Sarpedon, they are in harmony with the assumptions of all society.[10]

Only in the person of Achilles do we find so much as a hint that appearances may be misleading, and conception, in the form of words, a false and ruinous thing. When he answers Odysseus with his great speech in Book 9, he says he will speak out exactly what he thinks, and what will come to pass.[11] 'I hate that man like the gates of Hell who hides one thing in his heart, and says another,' he continues. Achilles' words ostensibly refer to himself: 'I will not mince words with you.' But the reader feels that they apply with a different force to Odysseus. Odysseus' elaborate and eloquent speech, spoken just before in the naïve confidence that Achilles, like himself, will consider the gifts as adequate symbols of honour, becomes a little hollow. Achilles' words here make it seem somehow dishonest at heart, and not in accordance with the essence of the situation.

Achilles' own speech that follows is of another sort. Passionate, confused, continually turning back on itself, it presents his own vision with a dreadful candour. And what this candour is concerned with is, precisely, the awful distance between appearance and reality; between what Achilles expected and what he got; be|tween the truth that society

---

[9] Particularly in Sophocles' *Philoctetes*. Cf. the lines referred to in n. 7, also lines 13–14 and 431–2.

[10] Cf. e.g. *Il.* 11. 404–10. Observe the connection of the word ἀριστεύειν in this passage with ἀριστεύς, of the word for prowess with that for social rank.

[11] Such certainty is godlike (and Achilles does hold the fate of the Greeks in his hand); in Book 1 (204), Achilles had said more doubtfully: ἀλλ' ἔκ τοι ἐρέω, τὸ δὲ καὶ τελέεσθαι ὀίω, but no action had followed on his words. Athene in her answer had said, like Achilles in Book 9: ὧδε γὰρ ἐξερέω, τὸ δὲ καὶ τετελεσμένον ἔσται (212).

imposes on men and what Achilles has seen to be true for himself.

I will not here discuss Achilles' speech in any detail. But few readers of it, I believe, would disagree that it is about such a cleavage between seeming and being as I have indicated. The disillusionment consequent on Achilles' awareness of this cleavage, the questions his awareness of it give rise to, and the results of all this in the events of the war, are possibly the real plot of the second half of the *Iliad*.

Achilles is thus the one Homeric hero who does not accept the common language, and feels that it does not correspond to reality. But what is characteristic of the *Iliad*, and makes it unique as a tragedy, is that this otherness of Achilles is nowhere stated in clear and precise terms. Achilles can only say, 'There was, after all, no grace in it', or ask questions that cannot really be answered: 'But why should the Argives be fighting against the Trojans?' or make demands that can never be satisfied: '. . . until he pays back all my heart-rending grief'.[12]

Homer, in fact, has no language, no terms, in which to express this kind of basic disillusionment with society and the external world. The reason lies in the nature of epic verse. The poet does not make a language of his own; he draws from a common store of poetic diction. This store is a product of bards and a reflection of society: for epic song had a clear social function.[13] Neither Homer, then, in his own person as narrator, nor the characters he dramatizes, can speak any language other than the one which reflects the assumptions of heroic society, those assumptions so beautifully and so serenely enunciated by Sarpedon in Book 12.

Achilles has no language with which to express his disillusionment. Yet he expresses it, and in a remarkable way. He does it by misusing the language he disposes of. He asks

---

[12] *Il.* 9. 316, 337-9, 387. The question in the second of these is answered by Achilles. But it seems, as we read the speech, of wider scope than the answer given. We feel that the justification of war itself is being called in question, as it is in 1. 152-7.

[13] The bard is held in great honour—witness Phemius' being spared in the Slaughter of the Suitors. But he is the servant of society. Both Phemius and Demodocus sing the songs they are asked to sing, and it is clear they sing in such a manner as to celebrate the kind of life their listeners lead. This is a great difference between the bard and the rhapsode.

questions that cannot be answered and makes demands that
cannot be met. He uses conventional expressions where we
least expect him to, as when | he speaks to Patroclus in Book 16
of a hope of being offered material gifts by the Greeks, when
we know that he has been offered these gifts and that they are
meaningless to him; or as when he says that he has won great
glory by slaying Hector, when we know that he is really
fighting to avenge his comrade, and that he sees no value in the
glory that society can confer.[14] All this is done with wonderful
subtlety: most readers feel it when they read the *Iliad*; few
understand how the poet is doing it. It is not a sign of artistic
weakness: Homer profits, by not availing himself of the
intellectual terminology of the fifth century. Achilles' tragedy,
his final isolation, is that he can in no sense, including that of
language (unlike, say, Hamlet), leave the society which has
become alien to him. And Homer uses the epic speech a long
poetic tradition gave him to transcend the limits of that speech.

[14] *Il.* 16. 84–6 and 22. 391–4. In the later passage, note how the conventional hymn
of triumph follows upon the strange but passionate reference to Patroclus in 387–90.
The passage from Book 16 has long been a stumbling-block. Because Achilles appears
to know nothing of Agamemnon's offer of gifts in Book 9, it has been held that there
is a hopeless inconsistency. Grote, following K. L. Kayser, was the first to discuss this
alleged inconsistency. In his *History of Greece* (4th edn. [London, 1872], ii. 112 ff.), he
argued forcefully that the whole of Book 9 is an interpolation, and his case finds its
best support from 16. 84–6. More recent critics have been unwilling to reject one of
the finest books of the *Iliad* in order to maintain the coherence of an hypothetical
Urilias. But, unable to answer Grote's arguments, they instead deny the authenticity
of 16. 84–6. See Willy Theiler, 'Die Dichter der Ilias' in *Festschrift für Edouard Tièche*
(Berne, 1947), 152, and Von der Mühll, *Kritisches Hypomnema zur Ilias* (Basle, 1952),
242. Wolfgang Schadewaldt, in his *Iliasstudien* (*AbhLeipzig*, 43 6 [1938], 128–9) seems
alone in defending the lines. Schadewaldt's intuitions, as so often, are right, but here
they are hazy. He does not perceive the depth of Achilles' dilemma here in the
beginning of Book 16. He wants to make Achilles' words reasonable, but of course
they are not reasonable. Neither are lines 97–100 in the same speech. They are the
expression of an impossible situation. Achilles cannot forsake the action which
manifests an insight that no available words will express directly. Lines 60–3 make
that clear. On the particle γε in line 61, I am tempted to say, hangs the whole tragic
decision.

   In a discerning, though somewhat legalistic, article, 'The Propitiation of Achilles'
(*AJP* 74 [1953], 137–48), David E. Eichholz has defended the lines on grounds similar
to Schadewaldt's: Achilles does want the gifts, but he 'must have [them] on his own
terms and in his own time' (p. 141). In view both of what Achilles says of the gifts in
Book 9, and of his supreme indifference to them when they are actually given him in
Book 19, it seems to me that the reader must coerce his imagination to believe that he
really wants them here. A few lines further down, Achilles gives an entirely different
reason for wanting Patroclus to restrain his attack.

2

Landscape in Greek Poetry

ONE searches in vain for a Greek word meaning *landscape*, just as for a single Greek word meaning *eloquence*. However, there is a disarmingly simple way of expressing the concept of *landscape* in a famous passage of Plato. Socrates and Phaedrus are in the Attic countryside, and Socrates says to his friend, 'Forgive me, excellent friend—for you know I am always trying to learn things; but the countryside and the trees have nothing to teach me: for that I must look to the men in town' (*Phaedr.* 230 D 3–5).

I believe that this statement is literally true of the historical Socrates: he did not learn from the beauties of nature; but I believe that in Plato's dialogue he in fact means the opposite of what he says. In the *Phaedrus*, the countryside and trees, or, better, simply 'places and trees'—χωρία καὶ δένδρα—have a good deal to teach him.

That, anyway, is the purpose of this essay: to examine what, within the artistic limits of certain Greek works, *landscape*— 'places and trees'—has to teach us. In what different ways is it used for dramatic effect?[1]

We might start with some general distinctions: Literature is essentially concerned with human feelings, actions, and judgements. Natural scenes, when they are introduced, have a metaphorical value: they are ultimately figures of something human. Very broadly speaking, there are two modes of use of natural scenes in poetry, and these modes seem to belong

[1] Cf. Ruskin's *Of Classical Landscape*; by no means outdated, but an intelligent discussion of how the (Homeric) Greeks conceived of the gods and the physical world, and how this conception affected their *choice* of landscape. It says little, however, of the *function* of landscape within the dramatic situation of a single work.

Yale Classical Studies 15 (1957), 3–29. Reprinted by permission of Yale University Press.

generally, one to the earlier and one to the later stages of a culture.

In the first, a natural phenomenon is made directly a simile of something human:|

He fell as an oak tree falls, or a poplar, or a soaring pine.

> ἤριπε δ᾽ ὡς ὅτε τις δρῦς ἤριπεν· ἢ ἀχερωίς·
> ἠὲ πίτυς βλωθρή·   (*Il*. 13. 389–90)

or:

Like to the sweet-apple that reddens on the top of the branch.

> οἶον τὸ γλυκύμαλον ἐρεύθεται ἄκρῳ ἐπ᾽ ὕσδῳ
>   (Sappho, 93 Bergk)

However conscious of the bitterness of life, man in the youth of a culture possesses a kind of confidence which does not allow him to feel alien from the world about him. The flowers, trees, stars, and animals feed his spiritual self even as the herbs and beasts in Eden Garden were said to be created for his use. They are the images of the state of his soul and of his body. The song of the nightingale may have been a song of unquenchable grief, but the nightingale had been the daughter of Pandareus, and it was human grief she sang:

But dearer to *my* heart by far is she who weeps for Itys, evermore for Itys, that bird distraught with grief, the messenger of Zeus.

> Ἀλλ᾽ ἐμέ γ᾽ ἁ στονόεσσ᾽ ἄραρεν φρένας,
> ἃ Ἴτυν αἰὲν Ἴτυν ὀλοφύρεται
> ὄρνις ἀτυζομένα, Διὸς ἄγγελος.
>   (Soph. *El*. 147–9. Adapted from Jebb's translation)

Earlier European poetry seems in general to show this same closeness to human experience in the use of natural things, and the same inner certainty that the intimate connection of human and nonhuman life is such that the latter can appropriately figure the former. here are a few random examples:

This from Dante:

> Al poco giorno, ed al gran cerchio d'ombra
>   Son giunto, lasso! ed al bianchir de' colli,
>   Quando si perde lo color ne l'erba;
> E'l mio disio però non cangia il verde ...   (*Rime*)

Within the contrast, whereby the grass loses its green but
Dante's love does not, the reader feels that *si perde* applies
equally to the poet and to the verdure of the hills.

Or this of De Góngora y Argote:

> Desata como nieblas
> todo lo que no ves,|
> que sospechas de amantes
> y querellas después,
> *hoy son flores azules,*
> *mañana serán miel.*   (*Romances*)

Or again, Ronsard:

> Mignonne, allons voir si la rose
> Qui ce matin avoit desclose
> Sa robe de pourpre au Soleil,
> A point perdu ceste vesprée
> Les plis de sa robe pourprée,
> Et son teint au vostre pareil.   (*Poésies*)

A less clear example of such identification is this, from Sidney:

> Me seemes I see the high and stately mountaines
> Transforme themselves to lowe dejected vallies:
> Me seemes I heare in these ill-changed forrests
> The nightingales doo learne of Owles their musique:
> Me seemes I feele the comfort of the morning
> Turnde to the mortal serene of an evening.   (*Arcadia*)

It is characteristic of the sestina form, as this of Sidney's and
Dante's above, to explore a great variety of relations between
human emotions and things of nature.[2] In the last quoted,
however, we already enter on the second mode of natural
metaphor indicated above; for the mountains, valleys, night-
ingales, and owls do not directly figure the courtier's emotion:
the verse is spoken by a shepherd; Sir Philip Sidney has found it
more to his purpose to speak through a pastoral mask; a new
dimension of profitable artificiality has been added to poetic
expression.

But first some evident corollaries might be drawn from the
simpler relation of man to nature we have held to be character-

---

[2] Cf. William Empson, *Seven Types of Ambiguity* (Norfolk, Conn., 1947), 36–7.

istic of earlier poetry. Though the relation is much closer than in the second mode which we shall describe, yet the 'pathetic fallacy' is, for that very reason, ruled out. There is no awful gap to be bridged between man and nature: they live in the same world, a world which can be called 'natural' in a wider, and older, sense. The poet does not have to *make contact* between himself and nature by attributing his emotions to the latter. Nature is so close to man that he can confidently maintain the sense of his own difference.[3] He does not have to forget | that he feels, where plants do not, that he is aware, where animals are not. There is no occasion for the desperate cry of Shelley:

> . . . Be thou, Spirit fierce,
> My spirit! Be thou me, impetuous one![4]

To sum up this aspect of the relation, the earlier poets seem instinctively not to have forgotten that in a comparison of the human and the non-human, the latter remains an image of the former, and this awareness seems to have been part of the familiarity between the two.

Another corollary is that as long as the natural world is thus close to hand, it is drawn from for poetic metaphor in small quantities. If there is to be what we call poetic resemblances, the symbol which is to confer lustre and meaning on the thing symbolized must be distinct and precise; else it will not help us to know the thing symbolized as the poet wants us to know it. As long as man, though different from the rest of nature (using the word in a wider sense), is not of another world from it, he will not choose nature as a whole, or nature in a multiple aspect, to figure something of himself. For this would involve his conceiving nature as something *other*. Moreover, earlier

[3] Cf. Ruskin's *Of the Pathetic Fallacy*: 'Thus, when Dante describes the spirits falling from the bank of Acheron "as dead leaves flutter from a bough", he gives the most perfect image possible of their utter lightness, feebleness, passiveness, and scattering agony of despair, without, however, for an instant losing his own clear perception that *these* are souls, and *those* are leaves: he makes no confusion of one with the other.'

[4] Indeed, the dominant impression of the 'Ode to the West Wind' is the hopeless estrangement of the speaker from the natural force he admires, an estrangement reflected in the inflation of the language as well as in the confusedly erotic overtones of many of the lines.

poetry is more chary of metaphor, and more precise in its choice. Therefore a single flash of natural splendour is chosen, and not dwelt on long. An example is the death of Gorgythion in the *Iliad*:

> Like a poppy he bent down his head: like a poppy
>   in a garden:
> Weighted with its seeds, and with the vernal dew:
> Even so he bowed down his head beneath the helmet's
>   weight.

> μήκων δ' ὡς ἑτέρωσε κάρη βάλεν· ἤ τ' ἐνὶ κήπῳ·
> καρπῷ βριθομένη νοτίῃσί τε εἰαρινῇσιν·
> ὡς ἑτέρωσ' ἤμυσε κάρη πήληκι βαρυνθέν.   (*Il.* 8. 306–8) |

Or again from Sappho:

> Evening that takest home all that splendid dawn has
>   scattered.
> Thou takest the lamb, thou takest the kid, thou
>   takest the mother her child.

> Ἔπρεπε πάντα φέρων ὅσα φαίνολις ἐσκέδασ' αὔως·
> φέρεις ὄιν· φέρεις αἶγα· φέρεις ἄπυ μάτερι παῖδα.   (95 Bergk)

Another consideration here is that in poets more concerned with passion than with manners, writers of πάθος rather than ἦθος, the author of the *Iliad* and the tragedians, prolonged description of any sort is avoided: the pace is too rapid to allow of it. But in any case, the poets of early Greece did not look to nature for something different from themselves.

This brings us to the time when they did, and to the second mode of natural metaphor. We might introduce this by taking up a question passed over above: when is landscape or natural scenery dealt with *for its own sake*? The phrase could be taken to mean that a poet describes something natural not merely to illustrate something else, but in order that his reader may delight in that natural thing itself. But the essential of a poetic metaphor is that it does not merely illustrate, no matter how brief it is: poetic language does not point to something outside of itself, and then lose its *raison d'être* when the outside thing is comprehended or carried out. In the Homeric lines quoted above, the poppy is as important as Gorgythion; and ultimately they are both metaphors of the sadness and beauty of death.

Interest in landscape, or nature, *for its own sake* could be best understood as applying to that literary art wherein man looks to nature for something which he has not within himself or which exists in an imperfect and adulterated manner in his daily life. This means a significantly different use of nature in poetry from what we described as the first mode. Nature no longer tells us what we are: it tells us what we are not but yearn to be.[5] Pastoral poetry—and romantic poetry, where it deals with nature|—fits into this category. A great poetic statement of such an attitude is Wordsworth's

> The world is too much with us . . .

This, and like poems, are not yet in the romantic tradition. They are the first shocked cry of a sensibility realizing that the bond between man and nature is snapped.

> Little we see in nature that is ours
>
> .   .   .   .   .
>
> For this, for everything, we are out of tune.

What Wordsworth expressed with amazement, his followers took for granted as the staple diet of their poetry. Shelley wept, and Keats delighted. For them all, nature was lovely and desirable, because it was a *thing apart*, containing a freshness and innocence not to be found in the sordid life of men. It is an attitude appearing again and again in Tennyson. Gerard Manley Hopkins turned it to religious account, often changing it little:

[5] The most exhaustive study of the philosophical implications of 'nature' and allied concepts is to be found in Lovejoy and Boas's *Primitivism and Related Ideas in Antiquity*, vol. 1 of *A Documentary History of Primitivism and Related Ideas* (Baltimore, 1935). See esp. the appendix, pp. 447 ff., where some sixty-six meanings of the term 'nature' are distinguished. Actually, it may be doubted that the meanings given there are all so discrete as the authors would have us believe. Many seem rather corollaries drawn from certain definitions than definitions themselves. In any case, the authors are concerned with philosophical expressions of 'primitivism' rather than with poetic or dramatic use of 'nature' (in any definition). Accordingly, in most of their examples, the *normative* use of 'nature' seems to be most prominent—nature as the basic order of things to which we *ought* to conform, if we are to realize our best potentialities.

Why do men then now not reck his rod?
Generations have trod, have trod, have trod;
And all is seared with trade; bleared, smeared with toil;
And wears man's smudge, and shares man's smell: the soil
Is bare now, nor can foot feel, being shod.

And for all this, nature is never spent;
There lives the dearest freshness deep down things . . .

(*God's Grandeur*)

An evident feature of such an attitude is that nature as a
whole now becomes a complex metaphor. It is apart from
man, and so he can conceive it by itself. Furthermore, true
vitality exists in nature, not in man, and nature's world is
relatively static. The puffing locomotive in the valley which
Thoreau wondered at was going somewhere, albeit to no
purpose; his pond simply *was*. *Its* virtue could be expressed
only by description. Nature, divorced from the human, is not
dramatic.

Now this sense of what nature is, and how it is to be used—
as a metaphor for man's unfulfilled aspirations—is a late and
extreme phenomenon of the Western European sensibility. It
found what was perhaps its highest expression in the poetry of
Wordsworth and Hölderlin,[6] dominated the nineteenth cen-
tury, | and now, at least in its more obvious manifestations, has
long since come to seem *passé*.

As evidence of its lateness, it is significant that the word
'nature' before the eighteenth century did not mean what it
usually means now. This present meaning is found under this
rather unclear definition in the *OED*, s.v. 'nature', meaning 13:
'The material world, or its collective objects and phenomena,
esp. those with which man is most directly in contact; freq. the
features and products of the earth itself, as contrasted with

---

[6] Hölderlin identifies his 'nature' at different times with childhood, with Grecian
antiquity, and with a kind of pantheism. His sense of alienation from it is both more
personal and more intense than that of Wordsworth:

Weh mir, wo nehm ich, wenn
Es Winter ist, die Blumen, und wo
Den Sonnenschein,
Und Schatten der Erde?
Die Mauern stehn
Sprachlos und kalt, im Winde
Klirren die Fahnen.

those of human civilization.' The first example given there which properly fits the definition[7] is from Dryden's translation of Virgil's *Georgics*:

> But time is lost, which never will renew,
> While we too far this pleasing path pursue,
> Surveying nature with too nice a view.
>
> (Virg. *Georg.* 3. 450)

The following quotation, from Cowper, shows clearly the anachoretic nuance 'nature' had at last assumed:

> To enjoy cool nature in a country seat.   (*Hope* 245)

It is significant that the word 'nature' does not appear in the Virgilian original of the Dryden passage:|

> Sed fugit interea, fugit inreparabile tempus,
> Singula dum capti circumvectamur amore.
>
> (*Georg.* 3. 284–5)

This is, in fact, a sense which φύσις, or *natura*, never assumed in antiquity; and the nineteenth-century attitude which went with such an acceptance of the word likewise does not appear in Greek and Roman writers. There is no Greek or Latin 'nature' poetry. There is, however, a typically classical poetic form which has a clear relation to it, being, like it, a manifestation of the second mode of natural metaphor, though a manifestation far less extreme: that is, the pastoral.[8]

The pastoral is a most complex form, and occupies a

---

[7] The definition, when you consider it closely, is hopelessly confused. It is the only one of those given under 'nature' in which nature and civilization are contrasted, which is what we want; but the first half, as far as one can make out, differs considerably from the second: a *material world* hardly seems to invite 'a sense of something far more deeply interfused', and one wonders what are the 'objects and phenomena with which man is most frequently in contact', the more so, in view of the examples from Dryden and Cowper. And if we insist on the 'especially' clause (which is needed and gives the best sense), it is difficult to make out how the two earlier examples (as one can see if one examines them) satisfy it: *nature*, there, is evidently that of the Greek philosophers—the *cosmos*, the material structure of the universe.

[8] The best modern discussion, though often hopelessly elusive, of the implications of pastoral poetry, and a book to which I here acknowledge a great debt, is William Empson's *Some Versions of Pastoral* (London, 1950). One has, it must be said, the impression in reading Mr Empson's book that almost anything can be a form of pastoral. He often uses the word, in a highly metaphorical sense, to mean any literary work in which the complex is expressed through the simple.

position roughly intermediate between the earlier use of natural metaphor and the extreme, nineteenth-century use. It is always the product of a sophisticated age, one whose civilization has made it conscious of a separation from a simpler and more innocent way of being.[9] But the device of pastoral poetry is not to recognize overtly this separation: it is to pretend it does not exist. To do this, it creates an impossible character who speaks for the poet and his audience: the poet-shepherd. He has affinities both to the poet and to nature: his speech is artful and polished but has the strength of rustic simplicity. One might say he is the legal fiction by which we return to the Golden Age. Through him, the poet can approach to nature. Pastoral poetry is a mask of innocence. The poet creates a setting of which the essential is that it be different from the urban life which has made the poet what he is. In this new scene, set at one remove from experience, with characters who live in two worlds, the poet can express himself with a directness which would otherwise have been impossible.

Such poetry, it should be noted, is always artificial. Some awareness of the artifice is necessary for its enjoyment. We must feel this world close enough for us to enter into it (therefore it is not really description of rustic life), and far enough from us | to be a metaphor of our lost innocence. Within these limits, considerable variation is allowed. The swain may wear a goatskin still smelling of rennet, and the goats may copulate while the rustics talk; or the shepherd's sorrow may be entertained by all the saints above. The fiction is always there, whether the setting be more or less realistic.

The greatest art is to live naturally, it has been said.[10] Indeed, it can be done only by art, and thus the props of pastoral poetry, far from being mere ornamentation, are essential. It is only by creating a special setting that the poet can say what he wants to say.

A good example of what we have been saying is Theocritus' Idyll 1. Far more than Virgil's and Milton's imitations of it, this

[9] Cf. the intelligent remarks of Kathleen Hartwell, 'Nature in Theocritus', *CJ* 17 (1922), 181–97, esp. 181–2.

[10] By Montaigne in *Sur l'expérience*. Montaigne means his epigram to be a paradox, for its point rests on the art vs. nature antithesis of late antiquity.

is a heroic song.[11] At least it represents with remarkable directness some of the essential feelings of epic and tragic poetry—the sense of death, of fate, of deliberate and dramatic resolution. So Daphnis, like Prometheus, or Ajax in the netherworld, disdains to reply to those who, in the situation of the poem, have become his inferiors:

> But the shepherd answered them not a word: he kept
> His bitter love, kept it till his doom's end.
>
> τὼς δ' οὐδὲν ποτελέξαθ' ὁ βουκόλος, ἀλλὰ τὸν αὐτῶ
> ἄνυε πικρὸν ἔρωτα, καὶ τέλος ἄνυε μοίρας.
>
> (Theoc. 1. 92–3)

and proclaims his triumph in the face of the goddess:

> Will you say now our last sun has set?
> Even in Hell Daphnis will be a bitter grief to Love.
>
> ἤδη γὰρ φράσδῃ πάνθ' ἄλιον ἄμμι δεδύκειν·
> Δάφνις κἠν Ἀίδα κακὸν ἔσσεται ἄλγος Ἔρωτι.   (102–3)

Daphnis, indeed, lives in a world where it is possible to be a hero. But this world is no longer quite the poet's own—or the audience's. Compare the *Iliad*: as we read it, we feel that the narrator is part of the society he describes: he would be a bard, who lives with heroes; and Achilles himself sang the deeds of | men. Nor is Homer's society here presented to us as anything other than simply the way things are. But in the poem of Theocritus, the heroic world is also the bucolic world. Thus in the first lines quoted above, there is a sharp finality of movement, and this, culminating in the word μοίρας, creates for us that sense of inflexible—and yet chosen—destiny which is part of the greatness of a hero. But the word βουκόλος, with the article, is so placed as to give it prominence and to insist on what Daphnis is. The implication 'Though but a shepherd, yet Daphnis behaved like a hero' is subtly modified into 'It was because he was a shepherd that he acted in this grand and simple way'.

The beauty of the poem lies in this fine balance of heroic

---

[11] Cf. Georges Łanowski, 'La Passion de Daphnis', *Eos* 42 1 (1947), 175–94, esp. 179, on the heroic character of Theocritus I. Łanowski's suggestions concerning the nature of Daphnis' passion, however, I do not find convincing.

austerity and pastoral delicacy. Daphnis dies a hero's death, but the whole dramatization of his death is modulated into a bucolic movement. When, in the *Iliad*, Achilles says to Lycaon:

> But upon me too is death and imperious fate . . .
>
> Ἀλλ' ἐπί τοι καὶ ἐμοὶ θάνατος καὶ μοῖρα κραταιή
>
> (*Il.* 21. 110)

this is the heroic mode, simple and unmatchable. It would hardly be Ἔρως that brings Achilles' death; but it is more than that word that makes Daphnis' death different from anything in the *Iliad*. The hypnotic repetition of ἄννε and the falling rhythm of Theocritus' line—ἄννε πικρὸν ἔρωτα καὶ ἐς τέλος ἄννε μοίρας—put the heroic gesture itself into the bucolic mode.

A corresponding complexity hinges on the same word βουκόλος when Daphnis answers Aphrodite. He reminds her of Anchises, himself a link between the heroic and pastoral worlds:

> What of that shepherd and Cypris? . . .
>
> Οὐ λέγεται τὰν Κύπριν ὁ βουκόλος . . .   (Theoc. 1. 105)

then he taunts her in a tone of impatience and contempt:

> Go off to Ida,
> Go see Anchises . . .
>
> ἕρπε ποτ' Ἴδαν
> ἕρπε ποτ' Ἀγχίσαν . . .   (105–6)

and then follow a line and a half of tranquillity and sensuous delight: |

> . . . There there are oaks, and galingale;
> There the bees buzz sweetly all about their hives.
>
> τηνεὶ δρύες ἠδὲ κύπειρος,
> αἱ δὲ καλὸν βομβεῦτι ποτὶ σμάνεσσι μέλισσαι.   (106–7)[12]

[12] The text and translation of these lines are not entirely free from doubt. Miss Hartwell, op. cit. (above, n. 9), p. 188, accepting the MS reading, translates: 'there you will find oak trees, but here grows the galingale, and the bees boom pleasantly about the hives', and draws conclusions about the kind of scenery Theocritus liked. But the opposition between oak trees in Phrygia and galingale and bees in Sicily is rather forced, and Miss Hartwell's interpretation, that Daphnis would be left alone in his softer Sicilian landscape, dissolves the dramatic tension of the lines. Meineke's

The effect of this is the counterpart of what we had before: the heroic tone there was blended with the pastoral; here pure pastoral is given an ironic edge and an epic sharpness. Daphnis' contempt still rings in the line describing the bees. He seems to be saying that this indulgence is good for Aphrodite; he, as a hero, is superior to it. Yet the beauty of the line is of his own making. We feel also in his words: 'I will show you how well I can sing. It is my gift of song that has made me great, and will cause universal nature to mourn my death.'

Everything in this part of the poem—Thyrsis' song—is such as to intensify the tragedy and the triumph of Daphnis' death through the appeal by the refrain, by every turn of language, to the simplicity of a pastoral setting. The point of all this is that *only in such a setting* can Daphnis, as a dramatic hero, exist. The poet first wins us over by the melody and the beauty of rustic landscape:

> Sweet is the whispering music of the pine-tree, goatherd,
> There, by the spring; sweet too is thy piping . . .
>
> Sweeter, O shepherd, is thy song that the clear
>   resounding
> Water yonder that falls high down from the rocks . . .
>
> Ἀδύ τι τὸ ψιθύρισμα καὶ ἁ πίτυς, αἰπόλε, τήνα
> ἁ ποτὶ ταῖς παγαῖσι μελίσδεται, ἁδὺ δὲ καὶ τύ
> συρίσδες . . .
>
> Ἄδιον, ὦ ποιμήν, τὸ τεὸν μέλος, ἢ τὸ καταχές
> τῆν' ἀπὸ τᾶς πέτρας καταλείβεται ὑψόθεν ὕδωρ . . .
>
> (1–3, 7, 8) |

These incantatory verses have as yet no heroic overtones. But they make for us the world of beauty, of simplicity, of rich potentialities, where the heroic mode has its place. If we compare, for example, Idyll 22, the Διόσκουροι, we are informed from the first line, 'We praise in song the twin sons of Leda and of aegis-bearing Zeus', that this is an Alexandrian poet writing a pastiche of a Homeric Hymn. And this feeling never entirely leaves us throughout the poem.

By a subtle device, it is the very magical unreality of the

emendations, on the other hand, based on an indirect tradition (cf. Gow ad loc.), give a satisfactory sense. They are in the texts of Wilamowitz and Gow, and I have translated accordingly.

poetic landscape in Idyll 1 that prevails upon us to accept Daphnis' words as straightforward heroic speech, and to take pleasure in them as spontaneous dramatic utterance; for this unreality wins us to itself. The unreality, be it noted, is not that of the exotic and far-away; rather it is the unreality of happy simplicity. Hence the common and the familiar, the animals, and all the scenes on the cup, have their place. We are meant, as we read the poem, partly to believe that rustic life is like this, and that we can share in it.

Pastoral might be described as a cover in an age of irony. It arose in Greece at a time when writers felt it impossible to deal with strong emotion directly and when literary works with reference to immediate experience were, for that reason, slight. Theocritus 15, the mime with the two Syracusan women, is a good example. It is an agreeable piece of genre-work, which reminds us of certain types of Hellenistic sculpture, but it has no poetic force. It is only within a contrived setting, where sophistication can transplant the reader into a world of nature, that poetic expression can achieve a kind of beauty and greatness. The mastery of Theocritus lies in his having been able to create this setting with more fullness and depth than any of his followers. There is more colour and richness in his landscape, and we are more steeped in it, than in that of Virgil or of Milton.

The pastoral, by finding the sources of its strength in a nature which no longer includes society, constitutes an implicit criticism of that society. In the fully developed pastoral, to be sure, the criticism is nowhere openly stated. It subsists more as an unconscious assumption than as part of the active artistic expression of the poem. And then there is a further dimension in Theocritus, in that he tends to address himself to a circle of friends forming a sort of society within a society, a circle which, | through its very refinement, is less irrevocably removed from nature than metropolitan life at large. The celebrated Idyll 7 and the distaff poem in Aeolic choriambs are the finest and clearest manifestation of this dimension of Theocritean art.

The beginnings of true pastoral are to be found, as we might expect, in the period when disillusion with society became a marked feature of literature: at the end of the fifth and the beginning of the fourth centuries—and in the works of the man who most clearly expressed that disillusion: Plato.[13] His image, in the *Republic*, of the philosopher standing in the shelter of a wall during the storms of the city-state, is an apt illustration of his attitude, one that would have been scarcely comprehensible in the day of Aeschylus. In him we see as yet but the adumbration of pastoral; it is only one element of his remarkable complex work.

Plato began by attempting to render the essence of his friend Socrates, who was still a citizen of Athens in the old sense, and almost a conservative, as Samuel Johnson was a conservative— one who set out to recall society to its traditional virtues in a time of decline. But the day when a writer could speak directly to society was passing; and Plato ended as a teacher of more or less esoteric doctrine to a small group of pupils. Tradition has it that he was also a poet; and if some of the poems ascribed to him were really his, there could be no doubt of his role as creator of pastoral. Thus:

> Let the Dryads' leaf-grown rock keep silence.
> Let the down-rushing springs be silent.
> Let the sheep with their sundry bleating
> Be heard no more. For Pan himself
> Pipes clear music; on his syrinx
> His lips form over the close-joined reeds.
> See dancing round him now
> The Nymphs of the Water: the Hamadryad Nymphs.

> Σιγάτω λάσιον Δρυάδων λέπας οἵ τ' ἀπὸ πέτρας
>   κρουνοὶ καὶ βληχὴ πουλυμιγὴς τοκάδων,
> αὐτὸς ἐπεὶ σύριγγι μελίζεται εὐκελάδῳ Πάν
>   ὑγρὸν ἱεὶς ζευκτῶν χεῖλος ὑπὲρ καλάρων. |
> αἱ δὲ πέριξ θαλεροῖσι χορὸν ποσὶν ἐστήσαντο
>   Ὑδριάδες νύμφαι· νύμφαι Ἀμαδρυάδες.   (24 Bergk)[14]

---

[13] It is to be hoped that no objection on formal grounds will be taken to the introduction of Plato in an essay on Greek poetry. In the Renaissance even a writer like Cicero was commonly called *poeta*; and is Menander more poetic than the author of the *Symposium*?

[14] Cf. Wilamowitz, *Platon* (Berlin, 1920), i. 456, n. 1. On the authenticity of the pastoral epigrams of Plato in general cf. Bergk, *Poetae Lyrici Graeci*, ii (Leipzig, 1914) 618: 'Quod si in tribus, quae primo loco memoravi (24, 25, 32), similitudo quaedam cum bucolicis carminibus deprehenditur, nihil est causae, cur propterea Platoni abiudices ...'

If he is not the author of this and of like poems, it may yet say something of the quality of his work that such poems were attributed to him. Pastoral, at any rate—I use the word in a slightly extended sense—is clearly discernible in one of his finest works, and in the form of both indulgence in natural beauty and criticism of society. The *Phaedrus* is a work of the middle period. The figure of Socrates in the dialogues of this period has not yet been abandoned in favour of an attempt to attain to truth through the exercise of logic, but he is handled with far more freedom than in the earlier dialogues, those concerned with his trial and death and with his daily life about Athens. Accordingly, they show more of Plato's own poetic genius than any other group of his works.

Socrates, you will remember, meets his young friend Phaedrus just inside the city walls. It emerges from their conversation that Phaedrus has spent the morning listening to a set speech of the celebrated orator Lysias which he admires extravagantly. Indeed, he has obtained a copy of the speech itself so that he can commit it to memory; but in his shyness he attempts—unsuccessfully—to conceal the document from Socrates. The two leave town together and walk barefoot along the bank of the Ilissus river, conversing by the way. Then they find a place to rest. Phaedrus reads the speech to Socrates, Socrates counters with a mock speech and then with a serious speech of his own; then they converse some more, and at last set off for home.

The *Phaedrus* is, among other things, an attack on one of the prime phenomena of the corrupt society—sophistical rhetoric. Lysias' speech on Love, so fondly guarded by the susceptible Phaedrus, is first and foremost characterized by perversity. It is an example of the arbitrary machinations of intellect and speech which had come to replace the traditional beliefs and feelings of earlier Athens as well as Periclean vision. The condemnation of this document (Plato employs his best wit in its construction) is carried out by having it read aloud in a setting removed from the orator's circle in town. It is Socrates' suggestion that they | go off into the country, wading through the Ilissus. The scenery is described in evocative detail:

PHAEDRUS.   Do you see that extraordinarily high plane tree?
SOCRATES.   I do indeed.

PHAEDRUS.   There is shade beneath it, and a nice breeze, and grass to sit on; or we could lie down.

SOCRATES.   Lead the way.   (*Phaedr.* 229 A 8–B 3)

The ambiance is one to suggest and reinforce that vision of natural truth with which Socrates wishes to counter Lysian rhetoric. Here the gods still live, who have no place in the sophistic milieu of the town.

It is at this point that the argument over the rationalistic interpretation of the Oreithyia legend takes place. Phaedrus' disbelief of the legend, and his desire to substitute a scientific explanation for it, is one with the perverse treatment of love in Lysias' speech. It is left to Socrates to state the rationalists' case and then dismiss it with contempt. Socrates' argument is reminiscent of Montaigne's rejection of the legal interpretation of his day in *Sur l'expérience*. Socrates points out that no greater knowledge is to be gained by such interpretation, and that there is no end to it. Then, like Montaigne, he appeals to a kind of natural simplicity and self-knowledge. Only his appeal to nature is more concerned with beauty of landscape than anything in Montaigne:

Dear Phaedrus, I have no time for such things; and I will tell you why. I am not yet able to obey the Delphic precept, and to *know myself*. How foolish then, it seems to me, when I am still in ignorance of this, to busy myself with matters that are not my own (τὰ ἀλλότρια σκοπεῖν). And so I am ready to leave such things in peace (χαίρειν ἐάσας ταῦτα), and to believe what people tell me about them; and to be concerned instead with my own self: Am I a beast more intricate and more wild than the monster Typhon, or am I rather a gentler and simpler animal, one privileged by nature with a portion that is divine and unmonstrous (θείας τινὸς καὶ ἀτύφου μοίρας φύσει μετέχον)? But by the way, dear friend, is this not the very tree to which you wished to lead us?   (229 E 3–230 A 7)

Socrates appeals to many emotions in this speech, to early fifth-century religious feeling not least, for his thought is re|markably close to that of Pindar's Pythian 8, where Τυφὼν ἑκατογκάρανος, the hundred-headed Typhon (we may compare the Τυφῶνος πολυπλοκώτερον of Plato), is made the symbol of the wildness and madness that alienates us from the gods. *Nature*, too, φύσει, in Socrates' sense, meaning *what I am by birth, by my race, by the dispensation of the gods*, is reminiscent of the strong emphasis on φυά, the synonymous word, in this and in other poems of Pindar. But φύσις is also the word meaning 'growth' and is close to the verb φῦναι, what a tree does, as, for example, in the line telling of the tree of which Odysseus fashioned his bedstead in the *Odyssey*:

An olive tree grew there long of leaf ...

θάμνος ἔφυ τανύφυλλος ἐλαίης ἕρκεος ἐντός

(*Od.* 23. 190)

and there seems to be a suggestion here that the natural and living beauty of the place is what makes the intricate ratiocinations of Lysias and the sophistic attempts to rationalize away the divine so peculiarly inappropriate. Socrates' 'by the way', then, μεταξὺ τῶν λόγων, with which he brings Phaedrus' and the reader's attention back to the fresh beauty of their surroundings, is Plato's way of seeming most casual when he is most in earnest. For this tree and the greenness of the landscape are the visible manifestation of that divine natural portion to which Plato, by his near-pastoral tone, gives such dramatic reality.

The succeeding speech of Socrates is a masterpiece of landscape description:

Yes, by Hera, this is a lovely place of rest (καλή γε ἡ καταγωγή). The plane tree is marvellously tall and outspreading; and the tallness and shadiness of the chaste tree is perfect. And see how it is in full flower, so as to fill the place with fragrance. And then the spring is most delightful as it flows from beneath the plane tree, and its water wonderfully cool, at least to judge by dipping my foot. The statues and sacred figures would seem to tell us it is a sanctuary of some of the nymphs and of Achelous. Or consider (Εἰ δ' αὖ βούλει) the freshness of the place (τὸ εὔπνουν τοῦ τόπου), how pleasant to all the senses! It has a sweet and summery echo of a chorus of cicadas (θερινόν τε καὶ λιγυρὸν ὑπηχεῖ τῷ τῶν τεττίγων χορῷ). But what is most elegant of all is the quality of the grass (πάντων δὲ κομψότατον τὸ τῆς πόας); how it lines in a gentle slope ideally suited to lie back and lean one's head upon. You have been a most admirable guide, dear Phaedrus.   (230 B 2—C 1)

Its grace and loveliness derive largely from the sophisticated and urbane turns of speech that are used to bring forth each descriptive touch. The use of the neuter singular of the adjective (τὸ εὔπνουν), the γες, the Εἰ δ' αὖ βούλει, and especially the word κομψότατον—these are modes of speech characteristic of the elegant society of the town. To understand natural beauty requires the greatest sophistication of all.

This is made clear by the irony that follows. Phaedrus remarks that Socrates behaves like a stranger, surprised that his friend does not, like himself, take the beauty of the place more for granted. For Phaedrus is less aware than the reader of the metaphorical richness of the landscape, the richness which is to make the abstract ingeniousness of Lysias seem so paltry a thing. But Socrates merely apologizes, professing to be really eager to hear Lysias' speech; and concludes with a perfect example of Socratic—or Platonic—irony:

Forgive me, excellent friend—you know I am always trying to learn things; but the countryside and the trees have nothing to teach me; for that I must look to the men in town.

Συγγίγνωσκέ μοι, ὦ ἄριστε· φιλομαθὴς γάρ εἰμι. Τὰ μὲν οὖν χωρία καὶ τὰ δένδρα οὐδέν μ' ἐθέλει διδάσκειν· οἱ δ' ἐν τῷ ἄστει ἄνθρωποι.

(230 D 3–5, cited above, p. 3)

The point is that in this context the men in town have nothing to teach Socrates; the countryside and the trees, with the gods and nymphs that dwell in them, everything.

Socrates' two counter-speeches, the burlesque and the straightforward, are made, one by the feigned, the other by the true, inspiration of the country nymphs. The second, of course, far transcends the rustic: but the rustic is its point of origin. After the sublime vision, and a discussion of rhetoric, the work ends with a philosophic prayer to the pastoral deities:

O belovèd Pan, and all you other gods who dwell in this place, grant that I may become beautiful *inwardly* . . .

δοίητέ μοι καλῷ γενέσθαι τἄνδοθεν . . .

It is, lest we lose perspective, part of the subtlety of the work that Socrates here is not a rustic philosopher; he has not lost connection with the questioner of the ἀγορά. His oneness with a religious-natural setting adds strength to the inheritor of

Athenian wit, and gives an unexpected turn to his criticism of his own | environment. It was not till Theocritus that a great poet devoted all his energies to the pastoral.

Our discussion of the use of natural metaphor in Greek poetry will gain a kind of fragmentary completeness with the introduction of one more literary form: the romantic comedy. This had to be a dramatic work, though not necessarily a play; a comedy, because it has a happy ending; romantic, because the action is transferred from immediate experience to a fabulous, exotic, or at least remote, setting. This last quality gives it obvious affinities with pastoral poetry, and it is a kind of precursor of it; but there are great differences. The romantic comedy is an earlier form, and its poetry has less softness and indulgence of the senses than pastoral. Its natural setting, then, is not Sicilian fields but the sea and distant islands. Moreover, the relation with society is different. The actual life of men, in the romantic comedy, does not seem wholly devoid of magic; therefore it is not totally excluded from the work. The notion is that society and nature are different but not hopelessly so. Society is a variable element. It can be corrupt and perverse; in fact, it usually is. But the same human excellence that finds its first and most apt expression in a natural setting can operate in society as well; it can, as it were, bring society round, make it recognize and conform to the virtues of nature. Nature and civilization, then, are not permanently estranged, but temporarily so. Civilization can be brought back to a natural way of being, and this is largely what such works are about.

The pattern of this is set by the *Odyssey*. Society is of course the court of Ithaca. Natural law is not respected there. The suitors, the agents of unfeeling and violence, are in control:

> They were before the doorway delighting their hearts
>   in games of draughts,
> Sitting upon hides of oxen they themselves had slain . . .
>
> . . . οἱ μὲν ἔπειτα
> πεσσοῖσι προπάροιθε θυράων θυμὸν ἔτερπον
> ἥμενοι ἐν ῥινοῖσι βοῶν, οὓς ἔκτανον αὐτοί . . .   (*Od.* 1. 106–8)

The αὐτοί, asserting their mastery, is a lovely stroke of quiet indignation.

Needless to say, there is never any doubt of the outcome of the story, just as there is no reason for the spectator to entertain doubt of the happy outcome of *Cymbeline* or *The Tempest*. The romantic comedy is, in this regard, a success story. In a tragedy, society and nature are not distinguished, but the hero | has ultimately no place in either. He can escape only by death; he cannot, like the hero of the romantic comedy, go to nature and then return, bringing its virtues with him. But tragedy assumes this: that society, the company of his fellow men, is the place for a man to live as long as he lives; a flight to a non-human nature is out of the question.

The *Iliad* is the model of all tragedy. It deals at all times with human experience directly; natural objects are straightforward images of things human: the stars in heaven show forth the vastness of the illumination of the Trojan campfires. It is a world where men are finally helpless but in the meantime, doers of deeds, the makers of the world. Their very closeness to death gives them a supreme importance: Athene and Apollo, in the guise of birds, sit delighting in *them*:

> Then did Athene and Apollo of the silver bow
> Assuming the forms of vultures seat themselves
> Upon a towering oak tree of Father Zeus
> Delighting in men. These sat in serried ranks,
> With shields and helms and spears all bristling.

> κὰδ δ' ἄρ' Ἀθηναίη τε καὶ ἀργυρότοξος Ἀπόλλων
> ἑζέσθην ὄρνισιν ἐοικότες αἰγυπιοῖσι
> φηγῷ ἐφ' ὑψηλῇ πατρὸς Διὸς αἰγιόχοιο
> ἀνδράσι τερπόμενοι· τῶν δὲ στίχες ἥατο πυκναί
> ἀσπίσι καὶ κορύθεσσι καὶ ἔγχεσι πεφρικυῖαι.

> (*Il.* 7. 58–62)

The *Iliad*, then, deals with its subject, human passion and death, with a striking immediacy. No descriptions give so strongly a sense of actuality, no speeches have so clearly the ring of the living voice in them, before Shakespeare. An exotic dimension in this sort of writing is unthinkable; the fabulous, as setting, has no place.

In the *Odyssey*, on the other hand, an alien setting is indispensable: for the Odysseus who is to return to clear out the suitors and set things right must be defined by the natural world he passes through; otherwise his return and his triumph will have no meaning. (His putting things once more in harmony with natural laws will be marvellously symbolized by the tree, of which we have heard nothing hitherto, out of which Odysseus' and Penelope's bedstead and the whole Ithacan palace grow.) This natural world, though filled with more or less anthropomorphic monsters, is non-human and static, in comparison with the human world of the *Iliad*. This fits the greater tendency to description in the *Odyssey*, with the emphasis on manners and | with the more relaxed tone. Longinus compares the two works well:

One might compare the Homer of the *Odyssey* to the setting sun: the greatness is there, but not the intensity. He does not maintain there, as in the *Iliad*, the concentration, the grandeur that never sinks to the trivial, the fire, the reciprocation of passions, the suddenness, the ring of speech, all that imagination condenses from reality itself. But rather, like the sea at ebb tide withdrawing into itself and to its natural level, there are the wandering rivulets and the shallows of the mythical and unbelievable. ... Thus the decline of passion in the greatest writers resolves itself into description and manners (εἰς ἦθος ἐκλύεται). (*De Subl.* 9. 13–15)

Before this last sentence he says:

I speak of old age, but the old age of a Homer. Only in all this writing the mythical takes precedence over the actual.

γῆρας δ᾽ ἡγοῦμαι, γῆρας δ᾽ ὅμως Ὁμήρου· πλήν ἐν ἅπασι τούτοις ἐξῆς κρατεῖ τοῦ πρακτικοῦ τὸ μυθικόν.

But of course the *mythical* is an essential of the work; and Longinus himself speaks of the *Storms* and the Cyclops episode as its glories. For in this poem it is, curiously, only in the non-human world that heroic virtue can be expressed; or at least it must first be expressed there. With the one exception of the Slaying of the Suitors—where there is, so to speak, the final return of nature to society—Odysseus' battles are all fought with monsters, not with other men. Odysseus is a hero, but his society does not offer scope for heroic action. As in Theocritus Daphnis must be in the pastoral, so in the *Odyssey* Odysseus must be in the fabulous, world. The strangely languid battle scene of Book 24 makes this evident. Or compare the provincial council of the nobles in Book 2 of the *Odyssey* with the tumultuous meeting of the Achaeans in *Iliad*, 2. In the former, nothing happens: it serves only to make clear the situation to the reader and to give Homer room to treat Telemachus with indulgent irony.

Society is treated with sophistication, interest in manners, love of description. Helen appears a gracious hostess, not the fatal beauty of the *Iliad* who accuses herself so bitterly. Menelaus and Nestor are country squires. They show a warm interest in | Telemachus and much affection for Odysseus, but do nothing. They speak of the Trojan War as of something *längst vergangenes*. As Longinus says: 'For what is the *Odyssey* after all but an epilogue to the *Iliad*?' (*De Subl.* 9. 12), and he quotes the nostalgic lines of Nestor himself:

> There warlike Ajax lies; there lies Achilles;
> There lies Patroclus, peer of the gods in counsel;
> And there lies my own dear son . . .   (*Od.* 3. 109–11)

Theirs is a show-piece society. It partakes of the fabulous itself. Menelaus' palace draws from the naïve Telemachus the comment:

Look there, son of Nestor, dearest to my heart, see the flash of bronze in the ringing halls, and the gleam of amber, and of silver, and of ivory. Surely the hall of Olympian Zeus is like this within!

(*Od.* 4. 71–4)

Neither here nor in the more real society of Ithaca can Odysseus become himself. The poet has created a fictitious world for him, in which he can prove himself against the asperities, at the same time that he enjoys the amenities, of nature.

Odysseus, however, is a heroic, and not a pastoral, figure. He belongs to the world of men; only in this world he has no equals, and no competition. Hence there is an ambiguity in his relation to the wondrous world of adventure through which he moves. He is in it, but not of it. He has a greater vividness than his surroundings. The grotesque shapes of Scylla and Charybdis lay hold on the imagination; but a firmer sense of reality, more like that of the *Iliad*, comes when we hear the almost ritual lines that tell of Odysseus' own feelings:

Then sailed we onward, grief in our hearts,
Happily free from death, leaving comrades behind.

"Ἔνθεν δὲ προτέρω πλέομεν ἀκαχήμενοι ἦτορ
ἄσμενοι ἐκ θανάτοιο, φίλους ὀλέσαντες ἑταίρους.

(*Od.* 9. 62–3; 9. 563–6; 10. 133–4)

Nature for him is not, on the surface, a refuge, but an exile.
It is the place where he can make good his mastery, his
awareness and his strength, his loyalty and his piety. But he
always desires to return. Hence comes the fundamental theme
of endurance. The same landscape that appears in so enchant-
ing a | light to the reader is for him a prison. The ambiguity is
expressed with particular point in the famous passage of Book
5. I quote from Chapman:

      . . . A Grove grew
   In endlesse spring about her Caverne round,
   With odorous Cypresse, Pines, and Poplars crown'd
   Where Haulks, Sea-owles, and long tongu'd Bittours bred,
   And other birds their shadie pinions spred.
   All Fowles maritimall; none roosted there,
   But those whose labours in the waters were.
   A vine did all the hollow Cave embrace;
   Still greene, yet still ripe bunches gave it grace.
   Four Fountaines, one against another powr'd
   Their silver streames; and medows all enflowrd
   With sweete Balme-gentle, and blue Violets hid.
   That deckt the soft brests of each fragrant Mead.
   Should anyone, (though he immortall were)
   Arrive and see the sacred objects there;
   He would admire them and be over-joyd;
   And so stood *Hermes* ravisht powres employd.
      But having all admir'd, he entered on
   The ample Cave; nor could be seene unknowne
   Of great *Calypso* (for all deities are
   Prompt in each other's knowledge; though so farre
   Severd in dwellings); but he could not see
   Ulysses there within. Without was he
   Set sad ashore; where 'twas his use to view
   Th'unquiet sea; sigh'd, wept, and emptie drew
   His heart of comfort.   (*Od.* 5. 63–83)

There is wit as well as delectation in such lines. The

admiration of the divinity is expressed as a potentiality to point
up the loveliness of the landscape; it is a hyperbole, like
Telemachus' words to Pisistratus, quoted above. But then the
hyperbole becomes actuality, for the god is there, and admire
and delight are exactly what he does.

Odysseus is in this landscape, or on the edge of it; but he will
not give in to its blandishments. Our notion of him must
include the awareness that it is he, of all men, who is chosen to
be on the goddess' island: that this is the sort of rich experience
that has made him the greatest of all travellers. But simultaneously we must know that for him this magic place is a test of
endurance; that his loyalty to his home appears but the
stronger for it. The dramatic shift in the last three lines: |

οὐδ' ἄρ' Ὀδυσσῆα μεγαλήτορα ἔνδον ἔτετμεν,
ἀλλ' ὅ γ' ἐπ' ἀκτῆς κλαῖε καθήμενος, ἔνθα πάρος περ,
δάκρυσι καὶ στοναχῇσι καὶ ἄλγεσι θυμὸν ἐρέχθων.   (81–3)

is a powerful one. It makes all the beauties of the preceding
lines appear a shade unreal, gives them a nuance of fantasy. The
cadence changes too from the effortless ease of the landscape
description with its falling rhythm and bucolic diaeresis—

σκῶπές τ' ἴρηκές τε τανύγλωσσοί τε κορῶναι·
εἰνάλιαι, τῇσίν τε θαλάσσια ἔργα μέμηλεν.   (66–7)

—to hexameters with more austerity and heroic impetus. The
last line is almost of the same rhythmic construction as the
description of the ranks of warriors in the *Iliad*, quoted above:

In the *Odyssey*:

δάκρυσι καὶ στοναχῇσι καὶ ἄλγεσι θυμὸν ἐρέχθων.   (5. 83)

In the *Iliad*:

ἀσπίσι καὶ κορύθεσσι καὶ ἔγχεσι πεφρικυῖαι.   (7.62)

This passage shows with unusual clarity the ambiguity of
the hero who is of society but necessarily from it, and in nature
but not of it. Landscape defines him, as it does Philoctetes
in Sophocles' play,[15] both affirmatively and negatively. Its

---

[15] Philoctetes cannot be said to have enjoyed Lemnos, as Odysseus enjoys his
landscape. But his nobility becomes firmly associated with the rugged beauty of the
land. Cf. *Phil.* 1453–68.

richness indicates his capacity for enjoyment: its remoteness exalts his endurance. It is in this light that we must understand the paradox of the hero who wishes ever to return home—

> longing to see even the smoke rising forth
> from his homeland   (*Od.* 1. 58–9)

—and is yet willing to spend a year as paramour of Circe.

His appreciation of the natural scene is more evident in the Cyclops episode, perhaps the finest of the poem. He approaches the Cyclopean land with the same formula of endurance we have mentioned above:

> Then sailed we onwards, grief in our hearts ...   (9. 105)

but he first describes in great detail the wild beauties of the island lying across the harbour: |

There lies an island stretched out before the harbour of Cyclops' land—not very far, nor yet very close. It is a wooded island. In it live goats in countless number, all *wild* (ἄγριαι). For no footfall of man disturbs them; hunters, those who toil in ranging the wooded mountain tops, never go there. Nor is it possessed by herdsmen, nor by the plough. Unsown, untilled perpetually, it knows not man (ἀνδρῶν χηρεύει), but nourishes bleating goats. ... It's not at all a *bad* land (οὐ μὲν γάρ τι κακή γε): it could bear everything in season. It has fertile watered meadows by the shore; vineyards could thrive there forever ...   (9. 116–24, 131–33)

With the word ἄγριαι, 'of the goats', the inhospitable note of the land is struck. The Cyclops is the type of the ἄγριος:[16]

> A savage, knowing not justice, nor the ordinances of heaven.
>
> ἄγριον, οὔτε δίκας εὖ εἰδότα οὔτε θέμιστας.   (9.215)

To counterbalance his pleasure, Odysseus must here exert not endurance, in unbroken resolve to return home, but all his courage and cunning, to overcome the monster who personifies the deadliness of the exotic world.

Cyclops himself reveals the ambiguity of the landscape. He

---

[16] This is the word in Greek to express the life of men or beasts outside society. Thus Athene uses it in her tactful lie to Telemachus in *Od.* 1. 198–9: χαλεποὶ δέ μιν ἄνδρες ἔχουσιν, ἄγριοι, οἵ που κεῖνον ἐρυκανόωσ' ἀέκοντα. In Sophocles' Philoctetes it becomes the point of the reader's judgement of the hero, for his ἀγριότης becomes identified with his heroic superiority to a corrupt society.

is first crude, monstrous, disgusting. Such is Odysseus' attitude when he offers him the wine:

> Here Cyclops, take some wine and drink, now that
>   you've supped on human flesh!

> Κύκλωψ, τῆ, πίε οἶνον, ἐπεὶ φάγες ἀνδρόμεα κρέα.   (9.347)

He sprawls drunk on the floor of the cave:

> So saying he stretched and fell over backwards . . .
> . . . Out of his gullet spewed wine
> And gobbets of man's flesh; heavy with wine he belched.
> (9. 371–4)

However, he is a monster in a fully pastoral setting. The | richness of the sheep, goats, full milk-pails, and cheese exerts such an attraction on the keen-eyed Odysseus that he commits one of his rare acts of imprudence in staying for their owner's homecoming:

> The wattles were laden with cheese . . .
>
> . . . all the vessels swam with whey . . .
>
> But I was not persuaded to leave—better if I had been.
> (9. 219, 222, 228)

There is a natural abundance of good things, as well as horror and death: and Odysseus takes up a characteristic attitude toward each.

What is sympathetic in Polyphemus, implicit in his setting, becomes explicit in his speech to his favourite ram, when he shows all the simple affection and tenderness of rusticity:

My good ram, why are you coming last like this out of the cave? You did not use to be left behind. No, you were always first to eat the tender blades of grass, as you went forward with great strides, and first to go to the river water, and first in your eagerness to return home in the evening. But now you are last of all. Is it that you grieve for your master's eye, which that evil man put out, with the help of his awful comrades, after he had vanquished my mind with wine?
(9. 447–54)

But lest our sympathy be too much with him, he straight-way turns to vindictiveness:

I mean Noman, who I say has not yet escaped destruction. O, that you could think as I do, and had a voice to speak, so that you could tell me where he is *now*, avoiding my might! His brain would be scattered from one end of the cave to the other as I smashed him against the rocky floor! (9. 455–9)

Odysseus' revealing of himself as his ship moves toward the island is his greatest imprudence, and, together with his revealing himself on another occasion, before the suitors in Book 22, his greatest assertion of heroic individuality. He is always the character who can have his cake and eat it too, as when he hears the Sirens: the natural world gives him glory and delight, but he rises above it to be himself. |

The price paid, if this is the right expression, in both the romantic comedy and the pastoral, for this device whereby the poet has a fabulous-natural setting in which to place his hero and to express himself without irony, is, as Longinus objects, the element of unreality that must attend it. The precision of observation, the grace of the language, and the keenness of intelligence in the *Odyssey* make this element never vexatious, usually not thought of. But the *Odyssey* nowhere engages our emotions as does the *Iliad*. The most horrible scenes in fact do not appal us. There is far more dread in Achilles' words to Lycaon:

> Come, friend, you too must die; why do you complain?
>
> Ἀλλά, φίλος, θάνε καὶ σύ· τίη ὀλοφύρεαι οὕτως; (*Il.* 21. 106)

than in all the anthropophagy of Polyphemus.

As regards the *Odyssey*, it should be remembered that the adventures of Odysseus proper occupy but a part of the work. The whole of it is fabulous, in varying degrees, until the end. The Phaeacian Isle, with its seamen with the nautical names and its ships that need no helmsman, appears now like a Greek feudal court, now like a fairyland: it is both. The long interval between Odysseus' return to Ithaca in Book 13 and his revealing himself in Book 22 has as its immediate purpose the prevention of any sympathy we might be inclined to feel for the suitors. But within it, Eumaeus is surely a figure of pastoral innocence. One has the sense, in reading the *Odyssey*, of a society which does not have a perfectly stable hold on reality. The beginning of the historical disintegration of a social structure (what we see in the movement to replace monarchy with oligarchy in Ithaca), one might suggest, is reflected in a certain lightness, a touch of the fantastic, which society possesses in the work of Homer's old age.

The *idyllic* has spread to all parts of the *Odyssey*. This makes for an extraordinary richness of sophistication and play of the fancy. The remarkable thing about the *Odyssey* is that it is of such early date. It could best be understood, one might almost say, as a work standing in the graceful decadence of the Mycenean, rather than at the austere beginning of the classical, phase of Hellenic culture.[17]

To sum up in perhaps overly formulaic terms: In the *Iliad*, in Sappho, in Pindar, and in most of Greek tragedy (though this is not true of, for example, the *Philoctetes*), landscape, as a distinct element, does not play a part; and natural description is used, sparingly and briefly, as a direct metaphor for things human. Nineteenth-century English nature poetry presents an idealized *nature* (and calls it that) with which the poet would like to identify himself, but cannot. Greek pastoral presents an idyllic nature in which the poet can move by proxy, and with which he is content. Plato, though never a pastoral writer in the strict sense, seems to have been its originator. The *Odyssey*, the first romantic comedy, presents a nature at once beautiful and savage through which the hero passes in order to return, at last, to human society.

All of these genres are largely determined by man's sense of the adequacy and vigour of human society. In those employing what was defined as the second mode of natural metaphor, society is never fully a fitting framework for poetic expression; and all these are characterized by the creation of a metaphorical dimension of nature. They are all represented by poetic masterpieces; but we read none of these with that final seriousness which the *Iliad* and the *Divine Comedy*, Sophocles and Shakespeare, demand of us.

[17] Wolfgang Schadewaldt, in *Aus Homers Welt und Werk* (Leipzig, 1944), brings the emphasis on ships and travel in both the *Odyssey* and the *Iliad* into relation with the era of colonization, during the beginnings of which they must have been written down. One might compare for this the influence of the early colonization of America on Shakespeare's *Tempest* (cf. Kittredge's introduction to that play). None the less, apart from the actuality of the figure of Odysseus, the dominant impression of the *Odyssey* (I believe) is of the fantastic and nostalgic, rather than of the fresh confidence of a commercial renaissance.

3

Thucydides, I. II. 2

περιουσίαν δὲ εἰ ἦλθον ἔχοντες τροφῆς καὶ ὄντες ἀθρόοι ἄνευ λῃστείας καὶ γεωργίας ξυνεχῶς τὸν πόλεμον διέφερον, ῥᾳδίως ἂν μάχῃ κρατοῦντες εἷλον, οἵ γε καὶ οὐχ ἀθρόοι, ἀλλὰ μέρει τῷ αἰεὶ παρόντι ἀντεῖχον, πολιορκίᾳ δ᾽ ἂν προσκαθεζόμενοι ἐν ἐλάσσονί τε χρόνῳ καὶ ἀπονώτερον τὴν Τροίαν εἷλον.

So the manuscripts and the Oxford text. The sentence has not yet been satisfactorily explained. None the less, Jones–Powell in the Oxford text, Miss Jacqueline de Romilly in the Budé, and Luschnat in the new Teubner retain the text as it stands, whether because they thought they understood it, or because the corrections proposed seemed to them no improvement. Luschnat in his apparatus remarks of the first εἷλον 'vix sanum', and notes Krüger's deletion of the word, as does Jones. Gomme, in his *Historical Commentary*, remarks, 'The repetition of εἷλον in §2 seems to me intolerable'. Then he adds: 'But we cannot just delete it after κρατοῦντες with Krüger . . .' He does not tell us why not.

The repetition of εἷλον is in fact intolerable, and there is no compelling reason to believe that Thucydides is responsible for it, particularly since its intrusion can be easily explained. Delachaux (*Notes critiques sur Thucydide* [Neuchâtel, 1925]) correctly says, 'La suppression de εἷλον . . . s'impose: la répétition fautive de ce mot à cette place . . . me paraît avoir été provoquée par la particule ἄν, à laquelle un lecteur peu attentif a cru nécessaire de joindre un verbe, dans l'idée que le second ἄν, au lieu d'être la simple répétition du premier, annonçait un nouveau verbe (le δέ qui suit πολιορκίᾳ et précède le second ἄν facilitait cette erreur).' The second δέ, in fact, more than the first ἄν, is the difficulty of the sentence, and is mainly respon-

American Journal of Philology 79, 3 (1958), 283–5. Reprinted by permission of the Johns Hopkins University Press.

sible for the intrusive first εἷλον: coming after the parenthetical clause οἵ γε ... ἀντεῖχον, it must have seemed to the scribe, as it has to modern editors, the introduction of an independent clause, in which case ῥᾳδίως ἂν μάχῃ κρατοῦντες would need a predicate.

Gomme, op. cit., p. 115, gives the necessary historical information. In an overseas expedition during the Peloponnesian War, the invading force had first to secure a mastery in the field; then, | if they could maintain this mastery, they proceeded to invest the town. It is clear that Thucydides regards the expedition against Troy as analogous to those he later describes in the *History*. This immediately disposes of any interpretation which sees in μάχῃ κρατοῦντες and πολιορκίᾳ προσκαθεζόμενοι two alternative methods of taking the city (Classen [1st ed.], Forbes). Moreover, it is not to Thucydides' purpose here to weigh different procedures by which the city might have been taken. His point is that the Greeks would easily have taken it in the proper efficient manner, had they come with an abundance of supplies. The single proper manner is what the two participial phrases together express.

This is clearly the sense we want; and to get it we have only to read the text aright, with the deletion of the first εἷλον. The hindrance has been the misunderstanding of the second δέ. Krüger wished to delete it along with the first εἷλον. This destroys the logical articulation of the sentence. Steup brackets it, and the words μάχῃ κρατοῦντες εἷλον as well. If we were to accept this, we should have the explanatory relative clause οἵ γε ... preceding the participle which it explains; which is awkward and un-Thucydidean. Moreover the sense would be incomplete. Stahl translates the δέ by *contra* and remarks: 'δέ autem ... in apodosi positum eam opponi superiori sententiae οἱ Τρῶες ... τὰ δέκα ἔτη ἀντεῖχον indicat'. This gives δέ no syntactical function within its own sentence. Delachaux, with Croiset, quoting 7. 33. 2 as a parallel, interprets the δέ as merely resumptive: '... it would have been easy, I say ...'

The δέ in fact is needed and has a precise syntactical function. It is the δέ which complicates or divides a subordinate clause or phrase, in particular the protasis of a conditional sentence. There is a parallel in the second sentence of the preceding chapter: Λακεδαιμονίων γὰρ εἰ ἡ πόλις ἐρημωθείη, λειφθείη δὲ τά τε ἱερὰ καὶ ... τὰ ἐδάφη ... If the combination of these things were to take place, then ... In 11. 2, there is a first protasis (εἰ ... ἦλθον), to which the apodosis is ῥᾳδίως ... εἷλον. This apodosis is further qualified by the two participles κρατοῦντες and προσκαθεζόμενοι, which are made by the second δέ into the complementary parts of a single action, an action which is both the result of the εἰ ἦλθον clause and a condition of the ῥᾳδίως εἷλον clause. We can then translate:

Whereas if they had come with | an abundance of supplies and all together without piracy and farming had prosecuted the war continuously, they would easily, by maintaining mastery in the field—obviously, since they held out with only whatever portion was on hand at the moment—, and by sustaining a siege, have taken Troy with less time and less trouble.

What makes the sentence somewhat hard to read in the Greek is actually the second ἄν, and it would be far better to delete it than the second δέ. But such a sentence as 2. 41. 1 shows how easily this particle can be repeated, and it is safer to leave it in the text.

Stahl translated πολιορκίᾳ προσκαθεζόμενοι as 'obsidioni ... assidue incumbentes'; and Steup says that this translation is proved wrong by the lack of the article before πολιορκίᾳ. Stahl's translation derives from an error of emphasis, and what further proves it wrong is the order of the Greek words. The emphatic points of the qualifying phrase are μάχῃ and πολιορκίᾳ; by battle and siege they would have taken the city.

The use of δέ we find in this sentence, though common enough, is not given a place in Denniston's *Greek Particles*. He divides connecting δέ into *continuative* and *adversative* uses, and says of the former that δέ 'is the normal equivalent of "and" at the beginning of a sentence'. From this, and from the examples he gives, it would seem that δέ connective always introduces a complete new thought. The use of δέ in Thucydides 1. 10. 2 and 1. 11. 2 is to continue and complicate a thought already begun, and which has not yet been completed.

4

What can we Do to Homer?

WHEN Matthew Arnold delivered his Oxford lectures *On Translating Homer* in 1861, he spoke as one who had settled the matter. With knowledge, taste, and system, he picked out the four qualities which Homer 'eminently' possesses: simplicity of thought, simplicity of expression, swiftness and nobility; and he found four illustrations of four modes of translation, each 'eminently' lacking one of these indispensable qualities. Chapman is complicated in thought, Pope complicated in expression, Cowper's Miltonic verse is slow, and the prize of ignobility was reserved for a contemporary named Newman, whom neither this honour nor his own indignant repudiation of it sufficed to save from oblivion. Newman, we note, was impressed by those who saw Homer as a folk-poet, and tried to render him in ballads, thereby becoming the forerunner of the poetical parts of the latest version of Homer, Mr Robert Graves's *Anger of Achilles*.

But Arnold's lectures did not settle the matter. There have been of course innumerable translations of the *Iliad* and *Odyssey* since 1861; but more than this, there have been modes of translation such as Arnold never conceived. A glance at some of these may help to put Mr Graves's curious offering in a clearer light.

The first radically new type of Homeric translation after Arnold was undoubtedly the Lang–Leaf–Myers *Iliad* and Butcher and Lang *Odyssey* of 1882. These versions are well-known to generations that grew up before the advent of the Penguin Book. Enshrined in the Modern Library edition, they were Homer pre-eminently, if not eminently, for decades. But when they appeared, they were, in their way, a fresh approach. They were the first serious version of Homer in English prose.

The Fat Abbot (Autumn 1960), 52–9. Review of Robert Graves, *The Anger of Achilles*, with illustrations by Ronald Searle (Doubleday, 1959).

It did not occur to Keats reading Chapman, any more than it had to Chapman, that Greek poetry in the high style could be properly rendered by prose in a modern language. Prose versions of course there were, in Latin and in the vernacular; but they were mere aids to the schoolboy, like Clark's interlinear *Iliad*—cribs, trots. And so strong was this sense that poetry is, | after all, poetry, that it was not till the 80s that the historicistic, naturalistic, prosaic nineteenth century brought out a version that tried to present not the poetry, but the 'historical truth' of Homer. For that is what the authors explicitly claimed to do. Others, they say in the preface to the *Odyssey*, have translated the spirit of Homer, we shall translate the facts. Without quarrelling over Andrew Lang's (of the four translators, he was the innovator and guide) definition of historical truth, we can observe that this was the first attempt to reduce Homer in translating him, to bring him down to something simpler, more pedestrian, more like us.

But the language in which Homer was turned to prose was far from ours. In creating a language for Homer, Lang created one that became standard for all translations from Greek, the famous 'Wardour Street' English, beloved by the Victorians: to the generation of Pound and Eliot, an object of loathing and contempt. Lang was following a precept of Matthew Arnold himself in making his archaizing style. Arnold had said that Malory and particularly the King James Bible were proper sources of good English poetic words. Archaizing works themselves, they were the sources the archaizers plundered indiscriminately for their prose Homer. Was it that they felt guilty over making Homer a prose writer? In any case, the poetry they took out of form and movement, they tried to put back in diction. The rules were simple: never use an ordinary word where a poetic or at least an obsolete word can be found (use *wain* for *wagon*, and *hest* for *command*), and always try to have one English word to match one Greek word. 'But when, or smitten with the spear or wounded with arrow shot, Agamemnon leapeth into his chariot, then will I give Hector strength to slay till he come even to the well-timbered ships . . .' As Professor Baird of Amherst has said, when the Bible became literature, literature became the Bible.

This style is now something for every modern translator to jeer at; yet it has more virtues than most of theirs. The Victorian translators made Homer stiff and self-consciously poetic, but they chose their words well, for they knew English poetry well. The Lang–Leaf–Myers *Iliad* has a genuine dignity, and the word-for-word, phrase-by-phrase translation does convey something of Homer no other translation has suc-ceeded in conveying: once we become accustomed to the archaistic patterns of speech (which make an initial impression quite unlike Homer), we can feel some of that formulary, at times almost ritual, quality which is so strong a feature of the original. The reader of Lang & Co., especially if he makes their acquaintance at an early age, will have phrases to remember. 'Then pale fear gat hold of him'—a little phoney by itself, but by dint of repetition, it assumes a meaning, it stands for the many scenes where it occurs, scenes where a hero confronts his death. It and a hundred like phrases have strength and are good to remember. No reader will find any words to remember at all in the smoothed-over versions in the Penguin series and now of Robert Graves.

Lang's basic mistake, of course, was in misunderstanding the nature of poetic language. He assumed that there there were just two kinds of literary | language, one consisting of spoken words, and one of poetic and artificial words. Homer's lan-guage was never spoken (which is certainly true), therefore he could be rendered by an English that was never spoken. But there are different ways in which the language of poetry can be living or dead. When Macduff says,

> Either thou, Macbeth,
> Or else my sword with an unbattered edge
> I sheathe again undeeded

he speaks words that never fell in colloquial speech. But they are not a dead language either. They have the ring of stress and battle in them. To the imagination of the audience, which is what counts here, they are not false and archaizing, but immediate, intense, dramatically more real than any naturalis-tic speech, because of the intensity of Shakespeare's imagina-tive grasp of the situation. The language of Shakespeare's

heroes is heightened, transcends the common speech to which
we (or a contemporary of his) are accustomed, and this strikes
us as right, because his characters show emotions of greater
force and clarity than those to which we are accustomed.
Given the dramatic assumptions which Shakespeare's language
itself imposes, his 'poetic' diction is realism.

The same is true of Homer's Greek. It is at once grand,
ornate, and immediately effective. When a Homeric hero
speaks in the words Homer created for him, we believe that he
is speaking, that these powerful words are torn from him by
the crisis in which he now finds himself. Neither realism, in the
nineteenth-century naturalistic sense, nor artificiality, in the
sense of the language of *The Idylls of the King*, have anything to
do with this. Both concepts are equally distortions of Homeric
speech, as much as they are of Shakespeare's. It was because he
saw only the artificiality that Myers could write: 'So perish all,
until we reach the citadel of sacred Ilios, ye flying and I behind
destroying. Nor ever the River, fair-flowing, silver-eddied,
shall avail you, to whom long time forsooth ye sacrifice many
bulls . . .' The intolerable *forsooth* only sums up the faults of this
speech. This is a costumed hero on a stage trying to sound the
way he imagines a Homeric warrior-prince would sound, no
killer on a battlefield.

The archaizing style met a sad end. Ezra Pound railed against
it in 'Some Translations from the Greek' in *Make it New*, T. S.
Eliot wrote that Gilbert Murray was almost single-handedly
responsible for the decline of the Classics. Fitts in his *Greek
Plays in Modern Translation* devoted almost the whole of his
introduction to holding it up to ridicule. The Victorian mode
of translation indeed seemed to bear the brunt of the whole
twentieth-century reaction against traditional English poetic
diction. And reacting against 'artificiality', translators turned
to 'realism'.

The model of the new style is Rieu's *Odyssey* of 1946. The
idea was not exactly his: such prose translators into commonplace idiom as T. E. Shaw (Lawrence) and Samuel Palmer had
anticipated him. But Rieu | was the first to demonstrate
beyond a doubt that Homer was really Anthony Trollope. As
the editor and master spirit of the whole series of Penguin

translations of Greek and Latin authors, Rieu can stand as the Lang of the new style. The simple principle of the new style is that Homer really talked like us. There is of course a problem in deciding who we are. But even if we neglect this problem, we may find the assumption a little foolish. It is in fact the same notion as that of the college student who said he liked *King Lear*, but wouldn't it be better translated into English? The world of Homer is in fact vastly different from ours, different from the worlds of all of us, and to pretend that Homer talked as we do leads to translation as unreal as to pretend that he spoke—or composed—like the Jacobean translators of the Bible.

Here is a random example. Odysseus at last stands before the suitors:

The unconquerable Odysseus looked down at them with a scowl. 'You curs!' he cried. 'You never thought to see me back from Troy. So you ate me out of house and home; you raped my maids: you wooed my wife on the sly though I was alive . . .'

Can Odysseus look at anyone 'with a scowl'? Can he say, in those mincing tones, 'you curs!'? Or is our sense of the appropriate really more offended by Butcher & Lang:

Then Odysseus of many counsels looked fiercely on them, and spake: 'Ye dogs, ye said in your hearts that I should never more come home from the land of the Trojans, in that ye wasted my house, and lay with the maidservants by force, and traitorously wooed my wife while I was yet alive . . .'?

It is worth noting that the Victorian translators have the virtue of accuracy. The ritual epithet πολύμητις, '*of many counsels*', is simply left out by Rieu, who frankly considers such epithets otiose unless they can be interpreted as jokes; and the suitors did not woo Penelope 'on the sly': they were quite open about it; and 'raped my maids' conjures up an image not quite fair to the suitors: what they did was make it hard for all but the most loyal of Odysseus' female servants to refuse them. But it is not on the point of accuracy that the Penguin translation should be judged. Rieu knows Greek well, if not so well as the Victorian scholars, and is clever at getting the meaning. The real point is whether Homeric heroes should be made to speak in the

bantering, slightly ironic tones of the Common Room. 'Telemachus now showed his good judgment. "Eurymachus," he
said, "and the rest of you who pay my mother your distinguished attentions ..."' The donnish sarcasm of 'pay ...
your distinguished attentions' is as untrue of the mood of
Telemachus as it is of the world of Homer. There is more
realism, or at least more reality, in the formal tones of
' "Eurymachus, and ye others, that are lordly wooers ..." '
But perhaps the pity is that we should have to make the choice.

Mr Graves's *Iliad* was hailed by *Time* magazine and others as
a work that at last freed Homer from the bondage of the
professors. The praise was misplaced. Graves in fact likes the
professors, and the faults of his work are quite professorial. His
prose is right in the Rieu–Penguin | donnish tradition. He cuts
down and simplifies, like all the modern professor-translators,
only a little more so, and his prose has exactly the same flat,
monotonous tones, the same dear, faint ironies of the English
schoolmaster's construe of the Homeric line. In the same way,
he makes Homer produce a language, that is not only unmemorable, but not really believable. Menelaus comes to Antilochus to announce the death of Patroclus:

Running towards him, he gasped: 'Prince, I bring bad news! The
Trojans are being assisted by Zeus, and I deeply regret to announce
that our greatest champion has been killed—Patroclus, son of
Menoetius—doubtless you guessed what had happened when they
counter-attacked? We are all heart-broken. Pray inform Achilles at
the camp!'

Deeply Regretted Announcements We Doubt Ever Got
Gasped On The Trojan Plain. It is hard to say quite who is
speaking these words, but it is clear that it is not any man
fighting with sword and spear for honour and life. The words
have no weight, no bite, no urgency to them. They do not
need to be brought up to date (do they speak like this even in
the Common Room now?) any more than they need to be put
into a picturesque past of *thees* and *tidings*. They need to be
something like real poetry.

Priam, on his desperate mission to Achilles' hut to redeem
the body of his son, meets Hermes, who, disguised as a

follower of Achilles, has come to guide him across the darkling
plain. When Priam offers him a gold cup for his services, ' "I
am your junior by two generations, my lord," Hermes
answered bashfully . . .' Hermes actually begins only by saying
that he is younger. The 'junior by two generations' is an
importation into the text, and so is the 'bashfully'. Graves
wants to make the scene cute, rather than sinister, and that is
just enough to make it lose its effect. The elegant young god
Hermes has his dark and chthonic associations: he is the god
who guides the souls of the dead to the world below, and
much of the impressiveness of the scene here comes from the
sense that Priam's visit is in a way like a journey to death. The
drawing-room tones and novelistic adverbs Graves puts in to
make the scene alive are such rather as to kill it.

What makes a man who is a good writer on his own
account, who loves Greece and Greek poetry, and who
moreover knows what war is, try to turn the *Iliad* into
something between an engraved calling card and a Redbook
novelette? The fault is theoretical, and not so different from
that of the Victorian translators. Graves does not believe in the
action of the *Iliad*. He does not believe, that is, that Homer
conceived his characters and what happened to them seriously.
Lang dresses his heroes up like a 'period' drama; Graves dresses
them down, like a performance of *Julius Caesar* in business
suits; neither is willing to take them seriously for what they
are.

Graves defends his theory in an odd introduction. The
stories of Homer, he argues, were originally told in prose. The
notion is so gratuitous and so contrary to reasonable scholarly
hypothesis that we must suspect that he is trying to justify his
own inability to use the poetic medium that Homer obviously
demands. Homer himself, we | learn, was a satirist. He was
cynical about the gods, cynical about the heroes. 'When I
"did" *Book 23* at my public school', he complains, 'the ancient
classroom curse forbade me to catch any of the concealed
comedy in the account of Patroclus' funeral games . . .' Graves
feels that there is concealed comedy everywhere in the *Iliad*.
To him this is what gives the *Iliad* its life, makes it like *Don
Quixote* and Shakespeare's later plays (how late?), 'tragedy

salted with humour', and unlike the *Aeneid*, the *Inferno*, and *Paradise Lost*, 'literary works of almost superhuman eloquence . . . seldom read except as a solemn intellectual task', Graves is like the schoolmaster who desperately tries to get his students' interest by insisting that a serious poetic work is 'really funny' when you go at it in the right way.

Now the *Iliad* is in fact not solemn in the way that *Paradise Lost* is. It has what Milton lacks, abundant humanity. And where there is humanity, there is a kind of humour. Nestor is garrulous, Ajax is simple, Agamemnon is tactless and self-pitying. And the gods, entirely awesome at some moments, are petty and almost grotesque at others. But the *Iliad* is in no sense satire. Its basic mood, the mood we have to get before we can take delight in personal eccentricities, is of uncompromising reality, of passions men pay for with their happiness and their lives, of glory and suffering and of *terribilità*. The satirical mode is quite foreign to it, because the fundamental note of satire is distortion, and satiric distortion is something we find in Euripides' *Cyclops* or in Aristophanes, not in the *Iliad*.

We can here draw a useful distinction between types of interpretation. There is critical interpretation and creative interpretation. Critical interpretation is bound by the intrinsic qualities of what it interprets. It is right or wrong according to whether it seizes and expresses the actual features of its objects. Creative interpretation uses its objects as a painter uses a landscape or a bowl of fruit: to make something new of it. It is not right or wrong, but good or bad art. Now it is perfectly possible to interpret the *Iliad* creatively as satire. The result, of course, will not be the *Iliad*, but something that takes off from the *Iliad*, and will probably be less like the *Iliad* the better it is. There is a great English literary work which interprets the *Iliad* as satire: Shakespeare's *Troilus and Cressida*. That Shakespeare's sources were not so much the *Iliad* as mediaeval and late ancient works ultimately deriving from Homeric tradition is not important. His play shows us what the *Iliad* is like in an Elizabethan, satiric version. The 'large Achilles' is an excellent creation of the imagination, and knowing the Homeric Achilles only enables us to appreciate him the more.

Robert Graves at his best is a good creative writer. Judged as a critical interpretation, his Introduction is nothing more than ignorant and pretentious. But what it could have been is a programme for a good satirical treatment of the story of the *Iliad*. There is a nice demonstration of this in the present edition. The illustrator, Ronald Searle, was clearly not bothered by the text either of Homer or even of Graves's translation, but | he did read the Introduction with some care. Those passages that Graves singles out as particularly juicy instances of 'concealed comedy' Searle illustrates as delightful unconcealed comedy. They are artistically the best things in the edition. As illustrations of Homer they are, not to put too fine a point on it, in dreadful taste; as illustrations of a satyr-play based on the *Iliad*, they are excellent.

Why was this satyr-play never written? Because Graves is too much of a professor to be an uninhibited artist. The very instances of comedy that he picks out for mention can be no more recognized from his translation than they ever were in 2,500 years before that translation was written. One good example will suffice. Machaon comes wounded to Nestor's tent with an arrow in his shoulder. Homer does not mention that Nestor took the arrow from Machaon's shoulder, but he does have Nestor indulge in a long reminiscing speech to the wounded man. Now no reader of the Greek or of Graves's translation would conclude from this that Nestor made 'no attempt to remove the arrow still protruding from Machaon's shoulder, though after fifty years of warfare he can hardly have avoided picking up a little simple surgery', as it is put in the Introduction. It is obvious that Homer, not a naturalistic artist, simply did not bother to narrate this detail, being more interested in Nestor's speech. Graves wants this to be broad satire, but he is too bound by the text to render it as such. But Searle is not, and the picture of Nestor, sunk in a pseudo-Homeric *fauteuil*, looking off into space and saying 'Some years previously, my father . . .' while Machaon sits in grim boredom with an arrow sticking out of his shoulder is very funny indeed.

Graves's bondage to the text comes out in other ways as well. He is aware, as Mr Rieu apparently is not, that Homer is genuine poetry, although he seems to think that the poetry comes off and on. His way of dealing with this insight is unique. When he comes to a moment he considers poetic, he suddenly breaks into song, or rather into doggerel. The effect of this is to make a curious sort of musical comedy. The quality of these 'lyrics' is surprising indeed from a man who is no mean poet in his own right. They tend to come when Homer introduces a simile, so that we usually have the additional discomfort of somehow having to slide back into prose:

> The mules put out enormous strength
> 　When down the mountainside,
> Some massive tree-trunk, or huge length
> 　Of timber, they make slide.
> And though their bodies gleam with sweat,
> 　Their spirit is not broken yet.

No mules ever worked so hard as did Menelaus and Meriones in carrying out that corpse!

Wow! |

　As one might have guessed, Graves is at his most pedantic when he touches on mythology. One of Homer's outstanding qualities is his sense of timing, his ability to end a scene or description at the right moment, and on the right mysterious, poignant note. A consequence of this instinctive art is that a good many mythological details are left out. A good example is the end of the τειχοσκοπία, the fine scene where Helen stands on the walls of Troy and names the great heroes of the Greek army to Priam standing beside her. The scene lasts about a hundred lines, and then Homer ends it swiftly and gracefully. Helen says that she cannot see her two brothers in the host, and wonders if they did not come on the expedition, or, if they came, whether they have withdrawn from the battle out of shame for her. But she did not know that they were already dead. Homer tells us this in two lines, and then abruptly shifts the scene. 'So she spoke; but the fruitful earth already possessed them, back in Lacedaemon, in their own native land.' So runs a literal translation. But Graves knows all about it, and will tell us what Homer does not: 'The truth was that both her brothers were long since dead and buried in their native country, after they had ambushed a pair of rival twins from Messene: Idas

and Lynceus. None of the four survived that fight.' Thank
you, Mr Graves! A piece of poetry has been neatly turned into
a page from a Classical handbook.

Robert Graves's new translation falls very much between
two stools, or Homeric *fauteuils*. He lacks the imagination and
the knowledge and the poetic gift to give us very much of
Homer as he is; and he is too faithful a student of the Greek text
to give us a new Homer of his own making. He should have
written another novel.

It is of course easier to see what a good translation of Homer
is not, than to see what it is. Certain assumptions will make
such a translation impossible to begin with. The notion that
Homer is 'artificial' is one of these. The notion that he speaks
like us, or like our schoolmasters, or like our maiden aunts, is
certainly another. The notion that Homer took his story as a
joke, and wrote like a drawing-room comedy at one moment,
and like a commercial jingle-writer the next, with lessons of
mythology added where they are needed, is sheer ruin. The
right translator needs first to regard the work as something like
what it is: a serious and complex work of dramatic poetry,
composed in a language that is not conversational English, but
is poetic diction that fits so tightly and so naturally the
imaginative grasp of the poet that it will seem the only possible
language that could have been used to express these gestures,
these emotions, these poetic facts. Only a great poet could
translate Homer properly, and such a poet, if there were one
now, would be writing his own poetry. But a lesser poet, by
facing the work squarely and not trying to make it into
something smaller than it is, can have some success. Richmond
Lattimore's *Iliad*, which Graves refers to as a good crib, does
more to save Homer from the professors, or from ignorance
and incompetence, than either the archaizing or the moderniz-
ing prose translators.

5

Shakespeare and History

THERE is an *éminence grise* behind much of the theoretical portion of this interesting book who receives no mention: that influential, provocative, slightly mad, and now so unfashionable speculator in history, Oswald Spengler. He told us in the 20s, as Mr Driver tells us now, that modern thought (Mr Driver says Christian thought) is deeply historical, and moves in a straight line, and is dominated by a sense of the boundless; whereas ancient thought (Mr Driver confines it to Greek thought) is obsessed by limits, moves in an Aristotelian circle, and is essentially unhistorical. From Spengler we learned, moreover, that the ancients had no sense of the supreme importance of the irreducible individual event, or of the psychology of particular men. They had instead a sense of types, of principles, and of binding measure and proportion. Fate for the ancient Greek was a timeless, spatial entity imposed on man (who could only suffer it) from without; for the Faustian Soul (an irreligious coun|terpart to Mr Driver's Christian Mind) we have an active fate or Providence, expressing itself in history and through the free will of man and the particular events, occurring at particular times, that man shapes. The contrast of Shakespeare with the Classical Greek dramatists is one of Spengler's favourite illustrations of his thesis. Shakespeare is 'inward' (*innerlich*); they are 'external'. His characters have psychology, theirs are types. Shakespeare writes 'character drama', the Greeks write 'situation drama'.

The following of Spengler is a good example:

What fullness of the *characteristic* is already to be found in Nordic mythology with its sly dwarfs, doltish giants, mischievous elves, with Loki, Baldur and the other figures, and how *typically* in contrast

Yale Review (Summer 1961), 603–10. Review of Tom F. Driver, *The Sense of History in Greek and Shakespearean Drama* (Columbia University Press). Reprinted by permission of Yale University Press.

does Homer's Olympus affect us! Zeus, Apollo, Poseidon, Ares, are simply 'men'; Hermes is 'the youth', Athene a more mature Aphrodite [!], the lesser gods, as later sculpture demonstrates, distinguishable only by their names. And all this is fully as true of the figures of the Attic stage. In Wolfram von Eschenbach, Cervantes, Shakespeare, Goethe, the tragedy [*das Tragische*] of the individual life develops from within outwards, dynamic, functional, and the course of human lives is fully comprehensible only against the historical backgrounds of their centuries; in the three great tragedians of Athens, tragedy comes from without, is static, Euclidian [= spatial].

The reference to Nordic mythology reminds us of Spengler's own roots; they go back to late eighteenth-century and early nineteenth-century German romanticism, the movement of ideas popularized by Mme de Staël: from the northern meres the strange figures of Germanic and Mediaeval myth reappear after their long oblivion, challenging and overcoming the Classical gods and heroes, and with them a new kind of poetry, deep, inward, expressing 'Geist', to be set against the cold formalism of the Classical and Neo-classical. What Spengler did was to expand these ideas of German romanticism into an historical framework; historical first, in that he identified the Germanic and Mediaeval spirit with the whole European civilization from AD 1000 to the present, and Classicism with ancient civilization both Greek and Roman; and second, in that he saw a 'sense of history' as one of the characteristic features of our 'Faustian' civilization, and a complete lack of it ('the utterly unhistorical sense of Thucydides') as inseparable from the closed, spatial vision of ancient man. Mr Tom Driver, writing from the halls of the Union Theological Seminary, gives another twist to romantic dogma: it is Judaeo-Christian, and perhaps Roman, *Weltanschauung* that is historical, particular, psychological, inward; whereas Greek vision—and dramatic art—is closed, timeless, typical, external, etc. Wherever these eras, or complexes of thought, are historically | placed, however, the comparison of Shakespeare with the Attic tragedians remains roughly a constant.

If Oswald Spengler does not appear in *The Sense of History* . . ., that is because Mr Driver either is not acquainted with him, or has forgotten him. Spengler created a small tradition, and has largely been absorbed in it. His theories, in various transformations, theological and otherwise, are a common stock now in writers and lecturers on the History of Ideas, and Mr Driver could find all of them in the sources he quotes: Oscar Cullman, Reinhold Niebuhr, Thorlief Boman, Mircea Eliade, Paul Tillich, even W. H. Auden (for Classical = external, situation drama; Shakespearean = individual, introspective drama). But it is time that the historical origin of these concepts be made somewhat clear.

They are German–Romantic myth. And how far a writer like Mr Driver is from an impartial survey of the material, how much his distinctions derive from preconceptions, can be seen if we compare his second chapter, on Hellenic Historical Consciousness, with his third, on Judaeo-Christian Historical Consciousness. In the third chapter, he works largely from the text of the Bible; he makes points and confirms them by quoting Scripture—though even here he accepts without reservations Erich Auerbach's hopelessly exaggerated contrast between the Scar of Odysseus episode in Homer and the Sacrifice of Isaac in Genesis: anyone who says, or follows anyone who says, '[Abraham's] soul is torn between desperate rebellion and hopeful expectation', has forgotten the words of the Bible in favour of an *a priori* notion of what the Bible is like. But in the chapter on Greek historical sense, the quotations are almost entirely from secondary sources. Collingwood, Jaeger, Tillich, and others assure Mr Driver that the Greeks had no sense of history. From Thucydides there is quoted just the one sentence in his entire work which might (though not accurately) be taken as evidence of a cyclical theory of history; from Herodotus, nothing. Instead we get a mishmash of *Scheinwissenschaft*, including the statements that Thucydides 'had no successors', whereas in fact his work was directly continued by Xenophon, Ephorus, and Theopompus; that Herodotus, 'describing . . . phenomena from the outside', was not historical because he was *scientific*; that Thucydides, who in fact saw the Peloponnesian War as the convergence of all Greek history, was only interested in 'illustrations of . . . political wisdom', and that present time to the Greeks 'cannot be, in the final count, of any fundamental consequence', although, at the same time, in Homer (here he quotes C. S. Lewis), 'nothing has a significance beyond the moment'.

Enough has perhaps been said to make the point that the theoretical structure of Mr Driver's book is both tralatitious and unsound. All the | more credit is due him, therefore, as a careful and sensitive reader of poetry, for the excellence of parts of the four succeeding chapters (v–viii), in each of which he makes an extended comparison of one Attic and one Shakespearean play. It is true that regarding all these plays solely from the viewpoint of 'time', and constantly overplaying Shakespearean and underplaying Attic dramatic uses of time, produces some distortion. None the less, several of these essays, especially as they deal with Shakespeare, are very fine; and even where we do not accept their conclusions, they reveal new aspects of Shakespearean and fifth-century tragic art.

The great drawback of these essays, which the reader must force himself to overlook as he reads them, is the pervasive bias in favour of Shakespeare. Shakespeare (p. 199) cannot 'be made to wear the flowing robe of Athens', the implication being that Shakespeare has something much more real and human to offer. The use of time in the *Oedipus Tyrannus* is blamed as 'schematic' and 'structural', the latter word (see p. 159) strangely acquiring a pejorative tone. 'The result [of the balance of expression in *Alcestis*] is brilliant theatre; but however much we may enjoy it technically or because of its emotion [?], it can never represent to us what the tragi-comedies of Shakespeare do.' The Attic dramàtists (p. 201) were led into 'questions about man as a historical creature which the culture could not help them to formulate'. Aeschylus, Sophocles, and Euripides could only anticipate their own fulfilment in Shakespeare.

This is what Northrop Frye in the first chapter of his *Anatomy of Criticism* calls Selective Criticism, and its weakness, as Frye brilliantly argues, is that it is not properly literary criticism at all, because it derives its standards from something outside the body of literature. In this case, Mr Driver likes Shakespeare because he sees him as a Christian poet concerned with man's relation to Providence and Judgement. It is true that Shakespeare makes some use, particularly in a play like *Richard III*, of something like a Christian time-scheme; but that in itself makes him neither a better nor a worse poet than Sophocles.

To put the matter in another way, a reader, critical or otherwise, may feel that in some ultimate sense Shakespeare is a yet greater dramatic poet than any of the great Attic tragedians. The present reviewer, in point of fact, shares that feeling. But once this has been said, there is nothing more to say. A preference, or a fuller sense of admiration or satisfaction, cannot be made the basis of a literary theory.

Of the individual essays, the best is that on *Richard III*, and undoubtedly because it is more amenable to Mr Driver's theoretical structure than any other. Shakespeare is intensely concerned with *time* in this play, and a providential and judgemental view of history, as Mr Driver ably points | out, is one of the meanings that the time references impose on us. Richard is 'untimely' in several ways: 'Deform'd, unfinish'd, sent before my time'; and Mr Driver stresses the significance of the sinister repartee in II. v. between the Duchess and the young Duke of York:

> Marry, they say my uncle grew so fast
> That he could gnaw a crust at two hours old.

Richard is untimely in that he interrupts the providential flow of history that eventually puts the Tudors on the English throne. At the same time, his very interruption makes the triumph of Henry VII possible. Speedy as he tries to make his work (improperly hastening his coronation and being swift to murder his rivals), time and destiny catch up with him, and he is left behind, damned by his own misuse of time—as Bosworth Field, reminiscent, as Mr Driver shows, of Armageddon, brings history up to date. This fine essay brings out the full meaning of crucial passages such as that in which Elizabeth replies to Richard's oath by 'time to come':

> That thou hast wrongèd in the time o'erpast;
> For I myself have many tears to wash
> Hereafter time, for time past wronged by thee.
>
> .　　.　　.　　.　　.
>
> Swear not by time to come; for that thou hast
> Misused ere used, by times ill-used o'erpast.

Turning to the *Persians* for comparison, Mr Driver, reasonably enough, does not find any like treatment of time. Like the other Greek plays, the *Persians* is treated in negative fashion as failing to exhibit the temporal features of Shakespearean drama. It is less reasonable for Mr Driver to be surprised by the absence of a long perspective of history in the *Persians*. Aeschylus himself, he knows, fought at Marathon, and probably at Salamis and Plataea: why is he then interested in the eternal rather than the historical aspects of the conflict? Surely he has himself given the answer. Aeschylus knew the single event from his own experience, and wants to make drama of its moral and pathetic import. Shakespeare draws on what is already history, in Holinshed, and naturally emphasizes the dramatic aspects of historical movement. The simple fact that historiography, and Greek prose in general, were only beginning to take shape in the lifetimes of the three Attic tragedians, does not affect Mr Driver's presentation as it should. Herodotus finished his work in the 420s, when Sophocles and Euripides were old men and Aeschylus long since dead, and Thucydides left his unfinished work behind him at his death around 400, when the great age of Greek Tragedy was ended. Shakespeare, on the other hand, was born into a world of chronicles. He had not the | present body of myth available to the Greek dramatists; he had instead history, and he used that.

In dealing with the three other Shakespearean plays he has chosen for his comparisons—*Hamlet* with the *Oresteia*, *Macbeth* with *Oedipus Tyrannus*, and *The Winter's Tale* with *Alcestis*—Mr Driver tries to find in each a Christian-historical pattern resembling the one he established for *Richard III*. *Hamlet* comes out a play of providence, *Macbeth* a deeper play of personal damnation set against a providential background, and *The Winter's Tale* appears a play of repentance and reconciliation through time. In each of these expositions he has variable success. None of them conforms with the ease of *Richard III*.

To take one of them: *Hamlet*, Mr Driver argues, is a play about a man on whom is imposed, and who at last accepts, a providential mission. The idea of the will of heaven is of great importance: Mr Driver leans very heavily for this on the Biblical associations in v. ii. 217: 'There is special providence in the fall of a sparrow'. The great question in the play is Hamlet's *vocation*, a word clearly chosen for its religious connotation. *Vocation* means *acting* in accordance with the will of heaven, and Mr Driver makes this the basis of a distinction between *Hamlet* and the *Oresteia*: in the latter, event leads to knowledge; e.g., the death of Agamemnon, offstage, and interpreted for us by Cassandra, leads to further knowledge of the forces of justice and blood and the will of Zeus. But in *Hamlet*, knowledge leads to event.

Mr Driver gives a pretty example of the way, he says, in which actual event impinges on reflection in Shakespeare. In the first scene of the first act, Barnardo's description of the appearance of the ghost on previous nights is broken off midway by the actual apparition of the ghost. Mr Driver comments: 'The narrative he began, and which in any Greek play he would have been allowed to finish, would have imparted a static quality to the scene. Shakespeare sees to it that rhetorical description is broken off by a new event demanding a new reaction . . . The life of reflection is continually disrupted by the demands of historical existence.'

That is very good observation. He then tries to extend this principle to the Gonzago play at the centre of the work. Here Hamlet acquires that certain knowledge that leads to his final acceptance of his mission and to the act that resolves the whole action of the work, carries out Heaven's will, and establishes right kingship once more, even though not the kingship Hamlet had ever had in mind.

The trouble with this analysis is that the Gonzago play does not lead to any action other than Hamlet's futile scolding of his mother. What Mr Driver seems to miss is the essential detachment of Hamlet. | He can absorb knowledge and reflect on it, but he is always removed from direct action. And this quality of being removed, this transcendence of immediate action despite all knowledge and prompting of the supernatural, is precisely what gives Hamlet his greatness. Like all truly detached persons, he can reflect on his very detachment, blame himself, tell himself at one moment what he must do and wonder at the next if he should do it. A unique dramatic figure, he is somehow in the dramatic action and yet not of it, standing, one might say, almost closer to the mind of the creative artist than any character in a play has a right to. This gives him from the very start that tragic knowledge, that perspective that makes all actions appear both uncertain and unimportant even when they involve him most, which other tragic heroes (like Macbeth) only attain to at the end of their dramatic journey. Hence he can give lessons on the art of acting, playing the recorder, on the ideal man, with a kind of final authority, and at the same time torment himself for not being sufficiently *engagé* to act in a simple and obvious fashion.

> Now whether it be
> Bestial oblivion or some craven scruple
> Of thinking too precisely on th'event—
>
> . . . . .
>
> . . . I do not know
> Why yet I live to say 'This thing's to do' . . .

All this, as he sees the simple Fortinbras, long after the Gonzago play, and after he has declined his chance to kill the king.

Hamlet is the man whose intelligence, or *thought* (with the usual Elizabethan connotation of *melancholy*), goes beyond and misses the *event*. He is tormented by what he feels, with unequalled intensity, to be his inability to feel, and to act directly on his feeling. Hence his fury at Laertes' display of grief over his sister's death in v. i.

What he finally accepts is not so much his mission as his mortality. The speech on providence in the fall of a sparrow, and 'If it be now, 'tis not to come—if it be not to come, it will be now—if it be not now, yet it will come—the readiness is all', must be understood in the light of the preceding graveyard scene. It is more a litany of mortality than a confession of faith in providence. Now the stage is set for the final play, in which he must be an actor, no longer a spectator (cf. the stage direction following: *Attendants enter to set benches and carry in cushions for the spectators*); and what Hamlet rises to is a confrontation of death.

Mr Driver sees the final triumph of Fortinbras as analogous to the triumphs of the Duke of Richmond and of Malcolm. He blames Aeschylus (p. 137) for being uninterested in the future of Orestes or the house of Atreus, implying that Shakespeare is interested in the future of Denmark, | an example of how he drives a valid principle too far. Aeschylus was 'interested in the future of Orestes' to the extent that he saw the political friendship of Argos as deriving, at least metaphorically, from the absolution of Orestes in Athens, and as vital to present Athenian prosperity. That 'myth' was for Aeschylus more historical than the 'history' of Hamlet was for Shakespeare. For the government of Fortinbras is quite unimportant to Shakespeare except as a foil to the solitary resolution of Hamlet. The history of *Hamlet* is a myth.

The essays on the four Shakespearean plays are by far the most valuable portion of *The Sense of History*. They will not command universal assent, and that on *Richard III*, and after it that on *The Winter's Tale*, do more justice to their plays than do those on *Hamlet* and *Macbeth*. But all four are honest, careful, and provocative essays. Of the Greek plays compared, *Alcestis* receives the most interesting treatment. But all four Greek plays suffer from the negative light in which they are considered. Moreover it should be said that the author does not know the Greek plays as well, and that his knowledge of Greek falls short of what he needs, as witness his translation (p. 131) of καιρόν χάριτος (*Agamemnon* 787) as 'grace of this time', when it clearly means 'proper degree of gratitude'. None the less, it is a pleasure to find a Shakespearean critic who can read Greek and show an intelligent interest in Greek works of literature.

6

Bacchylides: An Introduction

THE Greek poet Bacchylides was a contemporary, perhaps slightly younger, of a greater and more famous poet, Pindar, who lived from 518 to 438 BC. Already in late antiquity, in the critical treatise *On the Sublime*, Bacchylides was compared to Pindar as careful mediocrity to occasionally lapsing genius. And this judgement has persisted to our day. Bacchylides has been considered a dull and slight, or, more favourably, a sweet and sometimes charming, practitioner of the kind of poetry which Pindar created with profundity and magnificence. At best, Pindar is a soaring eagle to Bacchylides, 'the nightingale of Ceos', in the poet's own words, too often quoted in this tone of condescension.

Now Pindar is by far the greater poet; and he and Bacchylides do work in the same genre. But they handle the genre differently; they have distinct virtues, and are interested in different things. To blame Bacchylides for not being Pindar is as childish a judgement as to condemn Vermeer for falling short of Rembrandt, or Marvell for missing the grandeur of Milton.

Pindar's movement is deliberate and large. He is a master of poetic phrase at once monumental and supple. He describes the mountainous coast of north-west Greece:

βουβόται τόθι πρῶνες ἔξοχοι κατάκεινται
Δωδώναθεν ἀρχόμενοι πρὸς Ἰόνιον πόρον

'where the pasturing headlands lie high and clear/ from Dodona down on to the Ionian Sea.' Bacchylides' rhythm, even where he evokes a broad and static image, is quick and nervous:

Bacchylides, tr. R. Feigles, Yale University Press (1961), introduction (pp. xv–xxiv). Reprinted by permission of Yale University Press.

οἶά τε φύλλ' ἄνεμος
Ἴδας ἀνὰ μηλοβότους
πρῶνας ἀργηστὰς δονεῖ |

Flitting like leaves
That rustle to breezes
On Ida, grazer of sheep,
And her headlands flashed in sun.

Pindar owes much of his power to an intense and openly dramatized personal commitment to his subject. He reveals himself and imposes his vatic presence with a freedom enhanced by the ceremonial rhythm of choral verse. In the most moving of all his poems, Pythian 8, a work of 100 lines, he uses the first person singular seven times and the second person singular to address a god six times. He creates a goddess to please himself, Hesychia, 'kind goddess of Peace and Harmony', and speaks to her as to a friend. He talks of the 'winged device of my art' that will bear the victor's glory. He quotes a vision by the hero-prophet Amphiaraus of his son Alcmaeon, then recounts, as a simple fact, how Alcmaeon 'guardian of my possessions, met me as I was going by Delphi, and laid hold on prophecy with the arts that were his heritage'. Then with a sudden movement he speaks to Apollo who keeps the 'welcoming' temple of Delphi, and has 'at home' given the victor the sought-for prize: 'O my king, I beg you, may you and I look with one mind on all that I encounter.'[1]

It is clear that Pindar's closeness to the god of the lyre and of prophetic vision goes far beyond any rhetorical stance. This poem is a personal statement of a special role to which he feels he has been elected. The victory is the occasion for him to speak what is closest to his heart: the right and necessary relations of gods and men. These relations can be intimate and beautiful when they are informed, on the side of men, by a deep generosity like his own. Or they can be harsh and terrible. The life of man can be destroyed by force, it can be | loneliness and loss when the gods are not our companions. It is because the whole poem has been about Pindar and Pindar's vision that the famous lines at the end have so peculiarly strong an effect:

---

[1] The reference of ἑκόντι νόῳ and βλέπειν is deliberately unfixed to encompass both himself and the god.

Our life is a day. What is it? What is it not? Man is a shadow, and the dream of a shadow. But when the god-sent brightness comes, glory of shining lies all about us, and our time is sweet.

This is the sort of thing Bacchylides does not do. But it is important to observe that he does not try to do it either. He is an Ionian poet, much closer to Homer than Pindar is, and very close to his own uncle Simonides. He is neither cold nor perfunctory, and these translations and notes will demonstrate that his moral reflections are neither shallow nor extrinsic to the structure of his poems. But he stands back from his work, and prefers to consider himself a craftsman rather than a prophet. When he speaks of himself, it is within the clear limits of convention:

> By the earth I press
> Not once has a rival . . .

or

> . . . resound too the rolling finesse
> Of the nightingale of Ceos.

The relevance of poetic statement and story to the experience of his audience is important to Bacchylides. There is not one of his longer poems that does not have some central didactic focus to which the myth, the images, and the reflections refer. But he is more likely to tell us about the laws of the gods than to become the gods' spokesman. Hence the element of narrative, the Homeric art, is more central in Bacchylides than in Pindar. What the heroes of his poems say and do is more vital to his statement than what he is and says. | When a god speaks in Bacchylides, he speaks not to the poet, but to a hero:

> So Lord Apollo, true to his mark,
> Gave warning to Admetus . . .

Bacchylides lacks the inwardness of Pindar, and his poetry is never so personally compelling. He is cooler, brighter, more objective. In narrative grace and crisp elegance he is his superior; and, less overwhelming, he is more immediately accessible to readers from whom the Greek gods are far.

Both Pindar and Bacchylides wrote poetry of a kind which

has no close parallel in the modern world: it can be described by the generic term *choral lyric*, or simply *choral poetry*. It consists of songs varying in length from about 20 to about 500 lines,[2] sung to music by a trained chorus which danced as it sang. Poetry of this kind is best known to us from the choruses of Greek drama. The only immediate parallel is the modern opera, and Monteverdi, the chief inventor of the opera, in fact derived the new form in an attempt to reproduce the choral song of ancient Greece.

The opera, however, is a misleading parallel. Despite the innovations of Wagner, it is clear that everything in an opera is subordinate to the music. Few libretti, if any, make interesting reading in themselves. Greek choral poems are written in exact and intricate metres, in rich and self-sufficient poetic language. We have lost both the music and choreography for these works. But it is certain that the poetry was always, in the great period of choral song, the dominant element.

The form of these songs was usually *triadic*. There is a *strophe* of from about three to a dozen lines, comprising a complete metrical pattern based, as is all ancient Greek poetry, on | an alternation of long and short syllables. The metre, and probably the dance steps and the music, of the strophe, were matched in the *antistrophe*. The *triad* is then completed by an *epode*, in metre, dance, and music similar to, but not identical with, strophe and antistrophe. A song may consist of one triad, or of several, each repeating the other almost exactly. A few shorter poems have no epodes, but only matching strophes and antistrophes.

The subjects of these poems were both occasional and eternal. They were occasional in that they were written for particular festivals or celebrations, sometimes as entries for poetic competitions (like Attic drama), sometimes on commission from individuals. And they celebrated the particular. They were eternal in subject matter because they dealt always with relations between gods, heroes, and living men. They

---

[2] It might be more accurate, since these 'lines' themselves vary in length, to say that the choral poems of Pindar and Bacchylides take from about two to about forty-five minutes to read aloud, and the time of ancient performance would not have been much greater.

present, we could almost say create, a world in which the gods of Olympus, the heroes of myth and their descendants and analogues, the cities and men the poets knew, all exist together. The actions of a hero like Heracles or Achilles are like those of a victor in the great athletic games. The contemporary victor had a hero for his ancestor; or if he did not, his city did, and a city is like a large family. The heroes are like Plato's Demon of Love, more than man, less than god. They make the essential unity of the world of choral song; and hence myth, and the lasting moral and philosophic truths associated with myth, are the essence of choral poetry.

Almost all the extant works of Pindar, and the greater part of those of Bacchylides, are *epinician odes*. These are songs commissioned by men who have won victories in the games, or by their kinsmen. Most celebrate great victories, that is, those won at the Panhellenic games at Olympia, Delphi (the Pythian Games), Nemea, or the Isthmus of Corinth. But there were innumerable lesser and more local games as well, and we have a few examples of songs written for these by both Bacchylides and Pindar. |

The most sumptuous epinician odes are those written for victories in the chariot or horse race, and the reason for this is simple: a victor in these games had to be a rich and powerful man in order to provide so dear an entry in the contest. He himself would not ride in the race: it was a detached exhibition of glory. To win the prize for boxing, wrestling, or running, a man had himself to be an athlete. We can see in this collection the differences between the songs written for the two kinds of victory.

Choral poetry was in origin a Dorian art. The earliest example we have is the maiden-song of Alcman, a poet of Sparta who lived in the seventh century BC. Hence the language of the choral ode was always a modified form of the Dorian dialect, which was chiefly distinguished from the Ionian dialect of Asia Minor, the Greek islands, and Athens by a predominance of the long ā sound as opposed to the Ionian long ē. We can see at once how much this was conventional literary language when we observe that two of three great choral poets of antiquity, Simonides and Bacchylides, were themselves Ionian Greeks. The inescapable stage pronunciation of Shakespeare, in America as well as England, with its clipped vowels and frontal consonants, is some kind of analogy to the assumption of an artificial language. But in choral poetry the difference was not merely in performance: it was written into the very form of the words.

Along with Dorian words and forms, the other great element in the language of choral poetry was Homer. The stories of Homer, his colour and heroic gesture, his elaborate epithets and the movement of his syntax, persist in the works of the choral poets, and in a more evident and less transmuted fashion in Bacchylides than in Pindar. For Homer was himself an Ionian poet, and the tradition from which he derived ignored those late-comers, the Dorian Greeks, entirely.

The choral song in its origin was expressive of stability and locality. It was sung at a city festival by a chorus of singers | from the city; it celebrated local gods, and heroes who were buried nearby, and local religious practice. Homer, on the other hand, is the poet of Greeks who, forced from their homes by the invading Dorians, had gathered in Attica, and migrated to the Aegean islands and the Asian coast. His heroes are not bound by time and place. They are men of the vanished kingdoms of the Mycenean Age, and his long epic narratives of their lives and deaths, and of the universal gods who move in their midst, were sung by bards scattered all over Ionian lands. The fully developed choral poetry of Simonides, his nephew Bacchylides, and Pindar is an artistic merging of two lines of Greek tradition.

It is through Homer that we can best understand the achievement of the great choral poets. They took his stories, and stories like his, and attached them to local and contemporary scenes. Ajax and Achilles in the *Iliad* are great heroes who have left their homes to fight—and die—at Troy. In Bacchylides 13, we see them again at Troy; the occasion of our seeing them is their grandfather, Aeacus, who had been the great king of Aegina; and the ode is written to an Aeginetan victor in the παγκράτιον at Nemea.

The mythical narrative, which Homer presents in detailed and linear sequence, is condensed in Bacchylides and presented as separate moments of glory. The dramatically ordered world of the *Iliad* and *Odyssey*, where gods and men and meanings of actions are connected by events, is resolved in choral poetry into a more purely poetic (in our sense) nexus of associations and correspondences.

The later choral poets also reflect the growth of abstract language in Greek. Homer is didactic, in his way, more so than might at first appear. But the general significance of what happens in the *Iliad* and *Odyssey* is always bound up with particular actions or speeches. Bacchylides and Pindar have a natural tendency to draw from myth and situations reflections of a general character, and their store of language | contains a far greater number of abstract nouns to do this. Homer will speak of gold, and describe the delight men take in it. Bacchylides will say simply that gold is joyousness.

The fixed epithets of Homer are perhaps the most distinctive feature of his style. We are all familiar with them: the 'wine-dark sea', the 'glancing-eyed' Achaeans, the 'swift' or 'hollow' or 'black' ships. We know now that these epithets were an indispensable element of the formulae which made it possible for an epic minstrel to improvise his song as he performed it. But we do not need this knowledge to be aware that Homeric epithets rarely have a specific meaning in the contexts in which they occur. 'For Agamemnon king of men, the son of Atreus, had himself given the Arcadians well-benched ships to cross the wine-dark sea' is no more than the normal epic expression of 'Agamemnon had provided the Arcadians with ships'. The value of Homeric epithets is not specific but generic. As they fall on the ear over and again in almost every line of the *Iliad* and *Odyssey*, they tell us nothing special about what is happening at the moment; they rather evoke the whole heroic world of the epic.

Bacchylides adopts many of the *epithets* of Homer, and fashions others after their model. A profusion of individually selected epithets is as characteristic of his style as the fixed epithets are of Homer's. But Bacchylides' epithets are more than either ritual expressions or deliberate adornments, for they are images. The epithets of Homer function as kennings. The first word of Ode 17 is κυανόπρωρα, *dark-* (literally *dark-blue-) prowed*. This word occurs frequently in Homer as a fixed epithet of ships, and always in the same phrase, νεὸς κυανό-πρῴροιο, in the unstressed position at the end of a line. The word in Bacchylides stands emphatically at the prow of his poem, and is used deliberately to give the right sinister tone of the voyage to Crete.[3] Or again, consider the | words with which Meleager describes his sister in 5: she is χλωραύχενα ('her neck glows with the gloss of youth'); and Aphrodite is θελξιμβρότου ('Cypris the gold whose magic strikes'). Neither of these adventurous epithets are in Homer. The first Bacchylides took from Simonides, the second he probably made himself. But they are formed exactly like Homeric epithets. The striking difference is in their use: Bacchylides' epithets tell a story. The innocence and beauty of Deianeira is the deadly lure with which Aphrodite will strike unsuspecting Heracles.

By his constant use of epithets of Homeric form, sometimes as many as three different epithets gracing a single noun, Bacchylides wants to recall the voice and vision of Homer. But Bacchylides' epithets are new-fashioned in great part, and meant to be felt as such. More important, in his use of them he, like the other choral poets, goes totally against the Homeric tradition. He sets the epithets free from the fixed formulary pattern of Homeric tradition, gives them independent life, and makes them carry much of the bright meaning of his poetry.

This deliberateness of glorification and artful distance from direct experience of life may surprise the modern reader, though he may reflect too that it is a mode of expression in some respects not far from Mallarmé or Lorca. W. H. Auden, in his inaugural lecture as Professor of Poetry at Oxford, lays down several 'touchstones for critics'. The critic who really enjoys poetry will like 'long lists of proper names, such as the Old Testament genealogies or the Catalogue of Ships in the *Iliad*; riddles and all other ways of not calling a spade a spade; complicated verse-forms of great technical difficulty . . .; conscious theatrical exaggeration; [and] pieces of baroque flattery, like Dryden's welcome to the Duchess of Ormond'.

The standards Mr Auden so ingeniously sets up to dis-

---

[3] The adjective κυάνεον modifies νέφος to mean the dark cloud of death in Simonides and in Bacchylides 13.

tinguish the lover of poetry from the Philistine are standards of
pure enjoyment of words, of love of poetic brilliance and |
artfulness as such, apart from any prosaic value, or *meaning* that
might be valid in other contexts of speech and action. The
choral lyric of Bacchylides and Pindar is at first sight a perfect
illustration of poetry that could be used for such a test; and Mr
Auden might well trust the critic who likes a Greek epinician
ode. None the less, this category of pure poetry which he
establishes would not finally be the one to contain the odes of
Bacchylides. His poems are not something peculiar and artificial on the fringes of literary expression, to be enjoyed for their
very difference from plain statement of dramatic story and
moral reflection. The early fifth century was still an age in
which poetry was the fundamental mode of public utterance.
Bacchylides uses a traditional and artificial poetic language,
and his poems were commissioned by rich men and potentates.
But within this form, which was to him the natural and
traditional means of expression, he says what he thinks. Like
the plays of Aeschylus and Sophocles, Bacchylides' poems are
the proper voice of Greek civilization. There is a straightforward, positive, confident quality in them. They are meant to
exalt and to adorn, but at the same time they deal without
embarrassment or subterfuge with what the poet knows to be
the right laws of human society and the natural structure of our
lives. Wit, irony, word-play, and the oblique insight may
occasionally be present in these works, but they are peripheral
to their purpose. To borrow from Mr Auden in another vein,
Bacchylides speaks what we have almost come to consider
poetry no longer capable of expressing,

> All words like peace and love,
> All sane affirmative speech.

7

Homer: The Odyssey

THE magic of the *Odyssey* is that it makes beauty and strangeness real and immediate. A German scholar commented long ago that in reading Homer we find a world wonderful and unexpected and at the same time feel a nostalgia as for a familiar land we once knew. Here is a passage that illustrates well the suggestive paradox. Odysseus is furthest out in his fairy-tale adventures, in the land of the dead, and he sees the ghosts of famous women of the past. They come and tell him their stories:

First then I saw Tyro, of noble father, her who was said to be the daughter of splendid Salmoneus. And she said she was the wife of Cretheus son of Aeolus. She fell in love with a river, divine Enipeus, most beautiful of rivers that pour upon the earth. So she wandered by the fair streams of Enipeus. And looking like him, the holder of the earth, the earth-shaker, where the river pours out to sea, as it eddies, lay with her. The purple wave stood round them, like a hill, high-rounded, and hid the god and the mortal woman. He loosed her maiden girdle, and spread sleep upon her. But when the god had achieved the works of love, he took her by the hand, and spoke to her, and called her by name: 'Be happy, woman, in our love. In the course of a year you will bear shining sons, because the lovemaking of | the gods. is not vain. You must raise them and care for them. Now go home, and be silent and do not name me. But I tell you I am Poseidon shaker of earth.' When he had spoken so, he dived into the billowing sea. She conceived and bore Pelias and Neleus. These two became strong servants of Zeus, both of them. Pelias lived in spacious Iolcus, and the other in sandy Pylos. And she, queen of women, bore other sons to Cretheus, Aeson and Pheres, and Amythaon, delighter in horses (11. 235–59).

I have quoted, and literally translated, the passage at length, in the first place because it is a good example of the ease and grace

The Fat Abbot (Winter 1962), 48–57. Review of *Homer, The Odyssey*, trs. Robert Fitzgerald (Doubleday, 1961).

and steadiness with which Homer tells so improbable a story as Tyro's love for the river and her union with the god; and secondly because a comparison with Fitzgerald's translation will give a good picture of his art.

The subject is fantastic and mythical, like much of the *Odyssey*. The greatest critic of antiquity, in his essay *On the Sublime*, compared the *Odyssey* adversely with the *Iliad* for just this quality: writing the *Odyssey* in his old age, as Longinus metaphorically suggested, Homer constantly left reality behind for the unreal world of myth. But what Longinus did not give Homer credit for is the tone of ritual firmness that gives reality and substance to these airy flights of fancy. Cyclops is an ogre of purest fairy-tale, and Circe a witch from the same source. But the hero who encounters them has human weight and a hard eye like ours. Seen through his experience, the most incredible persons and most supernatural events become part of the sharp and familiar trials of life. He allows no doubt of their existence: he sees them too clearly, they exercise his wits too much, and cause him too much pleasure and pain.

The device in the *Odyssey* that most clearly manifests this confidence is the condition of Homeric art: the formula. The single autonomous line that introduces Tyro, serves with slight changes to introduce other spectacles of the nether world: 'And after her I saw Antiope, the wife of Amphitryon ...'; 'And I saw the mother of Œdipus, beautiful Epicaste ...'; 'And then I saw | Minos, shining son of Zeus ...' The epithets fill out the melodious curve of the line: they ennoble the persons to whom they are attached, and establish them in the remembered and canonical pattern of poetic legend. The whole formulary lines which could not be fashioned without these epithets provide a sense of regular and clearly recorded events. Into the framework of 'and then I saw' or 'after him/her I saw' could fit any name, and the epithet or descriptive phrase with which that name could be joined would make it fit the standard pattern. And alternating with this formula is another similar in function: 'then there came the soul of ...' to be followed by 'Achilles son of Peleus' or 'Agamemnon son of Atreus', 'Teiresias of Thebes', etc.

Odysseus introduces the sights of Hades' realm like entries in a log-book. The repeated phrases make it all straight record of fact, like Xenophon's account of the Greeks marching into Persia: 'Then he marched through Arabia, having the Euphrates river on his right, five days' journey and thirty-five parasangs'.

The heroic epithets, and a poetic language created for epic verse put the story into the splendid realm of epic legend, but the factual sobriety and the uncomplicated specificity of the formulae made from this language remove the sense of the fanciful or the exotic.

The formulary quality is most strongly felt in the introductory lines, but it is present in the whole texture of the passage. 'By the fair streams of', 'looking like him', 'holder of earth, earth-shaker', 'but when he had completed', 'in the course of a year', all these are familiar from other contexts in the song, some of them describing events and sights far more down to earth than these. So are whole lines: 'he took her by the hand and spoke to her and called her by name'. Our recognition of this as the proper way of introducing a serious and affectionate speech made by any one person to another is vital. If this is exactly what Antinous, a perfectly recognizable human type, does when he goes to speak to Telemachus in Book 2, why should we feel it any less real when a god looking like a river does it when he speaks to his mistress in a billow of the sea-wave? |

The dramatic hold, then, which is so much of the poetic power of the *Odyssey*, depends on a sustained and unfanciful realism; and this realism derives largely from a formulary language which extends the clear observation of experience to the richest flights of the imagination. A comparison made by Harry Levin will further illustrate this aspect of Homer: Odysseus and Don Quixote are two archetypal heroes of experience, journeying through the world and recording what they see. They both observe a sea of marvels, spectacles of wonder, and trials of every kind that can test the heart of man. But in the *Odyssey* all the marvellous things are real and out there in the world. The wonders of life are for Don Quixote in the true sense *figments* of the imagination, fictions imposed on a workaday world whose language has no place for them.

It is a consequence of this quality of the *Odyssey* that some of the very devices which are in the modern idiom most poetic are furthest from the beautiful, but even and direct, movement of Homer. Here is Fitzgerald's translation of the passage literally rendered above:

> Here was great loveliness of ghosts! I saw
> before them all, that princess of great ladies,
> Tyro, Salmoneus' daughter, as she told me,
> and queen to Krêtheus, a son of Aiolos.
> She had gone daft for the river Enipeus,
> most graceful of all running streams, and ranged
> all day by Enipeus' limpid side,
> whose form the foaming girdler of the islands,
> the god who makes earth tremble, took and so
> lay down with her where he went flooding seaward,
> their bower a purple billow, arching round
> to hide them in a sea-vale, god and lady.
> Now when his pleasure was complete, the god
> spoke to her softly, holding fast her hand:
>
> 'Dear mortal, go in joy! At the turn of seasons,
> winter to summer, you shall bear me sons;
> no lovemaking of gods can be in vain.
> Nurse our sweet children tenderly, and rear them. |
> Home with you now and hold your tongue, and tell
> no one your lover's name—though I am yours,
> Poseidon, lord of surf that makes earth tremble.'
>
> He plunged away into the deep sea swell,
> and she grew big with Pelias and Neleus,
> powerful vassals, in their time, of Zeus.
> Pelias lived on broad Iolkos seaboard
> rich in flocks, and Neleus at Pylos.
> As for the sons borne by that queen of women
> to Krêtheus, their names were Aison, Pherês,
> and Amytháon, expert charioteer.   (pp. 204–5)

Now this is a true piece of poetry, and it is enough to confirm the judgement made on all sides, that Fitzgerald's *Odyssey* is the best translation of Greek poetry, not excluding Lattimore's *Iliad*, to appear in many decades. The last two lines with the fine closing cadence of '. . . and Amytháon, expert charioteer' are truly Homeric: he has caught the poignant and unstressed feeling of the single epithet, the sense of manhood and glory now vanished. Homeric also is the word 'vassals' with which Fitzgerald translates θεράποντε, 'servants' or 'squires': he makes us feel the social reality behind the legend by choosing a word derived from European feudalism. 'Servant of Zeus' is the way a ruler identifies himself as one who draws his power from the king of the gods.

But in the handsomest lines of the passage there is a suffusion of the exquisite, a delicate incongruity of diction, that makes poetry at the expense of the true Homeric firmness. The introductory formula is lost in the exclamation 'Here was great loveliness of ghosts!' It is a literary exclamation, and reflects the pleasure not of the hero who values what he saw, but of the connoisseur who delights in that hero's words. And equally un-Homeric are phrases like 'gone daft' where Homer uses the simple word ἠράσσατο, 'fell in love', that appears in similar formulae; or the whole line 'lay down with her where he went flooding seaward': there Fitzgerald ingeniously reminds us of the double form of the god, human and fluvial at once. Homer, | once he introduces the god, treats him as anthropomorphic, and would never play with the ambiguity in this fine, metaphysical way. Then there is the consciously poetic note of '*limpid* side' or the sentimental hint of '*sweet* children', or the gallantry of 'I am yours', which, if Fitzgerald were less original or a less accurate reader of Greek, one would take to be a mistranslation of the particle τοι originally meaning 'to you' but in Homeric use no more than an emphatic adverb.

And finally, 'spoke to her softly, holding fast her hand'. The Greek word φῦ in the formula which this translates does imply grasping hard or clinging: and the formula is used in scenes of real or pretended affection. So the translation is not inaccurate; but, since it is made to fit the particular scene, the reader loses the sense of a gesture familiar to him from other scenes in the poem. Thus in Book 2 we have, for the same line

> Antinous came . . . over . . .
> and took him by the hand with a bold greeting　　(p. 39)

in Book 8

> and tenderly he pressed her hand and said　　(p. 145)

in Book 15

> called him apart and gripped his hand, whispering . . .
>
> (p. 296)

The conspiratorial gesture of the line in 15, the boldness of the line in 2, the tenderness and softness of the lines in 8 and 11 are all inherent in their contexts: but Homer uses a single functional phrase which acquires by its place the appropriate association and yet remains a firm element of construction unifying the experience of the poem. Fitzgerald elaborates the scenes with a novelist's technique, spelling out the emotions for us by rendering each phrase in unique language.

Similarly the epithets. It is in fact their constancy that makes them resonant for us. Thus Penelope is περίφρων, 'circumspect' or 'prudent'. It is important that she remain so, because this single quality, fixed in a single word, is the background against which | all her actions, including her perplexing and ambiguous relations with the suitors, take place. She is περίφρων in a ringing phrase that fills a whole line and comes out literally 'the daughter of Ikarius, circumspect Penelope'. This appears in the translation

1) careful Penelope / Ikários' daughter (p. 23)
2) She is . . . / to clear-eyed . . . / Penelope, Ikários' daughter (p. 211)
3) . . . the queen / Penelope, Ikários' daughter (p. 377)
4) the daughter of Ikários, Penelope (p. 399).

Nowhere does the expression appear in its entirety as the single line which musically affirms, over and again, one of the principal motifs of the story.

It would be wrong, and dully literalistic, simply to blame the translator for handling the formulary epithet in this fashion. On the contrary, it is his aliveness to what goes on in each scene and the variety of his expression that make the translation sparkle as it does. But it is important too to state clearly what is lost in Fitzgerald's inventiveness and conceit, as against those who have implied that the translation equals or surpasses the original.

The language of Homer is coherent, organic, generous, and

musical. The poet deals in traditional phrases, but his genius is revealed by the way he puts the phrases together, his juxtapositions and his timing. (That much the same epic stock of diction can be florid and tedious is evident from some of the less felicitous of the Homeric Hymns.) A simple example comes through in the translation: when Poseidon takes the form of the river, he is not named, but described by two epithets, a device Homer uses occasionally with major gods. His name is saved until the last line of his speech to Tyro, when in solemn revelation he announces himself to the future mother of his sons. In Homer the next to last line in the speech says 'Go home, be silent, and do not speak my name'. The revelation occupies the whole of the last line, and no more; and it begins with the predicate 'I tell you I am . . .' and ends with the familiar liturgically poetic phrase 'Poseidon, shaker of earth'. Fitzgerald has kept the dramatic | movement of the passage inherent in the postponement of the god's name. But even here we have lost some of the climactic effect, through the love-letter tone, through the spreading of distinct lines into sentences that cross from line to line, and finally by the translator's long epithetic description 'lord of surf that makes earth tremble' which he clearly chooses to render the majesty of the statement, but which puzzles by its elaboration, for we wonder a little at the relation in it of surf and earth.

This is in fact a very fine translation, and we are not likely soon to see another that can equal it. It almost never appears unworthy of its original, and often it renders the Greek with a hard and simple phrase that seems to be just what is wanted. But it has faults, though they are largely the faults of poetic creation, not of hackwork like the translations of Rouse, or Graves's *Iliad*. Its principal fault is the one we have examined: too much of the time we are given not the substance of the *Odyssey* with its plain magnificence of narrative, but a kind of literary commentary on it. The fault comes through most clearly if we try reading aloud, as a bard might recite, long passages. We find ourselves delighting in wit, at the flexibility with which the poet chooses among a range of tones and dictions (from the high poetry of 'where he went flooding seaward' to the cosiness of 'Home with you now'); but the weight and even tenor, the unexpressed conviction that this is a true and serious account of the world, the emotional truth that goes with an imposed familiarity of phrase, are diminished.

Three more passages will help fill out a proper estimate of Fitzgerald's work. The first is Odysseus building his raft on Calypso's island:

> A brazen axehead first she had to give him,
> two-bladed, and agreeable to the palm
> with a smooth-fitting haft of olive wood;
> next a well-polished adze; and then she led him
> to the island's tip where bigger timber grew—
> besides the alder and poplar, tall pine trees,
> long dead and seasoned, that would float him high. |
> Showing him in that place her stand of timber
> the loveliest of nymphs took her way home.
> Now the man fell to chopping; when he paused
> twenty tall trees were down. He lopped the branches,
> split the trunks, and trimmed his puncheons true.

Second, Odysseus' mother in the netherworld speaking of the condition of the dead (Odysseus wonders if they are hallucinations):

> All mortals meet this judgment when they die.
> No flesh and bone are here, none bound by sinew,
> since the bright-hearted pyre consumed them down—
> the white bones long exanimate—to ash;
> dreamlike the soul flies, insubstantial.

Third, Odysseus has told Penelope the secret of their bed, and she is at last certain it is her husband who sits before her:

> Their secret! as she heard it told, her knees
> grew tremulous and weak, her heart failed her.
> With eyes brimming tears she ran to him,
> her arms around his neck, and kissed him,
> murmuring: ...

The second of these is perhaps the most impressive bit of poetry, but the first is the one where we most truly feel Homer. Fitzgerald is not a master of the spoken word, and complex scenes too much tempt his wit. He is best at describing things or simple, satisfying actions; and as he describes the hero's carpentry his handling of the blank verse line with its crisp monosyllables ('lopped the branches / split the trunks', with a perfect rhythmic aphaeresis in the last line) shows him at his finest.

The netherworld passage is good, but too metaphysical for Homer: the puzzling condensation of syntax (Homer says in one line: 'their sinews no longer hold flesh and bones'), the ironic epithet *bright-hearted* (Homer has the traditional 'strong power of blazing fire'), and the remote Latinism *exanimate* make it a fine modern version of what Homer was telling, but miss the Homeric funereal familiarity with death. |

The third passage is Fitzgerald at his weakest. The attempt at interior monologue ('Their secret!'), though singled out for praise by at least one reviewer, is novelistic and cheap. And why should her eyes *brim* tears, and why should she *murmur*? At this moment of high emotion, the climax of the whole lovely story, Homer himself appears most traditional. 'So he spoke, and her knees and heart were loosened.' It is a line likely to be used for moments of extreme fear or dismay. So it appears in exactly this form when Penelope in Book 4 learns that her son has left Ithaca, and that the suitors are plotting his death. And with only the pronoun changed, in Book 22, when the suitors hear from Odysseus that he is going to kill them each and every one. By using the simple line here, without embellishment, at a moment of overwhelming joy, and following it by another perfectly traditional phrase 'she wept and ran to him', Homer achieves an emotion far truer than anything in the tremulousness of the translation.

The reader will notice other aspects of Fitzgerald's poem: the lack of large resonance in the verse, but now and then a line that does achieve some of it within the crisp bounds of a regular iambic pentameter:

> Then he ran up his rigging—halyards, braces—
> and hauled the boat on rollers to the water.

Or the clumsiness of some phrases in speech (Odysseus to Penelope: 'My strange one . . .'); and the fine visual perception of some lines:

> Skhería then came slowly into view
> like a rough shield of bull's hide on the sea

('rough ... hide' translated a single Greek word: Fitzgerald wants us to see the kind of shield Homer saw).

But what we have said is enough for a judgement. This is a noble work of translation. It cannot (what translation can?) embody the experience of Homer's words; but it has more of their delight and wisdom than one could have hoped.

8

The Two Voices of Virgil's *Aeneid*

I WANT to begin with the particular. Sometimes we come upon a short passage in a poetic work we know well, a passage we have never particularly noticed before, and all at once, as a kind of epiphany, the essential mood of the author seems to be contained in it. Here is a candidate for such a passage in the *Aeneid*.

There is at the end of Book 7 a kind of Catalogue of Ships, a muster or roll-call of the Latin leaders and their forces as they are arrayed against Aeneas in the long war which occupies the last books of the poem. One of these leaders is a quite obscure figure named Umbro. The Catalogue is a Homeric form, and Virgil here exploits it in the Homeric fashion, drawing out a ringing sense of power from place-names, epithets of landscape and valour, names of heroes. He endeavours further, again in the Homeric fashion, to give individuality within the sense of multitude by singling out some characteristic of each Latin warrior, a device the *Aeneid* has yet more need of than the *Iliad*, since Virgil has behind him no tradition of Latin song which could give the audience a previous familiarity with the heroes he names.

So Umbro comes from the Marruvian people, and he is most valiant: *fortissimus Umbro*. And he is a priest, who possesses the art of shedding sleep over fierce serpents. But—and here we catch the Homeric pathos—his herbs and his incantation could not save *him*, wounded by the Dardanian spear. Virgil then closes the brief scene with a beautiful lamentation:

> For you the grove of Angitia mourned, and Fucinus' glassy
> waters,
> And the clear lakes.

Arion (Winter 1963), 66–80.

Te nemus Angitiae, vitrea te Fucinus unda,
Te liquidi flevere lacus.

If we could understand wholly the reasons for this lamentation, so elaborate within its brevity, and what makes it so poignant, and why it is so Virgilian, we should, I think, have grasped much of Virgil's art. First, something I can talk about a little but not translate, there is the absolute mastery of rhetoric. We have a *tricolon*, three successive noun-phrases, here in *asyndeton*, that is, with no grammatical connectives, joined to one verb, *flevere*, 'mourned'; and this device combined with *apostrophe*: the dead warrior is suddenly addressed in the second person. The pronoun *te*, 'you', is repeated thrice, each time in the beginning of one of the three | elements of the tricolon, a repetition we call *anaphora*. So much is developed but standard rhetoric. Virgil's mastery consists not in that, but in the subtle variations of it we see here. The three nouns are all a little different. The first is a grove with the name of the goddess to whom it is sacred in the possessive singular: *nemus Angitiae*, 'the grove of Angitia'. The second is the name of a nearby lake, Fucinus, qualified by a noun and adjective: 'Fucinus with glassy wave', *vitrea Fucinus unda*. The third, beginning another hexameter line, is a common noun *lacus*, 'lakes', in the plural, with an adjective only: *liquidi lacus*, 'transparent lakes'. The first two nouns are opposed to the third by being names of places. The second and third are opposed to the first by having adjectives and by having adjective and noun separated, whereas 'the grove of Angitia', *nemus Angitiae*, comes together. But the first and third are also opposed to the second by the variation of the *anaphora*: *te*, 'you', embodying the directness of lamentation, begins the first phrase: *Te nemus Angitiae . . . Te* is repeated in the second phrase, but its directness is modulated, softened, by its coming second in the phrase, after the adjective *vitrea*, 'glassy': *vitrea te Fucinus unda*. Then in the third phrase, the tonic note is struck again: *Te liquidi flevere lacus*. And finally, the verse accent falls on the first *te* and the third, but not on the second:

Te nemus Angitiae, vitrea te Fucinus unda,
Te liquidi flevere lacus.

If this analysis seems too microcosmic, let me say that Virgil
may not be, surely is not, the greatest poet who ever lived; but
that in this mastery of the disposition of words within a formal
pattern, he has no rivals. The effect of the variation within the
symmetry is first to establish a rhythm, whose value might
finally have to be analysed in musical terms; and second, to add
emotion to the lines. The tricolon with anaphora is a strong
formal device, appropriate to the sounds of public lamentation.
The variations, like a gentle yielding within the firm tripartite
structure, add the note of genuine grief, invest the far-off place
names with something of what used to be called the lyric cry.

For it is the place-names in this passage that show us how
Virgil has departed from his Homeric model. The Homeric
lines the commentators cite here occur in the Catalogue of the
Trojans at the end of Book 2 of the *Iliad*:

The forces from Mysia were led by Chromius and by Ennomus, a
diviner of birds. But his birds did not keep *him* from black death. He
was to be slain by the hands of swift Achilles at the river, where
many another Trojan fell.

The moment of death, and the great slaughter of the Trojans
when Achilles returned to battle, is the picture we are left with.

Or again this, from Book 5 of the *Iliad*:

Menelaus son of Atreus caught with his sharp spear Strophius' son
Scamandrius, | a great hunter. Artemis herself had taught him how to
down all the wildlife that the woods nourish. But the huntress
goddess Artemis did him no good then, nor did his mastery with the
bow for which he was so famous. The son of Atreus, killer Menelaus,
struck him with the spear as he fled before him, right between the
shoulders, and drove the point out through his chest. And he fell
forward, and his armour rang as he fell.

Again, and more emphatically in a typical passage such as this,
the bitter irony of Homer leaves us with the image of the
instant death of the man: the glory of Scamandrius when he
lived and was famed as a hunter, then the uselessness of what he
was as death comes upon him.

Virgil in the lines about Umbro imitates these scenes. But
the image he leaves us with is not a fallen warrior, but a
mourning landscape. The dramatic preoccupation of Homer

with the single man and the single instant of time gives way to
an echoing appeal to the Italian countryside, and an appeal
strengthened in wholly un-Homeric fashion by historical
associations.

The place-names invoked by Virgil, Marruvium, Lake
Fucinus, the grove of the goddess Angitia, are from the
Marsian country, hill country to the east of Rome, where a few
generations earlier than Virgil a tough and warlike Italian
people, the Marsi, had lived in independence, as Roman allies.
In the Italian or Marsic war of 91–88 BC they had been defeated
by Rome, and though they had gained citizenship, they had
effectively lost their independence. To Virgil, this people
represented the original Italian stock. His feeling for them had
something in common with what Americans have felt for the
American Indian. They were somehow more Italian than the
Romans themselves. Proud, independent, with local traditions
hallowed by the names they had given to the countryside, they
succumbed inevitably to the expansion of Roman power. The
explicit message of the *Aeneid* claims that Rome was a happy
reconciliation of the natural virtues of the local Italian peoples
and the civilized might of the Trojans who came to found the
new city. But the tragic movement of the last books of the
poem carries a different suggestion: that the formation of
Rome's empire involved the loss of the pristine purity of Italy.
Thus the plot of the closing books of the poem centres on
Turnus, Aeneas' antagonist, who is made the embodiment of a
simple valour and love of honour which cannot survive the
complex forces of civilization.

In this light we can understand the form which the lamentation for Umbro takes. Umbro himself is not important. He is
no more than a made-up name. The real pathos is for the places
that mourn him. They are the true victims of Aeneas' war, and
in saying that they weep, Virgil calls on us to weep for what to
his mind made an earlier Italy fresh and true.

The lamentation of the ancient hallowed places of the Marsi
strikes a characteristic Virgilian note of melancholy and
nostalgia, a note produced by the personal accents of sorrow
over human and heroic values lost. But equally characteristic is
the aesthetic resolution of the lines. The lament is presented to

us as an object of artistic contemplation. By this I mean not simply that the lines are beautiful, for that is no distinguishing feature of Virgil. Nor do I refer to the vulgar concepts of 'word-painting' or 'scenic values', concepts often invoked in Virgilian criticism. The unexpected epithets *vitrea*, 'glassy', and *liquidi*, 'clear', do not, I think, 'paint a picture' for us. But they do create a sense of sublimation, a conscious feeling that the raw emotions of grief have been subsumed in an artistic finality of vision. Not only the death of Umbro but also the loss of Italy itself is at last replaced by an image of bright and clear waters. The word *vitrea* in the middle of the lamentation is particularly noteworthy, for its connotations of an artifact. It is as if Virgil were telling us that the way to resolve our personal sorrow over the losses of history is to regard these losses in the same mood as we would a beautifully wrought vessel of clear glass. The perfection of the lines itself imposes a kind of artistic detachment, and we are put in the position of Aeneas himself, as he sees, in Carthage, the destruction of Troy represented as paintings in a gallery of art.

These paintings remind Aeneas of all that has been, of the 'tears of human things' and at the same time, Virgil tells us, they fill him with hope. In a larger way, the whole poem is such a painting. It is about history, but its purpose is not to tell us that history is good, or for that matter that it is bad. Its purpose is rather to impose on us an attitude that can take into account all the history that is both good and bad, and can regard it with the purer emotions of artistic detachment, so that we are given a higher consolation, and sorrow itself becomes a thing to be desired.

Let us now consider the poem from a wider point of view. Here we take care not to let orthodox interpretations of the *Aeneid* obscure our sense of what it really is. The nostalgia for the heroic and Latin past, the pervasive sadness, the regretful sense of the limitations of human action in a world where you've got to end up on the right side or perish, the frequent elegiac note so apparently uncalled for in a panegyric of Roman greatness—like the passage at the end of Book 5 which describes the drowning of the good pilot Palinurus in dark and forgetful waters just before the Trojans reach Italy—the continual opposition of a personal voice which comes to us as if it were Virgil's own to the public voice of Roman success: all this I think is felt by every attentive reader of the poem. But most readers, in making a final judgement on the *Aeneid*, feel none the less constrained to put forth a hypothetical 'Roman reader' whose eyes were quite unused to the melting mood. *He* would have taken the poem ultimately as a great work | of Augustan propaganda, clapped his hands when Aeneas abandons the over-emotional Dido, and approved with little qualification the steady march of the Roman state to world dominion and the Principate of Augustus as we see these institutions mirrored in Anchises' speech in Book 6 and in Juno's renunciation in Book 12. This, we are told, is how we should read the poem. After all, what was Augustus giving Virgil all those gold-pieces for?

So Mr Kevin Guinach, the Rinehart translator, after putting forth these views, adds: 'From this it must not be inferred that Virgil was a hireling . . . it is fairer to suppose that he was an ardent admirer of the First Citizen and his policies, and sought to promote the reconstruction that August had in mind.' Apropos of Dido he says: 'The ancient Romans did not read this episode as tearfully as we do . . . From the Roman point of view, Dido was the aggressor in her marriage with Aeneas, an intolerable assumption of a male prerogative.' Moreover, he tells us, the Roman would have condemned her for breaking her vow to her first husband, dead these many years. Consider the case of Vestal Virgins . . .

But what, on the simple glorification of Rome interpretation, do we make of some of the finest passages of the *Aeneid*? What we find, again and again, is not a sense of triumph, but a sense of loss. Consider the three lines at the end of Book 2 which describe Aeneas' attempts to embrace the ghost of his wife:

> Three times I tried to put my arms around her
> And three times her image fled my arms' embrace,
> As light as the winds; as fleeting as a dream.

Like the lines about the fallen warrior, these lines derive from an earlier literary tradition. And again a comparison with this tradition will tell us something about the *Aeneid*. Virgil has two Homeric passages in mind, one in Book 23 of the *Iliad* where Achilles tries to embrace the hollow wraith of Patroclus:

> So spoke Achilles, and reached for him, but could not
> seize him, and the spirit went underground, like vapour,
> with a thin cry. And Achilles started awake, in amazement,
> and drove his hands together, and spoke, and his words were
>   sorrowful:
> Ah me! even in the house of Hades there is left something of
>   us,
> a soul and an image, but there is no real heart of life in it!

And a passage from Book 11 of the *Odyssey*, where Odysseus in the Underworld attempts to embrace the shade of his mother:

> ... I bit my lip
> rising perplexed, with longing to embrace her,
> and tried three times,
> but she went sifting through my arms, impalpable |
> as shadows are, and wavering like a dream.
> And this embittered all the pain I bore ...

So the Virgilian passage first of all serves to reinforce the identification, operative throughout the poem, of Aeneas with the heroes of Homer. But the identification only sets in relief the differences. Virgil's lines are characteristic of the whole mood of his poem, the sadness, the loss, the frustration, the sense of the insubstantiality of what could be palpable and satisfying. Virgil emphasizes the *image*—the word *imago* ends the second line; and we can think of countless like passages, such as the appearance of Aeneas' mother in Book 1, not recognized until after she has fled. The Homeric heroes are made angry by these signs of what lies beyond our physical existence. Achilles *drives* his hands together, Odysseus is *embittered* that this kind of frustration should be added to his troubles. The Homeric hero, however beleaguered by fate, loves and enjoys the warmth of life, and his course of action includes a protest against the evanescence of mortality. But the sense of emptiness is the very heart of the Virgilian mood. After the three lines I have quoted, Aeneas goes on simply:

'The night was over; I went back to my comrades.' And the third of the three lines, 'As light as the winds, as fleeting as a dream', receives a delicate emphasis, partly due to the two different words for *as* '*Par* levibus ventis, volucrique *simillima somno*', that blurs the contours of our waking senses and gives the line a force of poignant resignation absent from both Homeric passages.

One other passage here, which I will speak of later on. Aeneas comforts his men after the storm in Book 1 with a famous phrase:

Forsan et haec olim meminisse iuvabit,
Some day perhaps remembering this too will be a pleasure,

lifted again from the *Odyssey*. But the Homeric line is quite unmemorable. Odysseus says to his men that some day their troubles now will be a memory. He means only, they will be in the past, don't be overcome by them now. Virgil has made one clear change: the word *iuvabit*, 'it will be a pleasure', which makes a commonplace idea into a profoundly touching one. Not, I would insist, because Virgil is a greater poet, but because the kind of sentiment that stands out in the *Aeneid* is different from the kind that stands out in the *Odyssey*.

How much in general is Aeneas like the Greek heroes? We know from the first line that he is cast in the role of Achilles and Odysseus:

Arms and the man I sing ...

The *arms* are of course the *Iliad*, the man is the *Odyssey*. And the first six books of the *Aeneid* retrace the wanderings of Odysseus, the wars of the last six books follow the example of the *Iliad*. But the examples are not followed closely. The *Odyssey* goes on after | its first line to tell us about the single man Odysseus; the *Iliad* goes on to describe the quarrel that was the first step in the tragedy of Achilles. The *Aeneid* moves from Aeneas straightway to something larger than himself: Rome:

that man who was tossed about on land and sea
and suffered much in war until
he built his city, brought the gods to Latium
from whence the Alban Fathers, the towering walls of Rome...

Aeneas from the start is absorbed in his own destiny, a destiny which does not ultimately relate to him, but to something later, larger, and less personal: the high walls of Rome, stony and grand, the Augustan Empire. And throughout he has no choice. Aeneas never asserts himself like Odysseus. He is always the victim of forces greater than himself, and the one lesson he must learn is, not to resist them. The second book of the poem drills him thoroughly in this lesson. The word Aeneas keeps using, as he tells of the night Troy fell, is *obstipui*: I was *dumbfounded*, shocked into silence. Again and again he tries to assert himself, to act as a hero, and again and again he fails. He leads a band of desperate Trojans against the Greeks, but it all turns sour. The Trojans dress up as Greeks, an unheroic stratagem which works for a while, but then their own countrymen mistake them, and Trojans slaughter each other, while Aeneas himself ends up on the roof of Priam's palace, passive spectator of the terrible violations within. A key passage is the one in which Aeneas is about to kill Helen. At least the personal, if not entirely heroic, emotion of revenge can be satisfied. But his mother stops him, not with a personal plea, as Athena checks Achilles in the *Iliad*, but by revealing for an instant the gods at work destroying the city. Against such overwhelming forces as these, individual feeling has no place. Aeneas must do the *right* thing, the thing destiny demands, and sneak away from Troy.

One of the effects, then, of the epic identifications of Aeneas is ironic contrast: he is cast in a role which it is his tragedy not to be able to fulfil. Let us now consider another kind of identification: the historical ones. As well as being cast as Odysseus and Achilles, Aeneas has to be the emperor Augustus. Of many passages, this one in the third book particularly contributes to setting up the connection. Aeneas and his men coast along the western shore of Greece and stop at Actium, where there is a temple of Apollo. There they hold games and Aeneas fastens to the door of the temple spoils taken from the Greeks with the inscription THESE ARMS FROM THE GREEK VICTORS. The reason for this action in this place is that Augustus had won his great victory over Antony and Cleopatra a few years earlier at Actium. He had instituted

games in honour of his victory, and he liked to identify himself
with Apollo. Moreover THE GREEK VICTORS, who are |
now vanquished, represent the armies of Antony, who rec-
ruited his forces from the eastern Mediterranean, whereas
Augustus made himself the champion of Italy. So that the
victory Aeneas somewhat illogically claims here by dedicating
Greek spoils prefigures the victory that was to establish the
power of Augustus.

Some striking verbal parallels confirm the connection; and
give us as well insight into Virgil's technique. At the beginning
of Book 3, Aeneas sets sail from Troy.

> I am borne forth an exile on to the high seas
> With my comrades, my son, the Penates and the Great Gods
> Cum sociis natoque Penatibus et Magnis Dis.

The exact meaning of the phrase 'the Penates and the Great
Gods' is obscure. But it is clear that they are some sort of cult
statues of Troy, destined to become cult statues of the New
Troy, or Rome. The oddity of the phrase in fact helps us to
remember it—the Romans liked their religious language to be
obscure—and so does its remarkable thudding rhythm: *Penati-
bus et Magnis Dis.* This is Aeneas in his sacral character as bearer
of the divine character of Troy.

At the end of Book 8, Vulcan makes a shield for Aeneas, and
on it are engraved scenes from subsequent Roman history. One
of these scenes depicts the Battle of Actium:

> On one side stands Augustus Caesar leading Italians into
> battle,
> With the Fathers [i.e., the Senate], the People, the Penates
> and the Great Gods
> Cum patribus populoque, Penatibus et Magnis Dis.

Aeneas' shield shows the future version of himself.

But Aeneas is not just Augustus. There is also the possibility
of his being Augustus' bitter enemy, Mark Antony. Such is the
identification we are led to make when, in the fourth book, he
has become the consort of Dido, queen of Carthage. Thus the
contemptuous description of him by Iarbas, his rival for Dido's
love, 'that Paris with his effeminate retinue', closely matches

the image of Antony and Cleopatra with their corrupt eastern armies which Augustus created for Roman morale.

And Dido is Cleopatra. When she is about to die, she is said to be 'pale with imminent death', *pallida morte futura*. Cleopatra, in her own person, is described on Aeneas' shield in Book 8 as 'paling before imminent death', *pallentem morte futura*.

To understand the meaning in the poem of these historical identifications, we must first consider more fully the figure of Aeneas. We learn from the second line of the poem that he is a man 'exiled by fate', *fato profugus*, and we soon learn that fate has for | Aeneas implications that go beyond his personal journey through life. He is a man blessed—or is it cursed?—with a mission. The mission is no less than to be the founder of the most powerful state known to history; and so his every act and his every passion, all that he does, all that he feels, and all that happens to him is in the light or under the shadow of this immense prophetic future of which he, by no choice of his own, is the representative elected by the gods. Every experience he passes through, therefore, has a significance greater than the events of an ordinary man's life could possibly have. Every place he visits acquires an eternal fame of one kind or another. Every action he performs, every word he speaks, is fraught with consequences of which he himself can only dimly perceive the enormity.

This sense of pregnant greatness in every detail of experience is impressed on us too by the rhetorical exaggeration which pervades the *Aeneid*, and by the unrealism of many of its incidents. Juno's wrath in Book 1 is magnified far beyond Poseidon's resentment in the *Odyssey*; Athena's punishment of the lesser Ajax, which Juno would like to inflict upon Aeneas, is enlarged into a cosmic destruction. When there are storms, the waves rise up and lash the heavens. Dido is supposed to have arrived in Africa not long before with a small band of refugees; but already the construction of a tremendous city— the later Carthage, of course—can be seen, complete with temples and art-galleries. Aeneas is moving through a world where everything is a symbol of something larger than itself. The layers of literary and historical allusion reinforce this sense of expansion in space and time which every monumental hexameter verse imposes on the reader.

The potentialities of ages and empires are alive in the smallest details of the *Aeneid*, and Aeneas has been made into the keystone of it all. The inconceivable destiny of Rome rests upon his shoulders. The *Aeneid* can give a literal meaning to that cliché. So line 32 of Book 1:

Tantae molis erat Romanam condere gentem

It was a thing of so much *weight* to found the Roman race.

Aeneas can only leave Troy by carrying his aged father upon his shoulders. And Anchises is more than Aeneas' father. He is the burden of destiny itself. Thus in Book 6 it is he who unfolds the panorama of Roman history to his son who has descended to the Nether World to hear him. And at the end of Book 8, Virgil insists on Aeneas' role as bearer of destiny. The shield which Vulcan makes for him corresponds to the one he made for Achilles in the *Iliad*. Only Achilles' shield was adorned with generic pictures of life: a city at peace, a city at war, a scene of harvest, a scene of dancing, and so on. Aeneas' shield is adorned with scenes from Roman history, history which is future to him—it is here that we read of Augustus at the Battle of Actium—and as he puts it on, Virgil says: |

He marvels at the scenes, events unknown to him,
And lays upon his shoulder the fame and fate of his
   descendants

Attollens umero famamque et fata nepotum.

The burden may well be a heavy one to bear, particularly if the bearer of it himself is permitted only an occasional prophetic glimpse into its meaning. And when such a glimpse is permitted him, it is likely to be anything but reassuring. *Bella, horrida bella* . . . 'Wars, hideous wars!' the Sibyl shrieks at him when he questions her in Book 6. 'You will get to Latium, all right', she tells him, 'but you will wish you had never come!' *Sed non et venisse volent.* 'Go, seek your Italy!' Dido tells him, and then prophesies:

Let him beg for help in his own land, and when he has accepted the terms of a shameful peace, let him not enjoy his realm, or that light he has prayed for, but fall before his time, and lie unburied on the sands.

Sed cadat ante diem mediaque inhumatus harena,

whereby Aeneas is included in an almost obsessively recurrent series of images of disgraceful and nameless death.

Labour, ignorance, and suffering are Aeneas' most faithful companions on his journey to Rome. And at once to intensify his own suffering and lack of fulfilment and to magnify the destiny he is serving, Aeneas must witness the happiness and success of others. In the third book he visits his kinsman Helenus in Epirus, and there he sees a copy of Troy, laid out in miniature. Aeneas is at first hopeful as he asks the prophetic Helenus for advice:

Now tell me, for I have divine sanction for all I do, and the gods have promised me a happy course, tell me the labours I must undergo, and the dangers I must avoid.

But a little later, when Anchises enters, and he must set sail again, Aeneas falls into despair:

May you live happy, for your destiny is accomplished; but we are called from one fate to another . . . You have peace, you have no need to plow up the sea and follow forever the forever receding shores of Italy.

> Arva neque Ausoniae semper cedentia retro
> Quaerenda.

What this and other like passages impress upon us is something subtly at variance with the stated theme of the poem. Instead of an arduous but certain journey to a fixed and glorious goal, there arises, and gathers strength, a suggestion that the true end of the Trojan and Roman labours will never arrive. It is not that Aeneas will literally never arrive in Latium, found a city, and win his wars. That is as certain as it is that Odysseus will return to Ithaca. But everything in the *Odyssey* prepares us for a fuller end to | Odysseus' labours: we are made always to expect his reinstatement in kingship, home, honour, and happiness. In the *Aeneid* every prophecy and every episode prepares us for the contrary: Aeneas' end, it is suggested, will see him as far from his fulfilment as his beginning. This other Italy will never cease receding into the distance.

There is another dimension to Aeneas' suffering as the bearer of too vast a destiny. Aeneas cannot live his own life. An agent of powers at once high and impersonal, he is successively denied all the attributes of a hero, and even of a man. His every utterance perforce contains a note of history, rather than of individuality. He cannot be himself, because he is wired for sound for all the centuries to come, a fact that is reflected in the speeches of the *Aeneid*. The sonorous lines tend to come out as perfect epigrams, ready to be lifted out of their context and applied to an indefinite number of parallel situations. Aeneas arrives in Carthage and sees the busy construction of the city.

> O fortunate you, whose walls already rise!

he cries out.

> O fortunati, quorum iam moenia surgunt!

The line is memorable, too memorable perhaps for spontaneity. What Virgil has done is to turn to peculiar account what is at once the weakness and the glory of much of Latin verse: its monumentality, and its concomitant lack of dramatic illusion.

But Aeneas' failure as a hero goes deeper than the formality of his speech. As he makes his way through the first six books, we see him successively divested of every personal quality which makes a man into a hero. We have seen how the weight of his mission is made to overwhelm him at the very beginning of the poem. In the second book, he is in a situation which above all calls for self-sacrifice in the heat of battle. But this is precisely what he is kept from doing. Hector appears to him in a dream and tells him not to die for his country, but to flee. 'For if Troy could have been saved', the ghost says almost with condescension, 'my right arm would have saved it.' We understand that Aeneas' words in the first book, when he was overwhelmed by the storm, have a deeper meaning than the parallel lines of the *Odyssey*: 'O thrice and four times happy, you who fell at Troy!' Odysseus spoke out of a momentary despair. Aeneas' words are true for all his life. His personal ties too are not kept intact: in his haste to get his father and the state gods out of Troy, he leaves his wife behind; and when he returns to fetch her, she is an empty phantom, who can comfort him only with another prophecy.

But the most dramatic episode and the one in which Aeneas

most loses his claims to heroism is Book 4. The tragedy of
Dido is lucid and deeply moving. But the judgement it leads us
to make on Aeneas needs some comment. Generations of Latin
teachers have felt it necessary to defend Aeneas from the charge
| of having been a cad. Modern readers are romantic, but a
Roman reader would have known that Aeneas did the right
thing. So the student is asked to forsake his own experience of
the poem for that of a hypothetical Roman. Another theory is
that Virgil somehow fell in love with, and was carried away
by, his own heroine. But we cannot explain Virgil by assuming that he did not intend to write as he did. It is clear that on
the contrary Virgil deliberately presented Dido as a heroine,
and Aeneas as an inglorious deserter. Dido's speeches are
passionate, and, in their operatic way, ring utterly true. Aeneas
can apologize only by urging that his will is not his own. 'If I
had my way', he tells her, 'I would never have left Troy to
come here at all.' 'I would never have fallen in love with you in
the first place', he seems to mean. 'I follow Italy not of my own
choice.' *Italiam non sponte sequor.* Of course he is right. Aeneas'
will is not his own, and the episode in Carthage is his last
attempt to assert himself as an individual and not as the agent
of an institution. And in his failure, he loses his claim even to
the humbler of the heroic virtues. For piety, in the Roman
sense, meant devotion to persons as well as the state. 'Unhappy
Dido!' the queen about to die cries out, 'is it now his impious
deeds become clear to you? They should have before, when
you made him your partner in rule. See now his pledge of
faith, this man who carries about his gods, and his ancient
father on his back.' For pious Aeneas, as he is called, and calls
himself, throughout, cannot maintain even his piety in a
personal way.

Two later passages serve to emphasize this. At the beginning
of Book 5, the Trojans sail to Italy, troubled by the death-fires
they see back in Carthage. 'For they knew what a woman is
capable of, when insane with the grief of her love disho-
noured.' The Latin is perhaps more blunt. Dido's love was
literally defiled, *polluto amore*, and Aeneas is its defiler. Later, in
the Underworld in Book 6, Aeneas meets Dido. He wants
reconciliation now, and begs forgiveness. 'I did not know the

strength of your love for me', he says. Again the implication is
clear. Aeneas did not know, because he could not feel the same
love for her; because he is not master of himself, but the servant
of an abstract destiny. Dido, speechless in anger, turns away.
Aeneas is modelled on Odysseus here, and Dido's shade is the
shade of Ajax in Book 11 of the *Odyssey*. Virgil strengthens the
emotions this scene creates in us by recalling the one scene in
the *Odyssey* where Odysseus meets a hero greater than himself,
and is put to shame by his silence.

But Dido, we remember, is also Cleopatra, and we must
consider the meaning of that identification. Dido-Cleopatra is
the sworn enemy of Rome:

> Rise thou forth from my bones, some avenger!
>
> Exoriare aliquis nostris ex ossibus ultor! |

invoking the fell shades of Hannibal; but she is a tragic heroine.
Aeneas, on the other hand, could have been, and for a while
seemed to be, Antony, losing a world for love. Only he must
in the end be Augustus, losing love and honour for a dubious
world. The *Aeneid*, the supposed panegyric of Augustus and
great propaganda-piece of the new régime, has turned into
something quite different. The processes of history are presented as inevitable, as indeed they are, but the value of what
they achieve is cast into doubt. Virgil continually insists on the
public glory of the Roman achievement, the establishment of
peace and order and civilization, that 'dominion without end'
which Jupiter tells Venus he has given the Romans: *imperium
sine fine dedi*. But he insists equally on the terrible price one
must pay for this glory. More than blood, sweat, and tears,
something more precious is continually being lost by the
necessary process; human freedom, love, personal loyalty, all
the qualities which the heroes of Homer represent, are lost in
the service of what is grand, monumental, and impersonal: the
Roman State.

Book Six sets the seal on Aeneas' renunciation of himself.
What gives it a depth so much greater than the corresponding
book of the *Odyssey* is the unmistakable impression we have
that Aeneas has not only gone into the Underworld: he has in
some way himself died. He descends carrying the Golden

Bough, a symbol of splendour and lifelessness.[1] The bough glitters and it crackles in the wind: *sic leni crepitabat brattea vento.* It sheds, Virgil says, a strange discoloured aura of gold; and it is compared to the mistletoe, a parasitic plant, *quod non sua seminat arbos*, a plant with no vital connection to the tree to which it clings. A powerful contrast to the culminating image of the *Odyssey*, that great hidden rooted tree from which the bedchamber, the house, and the kingship of Odysseus draw continuous and organic life.

Aeneas moves through the world of the dead. He listens, again the passive spectator, to the famous Roman policy speech of Anchises, a speech full of eagles and trumpets and a speech renouncing the very things Virgil as a man prized most:

> Let others fashion the lifelike image from bronze and the marble;
> Let others have the palm of eloquence;
> Let others describe the wheeling constellations of heaven;
> Thy duty, O Roman, is to rule . . .
>
> Tu regere imperio populos, Romane, memento . . .

When he emerges, so strangely, from the ivory gate of false | dreams, he is no longer a living man, but one who has at last understood his mission, and become identified with it. Peace and order are to be had, but Aeneas will not enjoy them, for their price is life itself.

And yet there is something left which is deeper than all this. It is the capacity of the human being to suffer. We hear two distinct voices in the *Aeneid*, a public voice of triumph, and a private voice of regret. The private voice, the personal emotions of a man, is never allowed to motivate action. But it is none the less everywhere present. For Aeneas, after all, is something more than an Odysseus manqué, or a prototype of Augustus and myriads of Roman leaders. He is man himself; not man as the brilliant free agent of Homer's world, but man of a later stage in civilization, man in a metropolitan and imperial world, man in a world where the State is supreme. He cannot resist the forces of history, or even deny them; but he

---

[1] See *'Discolor Aura'*, by R. A. Brooks (*AJP* 74 [1958], 260–80), the best article on the *Aeneid* to date.

can be capable of human suffering, and this is where the personal voice asserts itself.

> Someday these things too will be pleasant to think back on,
>
> Forsan et haec olim meminisse iuvabit,

he tells his comrades in Book 1. The implication is that when the great abstract goal is finally somehow reached, these present sufferings, seen in retrospect, will be more precious than it.

And so this pleasure, the only true pleasure left to Aeneas in a life of betrayals of the self, is envisaged as art. The sufferings of the Trojans, as Aeneas sees them in Carthage, have become fixed in art, literally: they are paintings. And it is here first, Virgil tells us, that Aeneas began to hope for a kind of salvation. here he can look back on his own losses, and see them as made beautiful and given universal meaning because human art has transfigured them. 'Look here!' he cries. 'There is Priam; there are tears for suffering, and the limitations of life can touch the heart.'

> Sunt lacrimae rerum et mentem mortalia tangunt.

The pleasure felt here by Aeneas in the midst of his reawakened grief is the essential paradox and the great human insight of the *Aeneid*, a poem as much about the *imperium* of art as about the *imperium* of Rome. The images in Carthage make Aeneas feel Priam's death not less deeply, but more. At the same time they are a redemption of past suffering, partly because they remove one element of the nightmare: final obscurity and namelessness, partly because they mean that we have found a form in which we can see suffering itself clearly. The brightness of the image and the power of pleasurable vision it confers, consoles for the pain of what it represents.

The pleasure of art in fact gives value to the pain itself, because tragic experience is the content of this art. Virgil continues the | scene in the art-gallery: 'He spoke, and with deep sorrow, and many lamentations, fed his soul on the empty pictures.'

> Atque animum pictura pascit inani.

'Empty'—*inani*—is the key-word here. Consider again how many times Virgil creates his most touching scenes by dwelling on how something substantial becomes empty and insubstantial: the phantom of Creusa, old fallen Troy, the apparition of Venus in Book 1, the shade of Dido in the Underworld, the lost pledge to Evander, the outraged life of Turnus. *Inanis* is the very word that describes the tears Aeneas sheds upon leaving Carthage and Dido: 'His mind was unmoved; the tears he wept were empty.' That is, *of no avail.*

　　　　Mens immota manet; lacrimae volvuntur inanes.

Aeneas' tragedy is that he cannot be a hero, being in the service of an impersonal power. What saves him as a man is that all the glory of the solid achievement which he is serving, all the satisfaction of 'having arrived' in Italy means less to him than his own sense of personal loss. The *Aeneid* enforces the fine paradox that all the wonders of the most powerful institution the world has ever known are not necessarily of greater importance than the emptiness of human suffering.

9

A Note on the Origins of Teleology

In *The Liberal Temper in Greek Politics* (New Haven, 1957), Professor Havelock makes a distinction between *genetic* and *teleological* approaches to Greek anthropology and political theory. The genetic approach is flexible and empirical, and although necessarily theoretical rather than experimental, yet comes closer to the methods of modern experimental science. It regards human life and human society as engaged in a gradual and continuous development from the animals who are our forebears to society as it existed in the *polis*, and beyond the *polis* to horizons of larger and more complicated social structures which the Sophists, whom Havelock regards as the proponents of this genetic view, could as yet only dimly perceive. Philosophically, according to Havelock, the genetic view denies the existence of eternal values. Values are deter-- mined by response to given needs, and as these change, so do those; so that the moral and political values and sanctions at stage $n+1$ of human society will be a function of needs and pressures encountered at stage n. Politically, the tendency to regard human beings in large social groups and even as a species, and the assertion that values are variables dependent on group needs, inclined the champions of the genetic view to favour democracy rather than oligarchy or aristocracy; and Protagoras, to choose an eminent example, is represented in Plato's dialogue, the fullest source for his doctrine, as offering what is in effect a rationale of Athenian democracy in the fifth and fourth centuries BC.

To state it most abstractly, the genetic view finds the determinant principle for a process of development, in this case that of human society, within that process itself. The primary values for mankind are simply the desire to live, and as an

Journal of the History of Ideas 26, 2 (1965), 259–62. Reprinted by permission of University of Rochester.

extension of that, the search for pleasure, materialistic and otherwise. These primary values have undergone, and will further undergo, countless modifications, all dependent on the needs and circumstances prevalent at any given moment.

In this as in other respects, the teleological approach is opposed to the genetic. The determinant values of human life are fixed and absolute, and they exist outside the development of human society; and even, to borrow a late teleological metaphor, outside the flux of this sublunar world altogether. Our knowledge of these ultimate values may be incomplete or false, but they are not made thereby the less immutable; and they are, finally, the cause, the αἰτία, the οὗ ἕνεκα, of every striving and every movement of mankind. The purpose that men do, or at any rate should, keep before themselves is not the fundamental desire to live and its concomitant, the search for pleasure, but rather *excellence*. For the ultimate cause of life, the parent of *genesis*, is itself supremely excellent, and the teleological thinkers, if they could at times attain a degree of detachment from which they could refer to human affairs as of little moment, yet had enough faith in themselves and in their fellow man to hold that divine excellence has an irreducible counterpart in the human soul.

The political orientation of teleological theory was naturally oligarchic. The very existence of ultimate values firmly fixed outside the process of human life presented a pattern for authoritarian rule within political society. And whereas human needs for survival, comfort, and enjoyment are | common to all members of the social group, it is obvious that excellence is a differentiating quality, which some men have in far greater degree than others. Once it is established that the purpose of society is not to provide happiness and security to its members, but to conform as much as possible to the model of the Heavenly City, it seems to follow that that portion of society which has most nearly attained excellence must guide and rule the rest. Those values which it is the philosophical man's burden to impose on society as a whole are intransigent and admit of no *negotiation*, to borrow Professor Havelock's convenient term. Since everything is, or ought to be, for the best, and there is only one best, it is the duty of the best men to tell others what is best for themselves. It is accordingly not surprising that the grandest, and, for all its charm, the grimmest, statement of authoritarian political theory in the ancient world, should share a place, in Plato's *Republic*, with what is possibly the most brilliant and most extreme elaboration of metaphysical argument for a teleological theory of value.[1]

What are the Greek origins of teleology? Professor Werner Jaeger offered us an answer in the opening chapter of his *Paideia*. He found the essential moral conceptions of Plato and Aristotle already present, if in unselfconscious and unintellectualized form, in the aristocratic assumptions of the Homeric hero. The ideal of ἀρετή, he argued, is the dominant motive of the Homeric nobility. The Homeric hero is thus following the same exalted self-interest that is recommended in the ethical works of the fourth-century philosophers, in deliberately conforming his life to the pattern of a higher good. Accordingly, Professor Jaeger did not hesitate to use Aristotle's *Ethics* to explain Homer, for he held that the *Aretegedanke* of both these authors is fundamentally the same, and something essentially Hellenic. In this notion, he was in agreement with Wilamowitz, who said, for example, that 'Socratic thought [die Sokratik] is nothing but the unfolding of a blossom whose first bud came with the birth of the Hellenic race'.[2]

But is the teleological morality of Plato and Aristotle actually to be found in Homer? It may be true, as Professor Jaeger argues, that the classical philosophers of the fourth

---

[1] The central metaphysical argument in the *Republic* appears in several stages and in different forms: the assertion of the supremacy of the ideal over the practical (472 B–473 B), the Forms as objects of knowledge (475 C–500 C), the nature of the Good (503 D–509 B), then the Divided line (509 D–511 E) and The Allegory of the Cave. The argument as a whole might be considered primarily Realist rather than Teleological, but these categories in Plato are almost identical: the *goal* of the Guardians' study (504 D 3) must be something unchanging and real, and conversely, in the *Republic* at least, as in the *Phaedo*, the Forms, the essential predications of things, are seen from the perspective of this world as the end of a series of causes and aspirations. Thus, the moving sentence which first defines the Good (505 D 11), Ὅ δὴ διώκει μὲν ἅπασα ψυχὴ καὶ τούτου ἕνεκα πάντα πράττει ... is clearly teleological; while 507 B 2 ff. in the same section of the argument is a précis of the Theory of Forms and, taken alone, seems a statement of Realism.

[2] '... die Sokratik nichts ist als die Entfaltung einer Blüte, zu der der Keim zugleich mit dem hellenischen Volke entstanden ist.' Commentary on Euripides, *Heracles Furens* II², 109 (quoted by J. Stenzel, P.–W. s.v. Sokrates [Philosoph]).

century constructed their ethical and political doctrines on the foundations of the heroic and aristocratic code of behaviour; but is it not equally true that their teleological systems involved a thorough reinterpretation of that code of behaviour, a reinterpretation which amounts | to a new ideology, and one which does not go back in any complete form beyond the thought of Socrates?

Ἀρετή is not an especially common word in Homer. And when it appears, it does not connote an independent entity representing the final goal of human action, but rather something limited and practical, a natural and empirical value arising out of the conditions of the warrior's life. This value is not itself unchanging; ἀρετή is a flexible term which can stand for any particular excellence. Thus, Periphetes of Mycenae (*Iliad* 15. 638 ff.) was the son of Copreus, a better son of an inferior father, better in all kinds of ἀρετή, in speed of foot and in fighting. The plural ἀρεταί, in this passage, is hardly more than a grammatical variant of the adjective ἀμείνων, and this adjective is explained by abilities immediately useful to a fighting man. Or consider the speech in Book 13 of the *Iliad* in which Idomeneus assures his comrade Meriones of his high regard for him. 'I know your ἀρετή,' Idomeneus says, 'what kind of man you are.' οἶδ' ἀρετὴν οἷός ἐσσι … So far is ἀρετή here from a definition of a superhuman quality as we might have it in the ἡ ἀρετὴ τί ποτ' ἐστί of a Socratic dialogue, that it is grammatically proleptic of a detailed personal description. Idomeneus goes on with a concrete illustration of his friend's *sang-froid* in an ambush. This particular description of an empirical virtue constitutes what he means by speaking of Meriones' ἀρετή.

As in Plato, Homer's ἀρετή is connected with knowledge. But in Homer, this knowledge is not derived from contemplation and abstraction, but from experience. When Ajax mocks Hector before the two heroes fight their duel in Book 7 of the *Iliad*, Hector answers: 'Do not make trial of me as if I were a helpless child. I know well battles and slaughterings of men. I know how to swing my dried ox-hide shield to the right, and how to swing it to the left, and that is what I call being a tough-grained fighter. And I know how to leap into the tumult of rushing horses, and I know how, in standing combat, to dance the War-god's dance.' The entirely experiential nature of Hector's definition of the fighting man is evident. Hector knows these things because he has been born to them. And so, in a moral decision involving life and death, Hector appeals to social values deriving from circumstance and experience, not from theory; this is from Book 6: 'Nor does my spirit bid me [to play the coward], since I have learned to be brave, and to fight among the foremost of the Trojans.'

Similarly, Odysseus, in a revealing passage in Book 11, resolves an interior dialogue concerning the question of whether it is better to stand his ground though isolated in battle, or to flee: 'But why does my spirit present me with such an argument? I know that cowards withdraw from fighting, but that a man who behaves like an ἀριστεύς in battle has no choice but to stand his ground and strike down another or be himself struck down.' The decisive verb ἀριστεύει in this passage is the one to look at. It does not mean to follow, or to realize, τὸ ἄριστον, as it is so easy to think if you start from the moral preconceptions of Plato. It means rather, as I have translated, to be an ἀριστεύς; that is, to fulfil the obligations of a member of the social class of ἀριστῆες. You are born an ἀριστεύς. You learn, by experience and by precept, the values and the imperatives that maintain you in that position. These values certainly derive from needs, and recognised as such: the need of a ruling class to confirm its position in society, the | need of a member of that class to prove himself in competition with his peers, even the need of society as a whole as it was then organized to have a ruling class of fighting men who would obtain booty and land for the rest. Such considerations may somewhat change our reading of the famous line 'always to be best, and surpass all other men'. Its chief idea is: to play your part in the game as a member of the aristocracy.

All this is not to diminish the depth of commitment of a Homeric warrior-prince to an ideal of behaviour; a noble ideal, and one, moreover, extending beyond mere prowess in battle. It is rather to urge that the values in which this ideal consists are empirical and consciously felt to be derived from an existing social structure and its needs; that Homeric excellence is very far from the overmastering philosophical abstraction of the political and ethical theories of Plato and Aristotle.

For a number of reasons, among which was its interest in oligarchy, teleological theory attempted to usurp the aristocratic values of Homer. It may be that much of Plato's influence is due to the success of this attempt; Plato's attempt, that is, to present his philosophy as the heir and the natural continuation of Homeric values and attitudes. To understand the many ways in which Plato goes about this would be to uncover many of the secrets of Platonic art. Here it will suffice to point to one example, Socrates' identification of himself with Achilles in the *Apology* (28 B ff.).

Socrates as Achilles is in a way the central image of the *Apology*; and it develops naturally into the image of Socrates as hoplite soldier standing his ground under the command of the god. Both Socrates and Achilles choose to die; but this similarity is not enough; for this would be only a likeness of personal choice. Now Socrates' personal choice in not giving in to his accusers is stressed by Plato: his anger, his dignity, his courage. But for a philosopher there must be over and above this the imperative of an external value; Socrates cannot defy his accusers merely because he wants to, or because he is Socrates; he is serving a higher, superhuman cause, that of the god, that of φρόνησις and ἀρετή. Achilles therefore must also have been obeying an external value; and so Socrates is made to give a thoroughly Sophistic interpretation of his motives. 'Let me die then', Achilles now tells his mother, 'after I have brought justice upon the wrongdoer!'

Now a moment's recalling of Book 18 of the *Iliad* will reveal how little relevant such thoughts of establishing justice are to the mood of the Homeric Achilles. In order to make Socrates more like Achilles, Plato makes Achilles more like Socrates. And the purpose of all this is to convince us that Platonic philosophy is not radical, but rather a deeply traditional development. But the teleological tendency of Socrates as Plato presents him, his tendency to abstract values and to fix them as transcendent imperatives, was in fact radical enough; and it is a mistake on our part to interpret Homer as a manifestation of this tendency. If we look at the matter historically, we can say that the ethical theories of Protagoras, who saw moral values as inherent in the structure of society, are quite as relevant to a just appreciation of the Homeric world as are those of Plato.

10

Have we Homer's *Iliad*?

THE great historian of Greek literature, Albin Lesky, singles out as the Homeric Question of our time the relation of the orally composed song to our texts of the *Iliad* and the *Odyssey*.[1] Since the original assertion of Milman Parry, in the early 1930s, that modern improvised narrative poetry (such as that of Yugoslavia) is in fundamental respects similar to Homer, almost every Homeric scholar has discussed the validity of the analogy, and taken up a position in regard to it. The names of Sir Maurice Bowra, T. B. L. Webster, Cedric H. Whitman, Denys Page, A. B. Lord, and G. S. Kirk come immediately to mind. The last-named scholar in particular has directly confronted the question, first in two articles, 'Homer and Modern Oral Poetry: Some Confusions' (*CQ* NS 10 (1960), 271–81) and 'Dark Age and Oral Poet' (*PCPhS* NS 7 (1961), 34–48) and then in his comprehensive book on Homer, *The Songs of Homer* (Cambridge, 1962), esp. pp. 55–101.

It is with the arguments and the conclusions of Kirk's first article, and their amplifications in his book, that we shall be concerned here.[2] These arguments open the widest perspectives upon the traditional questions of the origin and the transmission of the Homeric poems. And they involve a more fundamental question: what are the *Iliad* and the *Odyssey*? |

By modern or living oral poetry, any student of Homer will now mean that of Yugoslavia, as it is known to us by the at

---

[1] In his *Geschichte der griechischen Literatur* (2nd edn., Berne, 1963), henceforward referred to as *Lesky*, p. 34. The whole section on oral composition has been added since the first edition of 1957/8.

[2] The book will henceforth be referred to as *Songs*, the first article as *Poetry*. The second article deals primarily with the question of oral poetry in the post-Mycenaean Age, and does not often enter this discussion. If the first article is referred to nearly as often as the book, it is because it states in explicit and argumentative form some relevant ideas which then become the theoretical basis of what is said in the book.

Yale Classical Studies 20 (1966), 177–216. Reprinted by permission of Yale University Press.

once scholarly and romantic field work of Milman Parry, carried on between 1933 and 1935, and then by its continuation under M. Parry's assistant and successor in this endeavour, A. B. Lord of Harvard University. This work is so far available in the form of (*a*) articles by Milman Parry himself after the Yugoslav work was begun, especially 'Whole Formulaic Verses in Greek and South Slavic Heroic Song', *TAPhA* 64 (1933), 179–97; (*b*) articles by A. B. Lord;[3] (*c*) in Lord's publication of the songs from Novi Pazar (English translations appear in vol. 1), first of a projected series of volumes of Yugoslav heroic narrative verse collected by Parry and Lord and now in the Milman Parry Collection at Harvard University;[3] and (*d*) in Lord's recent book, *The Singer of Tales* (Cambridge, Mass., 1960).

It cannot be said that the discovery and publication of this Yugoslav material put the whole question of Homer in a new light: Parry had already done that several years earlier by the elaborate and precise examination of the text of Homer in his French *thèses*, *L'Épithète traditionelle dans Homère* and *Les Formules et la Métrique d'Homère* (Paris, 1928), which showed that our *Iliad* and *Odyssey* are composed in a traditional style designed to enable illiterate singers to improvise heroic song.[4] But the exposition of the living tradition in Yugoslavia, which Parry conceived as an external confirmation of his analytic conclusions, following as it does Parry's demonstration of the nature of Homeric diction, has made a powerful impression of its own. It has seemed to some that we can now observe living bards who are, to be sure, less great than Homer and work in a less great tradition, but who, apart from the differences in aesthetic quality, of which little need be said after one has

---

[3] Articles: *TAPhA* 67 (1936), 106–13; 69 (1938), 439–45; 82 (1951), 71–80; 84 (1953), 124–34; *AJA* 52 (1948), 34–44. The initial volumes of *Serbocroatian Heroic Songs* (vol. 1, subtitled *Novi Pazar: English Translations*; vol. 2, subtitled *Novi Pazar: Srpskohrvatski Tekstovi*) were published in Cambridge, Mass., and Belgrade in 1954 and 1953 respectively. I have not dealt here through lack of space with the important work of J. A. Notopoulos in modern Greek oral poetry.

[4] That the style is traditional and therefore *oral* (for composition in performance and not dependent on the use of the written word) may be taken as proved: it is not necessarily proved that our *Iliad* and *Odyssey* were composed orally. See below, pp. 210 ff., and my forthcoming introduction to *The Making of Homeric Verse: the Collected Papers of Milman Parry* (Oxford: Clarendon Press) [ch. 15 in the present volume].

recognized that they exist, are in all essentials like Homer himself. It is, against all hope, our dream come true: we can see and hear Homer sing! There is no doubt that such a dream animated the ardent though precise mind of Parry himself; and I hope it is not wrong to say that almost all of what Lord says in his fascinating book assumes this essential equivalence of the Yugoslav bard with the author of the *Iliad*.

With this belief Kirk disagrees. He is not alone in doing so: there are those who have objected on general grounds, e.g. Wade-Gery in *The Poet of the Iliad* (Cambridge, 1952); and there are others who ignore the whole question, speaking of Homer in terms which any validity at all of Parry's and Lord's work would show to be wholly inapplicable to Homer.[5] And Lesky himself in his careful and magisterial survey (*Lesky*, 53–8) seems to bring up the analogy, only finally to reject it in favour of a Homer who wrote and cross-checked in writing, and was far more like a literary poet than like the minstrel in the Serbian coffee-house.[6]

It is Kirk, however, who confronts the matter directly. In his key article he is debating with Lord and to some extent with S. Dow, who has accepted and generalized some of Lord's conclusions. The purpose of this article is to continue the fruitful | debate. For we have here not merely a scholarly argument: what is in question is our whole conception of the *Iliad* and the *Odyssey*.

Parry avoided the old Homeric Question: was the *Iliad* (to concentrate for the sake of convenience on the greater poem), substantially as we now have it, the product of a single designer, or is our text some sort of composite to which many hands contributed?[7] The proof of the traditional character of Homeric diction seemed to Parry to make this question almost otiose: even if one singer did put together our *Iliad*, his debt to the tradition was so great that the song could still be said to be a

---

[5] e.g. S. Benardete's study of Homeric epithets, 'Achilles in the *Iliad*', *Hermes* 91 (1963), 1–16.

[6] This does not mean that Lesky rejects Parry's proof of the traditional nature of Homeric diction and much that the study of Yugoslav epic can tell us. He is only aware of the limitation of the proof as I stated it above (n. 4).

[7] Cf. Paul Shorey's review of M. Parry's *thèses*, *CP* 23 (1928), 305–6.

direct manifestation of the tradition and the work of the generations of bards who made and preserved that tradition. The important thing was the style, and above that, the mood, of heroic poetry. This belonged to all bards when the tradition was in its vigour. The particular responsibility for our *Iliad* was incidental. Such seem to have been Parry's feelings on the question. At any rate, the revelation of how thoroughly the language of the *Iliad* is controlled by a formulary system which it took generations of bards to form was, as Parry clearly saw, one more hopeless impediment to any analytic solution of the old Homeric Question: the style of both *Iliad* and *Odyssey* was so uniform in respect of formula and metre that chronological layers or different hands could not conceivably be detected. Parry therefore contented himself with defining *Homer* for practical purposes as 'either the text of the *Iliad* and *Odyssey* or the poet or poets of these poems',[8] and never entering into the question further, except to state that old-fashioned analysis was impossible and to imply that it was irrelevant.

Absolute dating interested him little more. He regarded the Homeric poems as historical documents only in so far as they attested and embodied an heroic world where bards kept alive the tradition of heroic narrative. The only *date* he was concerned | with was the point at which the traditional style loses its hold on the singers of tales.[9]

Lord has been less austere in relation to these points. Dating has not engaged his efforts, but he has generally, though without argument, assumed that the *Iliad* and the *Odyssey* have each one author or have the same author.[10] And on the matter of the formation of the written text and the date of composition he takes a clear, and original, stand. Two principles determine the answer to these problems: (1) an orally composed poem cannot be handed on by the tradition of oral song

[8] *L'Épithète*, 3 n. 1.

[9] e.g. Parry's discussion (*L'Épithète*, 163–4) of the Hesiodic line fr. 94. 21 Rzach, where he argues, from the oddity and possible particularity of the phrase, that the epic tradition no longer 'possédait toute sa vigueur'. The argument there depends on the assumption that the Hesiodic poet disposed of the same stock of formulae as Homer.

[10] Cf. *Singer*, ch. 7, 'Homer', 141 ff., where he constantly refers to, e.g., 'the composer of the Homeric poems'.

without fundamental change;[11] and (2) 'the [oral] poet's powers are destroyed if he learns to read and write'.[12]

Kirk rejects the first of these principles and accepts the second. It may be, as we shall consider later, that he has made the wrong choice. But first we must look at the consequences of | Lord's principles; for it is disagreement with what follows from them that leads Kirk to his criticisms of detail.

If an improvised heroic poem is, as Lord is convinced, a protean thing which can never be reproduced unless it is somehow recorded then and there, at the moment of its performance,[13] then our text of the *Iliad* can only have come into being at the point when it was first put into writing. Even if what Kirk calls the 'monumental' composition took place earlier, still that large-scale song would have so changed with each singing, that 'Homer' must have been the man who performed it when it became fixed with the help of the alphabet. This will necessarily give to the composition of the *Iliad* a *terminus post quem* in the later part of the eighth century, depending on when it is believed that the use of the alphabet achieved some sort of currency.[14]

[11] Kirk quotes Dow's formulation of the principle (*CW* 49 (1956), 117) as an example of the extreme view he wishes to combat: '*Verbatim* oral transmission of a poem composed orally and not written down is unknown.'

[12] The formulation is Dow's, from a mimeographed sheet accompanying his recent Sather Lectures. Cf. Lord, *Homer's Originality*, 129 ff., and *Singer*, 124 ff. Actually, Lord is rather more cautious: there are oral poets who can read and write (*Singer*, 129). He insists only that they cannot *use* their knowledge of writing to help them compose a song. The song itself is either a product of the unlettered tradition (though conceivably the singer might be able on another occasion to write something), or it is a literary poem, which, whatever its merits, will have lost the qualities of heroic song. On this last point cf. the quotations from Kačić, *Singer*, 132, and Lord's comments.

Parry could not make a statement on this point in his pre-Yugoslavian work, since there was no evidence. But his firm distinction between poets of traditional, and poets of individual, style (esp. *L'Épithète*, 146 ff.) favours such a conclusion on logical grounds.

[13] Cf. Bowra, *Homer and his Forerunners* (Edinburgh, 1955), 9; and *Lesky*, 33.

[14] On the date of the introduction of the alphabet, and the uses to which it was first put, cf. H. L. Lorimer, 'Homer and the Art of Writing', *AJA* 52 (1948), 11–23; Bowra, *Forerunners*, 5 ff.; Kirk, *Songs*, 68 ff.; L. H. Jeffery, 'Writing', in *A Companion to Homer* (London, 1962), 545–59.

On the question which concerns many writers—were writing materials in sufficient quantity available to Homer—L. H. Jeffery in *A Companion to Homer* is cautious, saying we know nothing except that the Greeks knew what papyrus was by the time of Homer, that they learned it from Phoenicia (*Βύβλος*), and that it may or

More important to our concept of what we have in the text of the *Iliad*, the application of Lord's first principle gives us the comforting sense that our *Iliad* is in its essentials a faithful transcript of the song the great poet sang. Admittedly there may be scribal errors and interpolations; but the thing must have been put into writing at the moment of composition, and there is good hope—though neither Lord nor Bowra nor Whitman enters much into this phase of the problem—that the written text survived at least two centuries or so until the | Panathenaic Recension (if such existed[15]), and thence down through antiquity and at last to our own day. It almost seems, by one of the many paradoxes that greet us as the problem unravels itself, that the very fluidity of oral transmission is what guarantees us the *ipsissima verba* of Homer.

Lord's second principle enables him to explain how the unique transcription from improvised song to established text took place. The oral poet—and this was Parry's great principle, long before the empirical evidence of Yugoslavia, a principle in which he was hardly anticipated except by the genial speculations of Robert Wood[16]—is fundamentally a different kind of artist from the literary poet. The two kinds cannot mix, and when they seem to,[17] the apparent exception proves the rule. Homer, then, cannot have used the newly introduced alphabet to record his own poem, so someone else—a genius, in his way, but one who had no oral style to lose—must have done so. Hence Homer dictated his text to a scribe, as the Yugoslav bards dictated theirs to Parry and Lord and their assistants twenty-eight centuries later. Again we note the romantic—but not necessarily for that reason less scholarly—strain: not only, as Lord would have it, can we see and hear Homer sing; we can even be his scribe, we can do what the first man who put the *Iliad* into writing once did.[18]

may not have been *writing* papyrus. She thinks on the whole that if the poems were written down before the sixth century, they were probably written on leather. H. L. Lorimer, dating the *Iliad* to 750 or earlier, sees no difficulty at all in there being an adequate supply of papyrus in the later part of the eighth century.

[15] Cf. J. A. Davison in *Companion to Homer*, 237 ff.
[16] *On the Original Genius of Homer*, 4th edn. (London, 1824), esp. 157 ff.
[17] See above, n. 12.
[18] Cf. Lord, 'Homer's Originality: Oral Dictated Texts', *TAPhA* 84 (1953), 124–34, esp. 132.

It is fairly obvious that this theory of the creation of our text involves some difficulties, although Lord presents it as what we are left with after the elimination of all impossible alternatives. Some of these difficulties have not been stated either by Lord, who champions the theory, or by Kirk, who doubts it. But we must now look at Kirk's own position.

He is unimpressed by the dictation theory (*Songs*, 98 ff.) but, as we indicated, he is 'prepared to accept absolutely' the | premise on which it is based: that literacy destroys the heroic singer's craft. Kirk can accept the premise but deny the conclusion because he holds that the other premise, Lord's first principle, 'that the poems must have been written down as soon as they were composed because otherwise they could not have been transmitted, is fallacious and must be absolutely rejected as it stands' (*Poetry*, 279).

Kirk in fact does want the 'monumental composition' of the *Iliad* to be a completely oral process, in accordance with Lord's second principle. But he does not want the work as a whole put into writing for at least 100 years after composition. His reasons for this assertion are negative, and he would evidently argue that they must necessarily be so. They are the difficulty of imagining the process of making a very large book at the probable time of composition, and the alleged interpolations such as the Doloneia and the end of the *Odyssey*, which presumably could have been added more easily to a written text than to one handed down by repeated singing. To these he would now add (in conversation with me) the argument that the written recording of a poem as long as the *Iliad* or the *Odyssey* in the late eighth century would have been an event of too great magnitude not to have left a memory of its own.

None of these reasons appears to be unambiguous or decisive. There is no evidence whatever that the act of writing on so large a scale would not have been possible at that date. It would, of course, have been a remarkable event. But the introduction of the alphabet was itself a remarkable event. And so was the composition of the *Iliad*. The interpolations are themselves questionable. The end of the *Odyssey* will seem to many an essential, if in some respects unsatisfactory, portion of that poem. And there will always be some who feel that the

tenth book is not extraneous to the *Iliad*. But if they, or other
passages, were added to the Homeric poems after the main
composition, this really tells us nothing of the state in which
the main compositions were transmitted. Interpolations, particularly if singing on a smaller scale continued, could always
be made in a | written text. As for the lack of any record of the
epoch-making (epic-making?) act of putting such long poems
into writing—is this any stranger than our virtually complete
lack of any record of the person Homer himself, on any
theory?

Denys Page, whose *History and the Homeric Iliad* has exerted
a strong influence on Homeric scholars, wanted an extraordinarily early date for Homer—around the end of the ninth
century. Since this would place the date of composition well
before the introduction of writing, it would make Kirk's
notion of *reasonably accurate oral reproduction* a necessity. But
Kirk himself, although he raises the possibility that Page's date
may be right,[19] is much more inclined to the now commonly
accepted late eighth-century date; and Page's own reasons
appear curiously casual in the texture of his brilliantly argued
book.[20]

Given the forcefulness with which Kirk argues the point, the
lack of genuinely strong arguments for an illiterate Homer
who once sang a song then reproduced accurately by word of
mouth for six generations (or three or four or five) is
disappointing. On the face of it, the idea seems hard to accept,
harder indeed than Lord's notion of a prototypical Parry or
Lord sitting down with Homer and a batch of appropriate
writing materials over a period of several weeks *c.* 725 BC. Had
Homer, Kirk can be asked, often sung the *Iliad* all the way
through? Given the experience of Avdo Međedović, the one
bard Parry and Lord found who sang a song comparable in
length to our *Iliad*, it was not a thing to be undertaken lightly.[21]
Did he say on some occasion: 'I'll never do it better than that:

---

[19] *Songs*, 287.
[20] *History and the Homeric Iliad* (Berkeley, 1959), 158. One wishes one knew who all
these 'less famous and interesting' persons flourishing *c.* 725 and well-remembered by
later generations were.
[21] Lord, 'Homer, Parry and Huso', *AJA* 52 (1948), 34–44, esp. 42. Avdo's song
may seem to an outsider a case of the effect of the observer on an experiment.

reproduce *that* version!'? Or were his versions so alike that his apprentices began to reproduce the whole song (each reproduction taking at | least a week), having learned it by repeated hearing? And were *their* successors so struck with this one version that they too reproduced it for all intents and purposes accurately, being willing to spend months or years working it into their repertories? The whole process by which a poem of that length (to say nothing of the *Odyssey* as well) becomes frozen into what anybody must admit was a fluid tradition before Homer, and then retains the stamp of its first singer over many generations of men, is very obscure; and I think the truth is that Kirk has not been able to imagine just how it did happen. Perhaps it is commendable that he has not tried to do so; but the difficulty may lead us to suspect the value of the theory itself.

In addition to the negative and uncertain reasons I have discussed above, Kirk seems to have been led to his theory by impatience with the arguments of Lord, or at least with the certainty with which these arguments were advanced by Lord and accepted by others. In *Songs* (99) he states: 'It is of itself improbable that writing and bookmaking techniques could cope with anything on this scale in this period', and continues 'and, as I think, no evidence . . . exists that such dictation was necessary for the composition of the monumental poems.'

We must at this point first realize that Kirk's hypothesis is unnecessary, even if we grant it possible; for it does not explain any facts which cannot be explained otherwise. But it may then be objected that any theory accounting for the transition of the Homeric poems from oral song to written text is conjectural, so that it becomes a matter of choosing according to our taste and our intuitive calculus of probabilities. This might be so if Kirk's theory of adequately accurate oral reproduction were not ruled out by what we know and can legitimately conclude of the processes of oral poetry. We must therefore consider those arguments which Kirk offers to show that his theory is not so ruled out.

We find at the outset a general difficulty. Kirk argues simultaneously that (*a*) the Yugoslav poetry itself shows the possibility of reasonably accurate oral transmission, and (*b*) accurate oral transmission of Homer was possible because Homeric poetry was significantly different from the Yugoslav. To be sure, (*a*) and (*b*) are not absolutely contradictory, but there is clearly some arbitrariness of choice here. Kirk selects certain material from the Yugoslav analogy, and rejects the rest, without setting forth clear and valid criteria of selection.

But the particular arguments will tell us most. Kirk has studied the Novi Pazar songs with sensitivity and care, and his fair report of the relevant data in this volume is one of the many valuable things in his own presentation. His interpretation is not, however, everywhere indisputable. He states a number of times that accuracy of reproduction is the aim of the Yugoslav bard (we shall see that this helps him to form his distinction of *reproductive* and *creative* bards, to be discussed below): '. . . we see clearly that it is the aim of the best singers in the Novi Pazar region to reproduce exactly each song that they hear' (*Poetry*, 275); 'It is their professed ideal to achieve *verbatim* precision' (276); 'Complete verbal accuracy is even now the ideal' (277); and so on.

Kirk leans heavily on these assertions of the Yugoslav poet's purpose. We must comment first that there is here a confusion of language: 'reproduce exactly', 'professed ideal', '*verbatim* precision' are not the terms in which bards speak. Nothing in what they say can be legitimately translated by such expressions; nor is it even possible to translate them back into language which they would understand. The following extract from an interview between Nikola Vujnović, Parry's assistant, and the singer Đemo Zogić of Stolać will make this most clear. Đemo was describing his success in singing in the place of another singer, Suljo Makić, the song that Makić was to sing.

N.   Was it the same song, word for word, and line for line?

Đ.   The same song, word for word, and line for line. I didn't add a single line, and I didn't make a single mistake . . .

N.   Tell me this, if two good singers listen to a third singer | who is even better, and they both boast that they can learn a song if they hear it only once, do you think that there would be any difference between the two versions? . . .

Đ.   There would . . . It couldn't be otherwise. I told you before that two singers won't sing the same song alike.

N.   Then what are the differences?

Đ.   They add, or they make mistakes, and they forget. They don't
sing every word, or they add words. Two singers can't recite a
song which they heard from a third singer and have the two songs
exactly the same as the third.

N.   Does a singer sing a song which he knows well (not with
rhymes, but one of these old Border songs), will he sing it twice
the same and sing every line?

Đ.   That is possible. If I were to live for twenty years, I would sing
the song which I sang for you here today just the same twenty
years from now, word for word.[22]

I have quoted this at length because it shows so strikingly
how far the bard is from any understanding of verbal accuracy
in our sense. (Kirk, by quoting only part of the above passage,
unwittingly gives a misleading impression, *Poetry*, 275.) The
fact is that a member of an illiterate culture not only does not
conceive of verbatim accuracy as we know it, but is also
psychologically incapable of grasping the abstract concepts
implied by such terms. The Serbo-Croat expression 'word for
word' (riječ za riječ) may seem at first sight to be equivalent to
Kirk's phrases: an examination of how the bard uses such a
phrase reveals that this is not the case.

Let us turn from what bards say to what they do. By far the
best example of accurate reproduction we have from the
Yugoslav material is Zogić's two versions of 'Alija Rescues
Alibey's Children': in other cases we find much greater
variation. Between Zogić's versions, sung at an interval of
seventeen years, we find, in Kirk's words, 'comparatively
minor differences, involving the occasional substitution of one
formula or line for another and the addition or subtraction of a
number | of incidental themes. Lord lists twenty-three alte-
rations of various degrees of importance in a song of which the
later version was something over 1,430 lines long and some
sixty lines longer than the earlier version' (275). Let us try to
assess as objectively as we can the *rate of change* here. A song of
1,370 lines becomes one of 1,430 lines, with changes in theme
and formula amounting to twenty-three distinct alterations.
The quantitative difference is sixty lines out of 1,370, which is

---

[22] Lord, *Singer*, 27.

about 4.4 per cent. This is a change taking place in one singer's performance in less than a generation. Over six generations, this would give us a quantitative difference of 6×4.4 per cent $= 26$ per cent. But this of course is far too simple a measure. Within the song there were twenty-three alterations. 6×23 (assuming but one minimal unit of change a generation) will give us 138 alterations in a song of about 1,400 lines. But this implies that a singer's successor will reproduce his song as accurately as the singer himself, which can hardly be the case. It also assumes that a song the length of the *Iliad* can be as accurately reproduced as a song the length of Zogić's song of Alija, which is again unlikely in the extreme. It is surely evident that an 'Iliad' transmitted in a series of oral perform-ances over six generations would end up as something vastly different: so different that it is surely the singer at the end of this process whom we should think of as the author of our text, rather than the hypothetical singer at the beginning who first put together the story of the *Iliad*.

This brings us to the vital question which no one has yet confronted clearly: what is the essence of the *Iliad*? How much would our vulgate text have to be changed before a reasonable student would have to say: 'This is no longer the *Iliad*, it is a song sung in much the same style, treating of similar themes'? To this question no precise answer can be given. But until we are ready to give it some kind of answer, we have no right, I submit, to talk about *accuracy of reproduction*; for to talk about such matters at all, we must have some clear and rational notion of what is, or is not, being reproduced. |

The whole Homeric Question,[23] from its beginnings with writers like d'Aubignac and Robert Wood to the present day, is, after all, a function of one thing: the overwhelming and universally acknowledged greatness of the Homeric poems *as we have them*. It is because these poems, and particularly the *Iliad*, are manifestly among the supreme creations of the

[23] For the history of the Homeric Question see Nilsson, *Homer and Mycenae* (London, 1933), 1–55; H. L. Lorimer, 'Homer and the Art of Writing', *AJA* 52 (1948), 11–23; J. A. Davison, 'The Homeric Question', in *Companion to Homer*, 234–66; *Lesky*, 49 ff.

human mind, that d'Aubignac wrote his attack on the concept
of a single author, that Wolf wrote his *Prolegomena*, that
Schliemann dug at Mycenae and Troy, that Parry carried
through his examination of Homeric diction, and that today
still the problem of Homer exerts so powerful a fascination
over the minds of scholars and all educated men.

The point hardly needs labouring; but it must be made here.
For the question, 'How elastic is our concept of the *Iliad*?' is not
other than the question, 'In what does the greatness of the *Iliad*
lie?' Neither of these questions can be answered categorically
here. But some answers are clearly wrong, or insufficient. The
greatness of the *Iliad* cannot be that of any poem composed in
the formulary style examined and described by M. Parry. The
Odyssey for all its excellence is evidently a less great poem than
the *Iliad*. The Homeric Hymns are composed in a style
sufficiently like that of the *Iliad* and the *Odyssey* to demonstrate
that the formulary style can be used to create poems unquestio-
nably inferior to them both. And within the *Iliad*, some
passages are so much more powerful and moving than others
that we must agree without hesitation that the Homeric style
can be employed with as much final variety of quality and
effect as, say, the individual styles of Elizabethan blank verse.

But it is also not true that the story of the *Iliad*, that is, its
succession of themes,[24] what makes it great. Its excellence is | no
more mere μῦθος than is that of Sophocles' *Oedipus*.[25] For in
that case a translation, and even a paraphrase or extended
summary, would have the same greatness.

Nor is the greatness the negative one of the consistent
avoidance of inept or obscure expressions over a considerable
number of lines. We may or may not agree with Longinus'
contention that negligence is a mark of artistic genius as it is of
great wealth,[26] but we must surely agree with him that purity
from definable faults, even if the *Iliad* showed this quality,
cannot confer greatness.

If these qualities, the proper use of the formulary style, the
plot, and freedom from awkwardness of expression, cannot

[24] Using *themes* more or less as Lord uses the term, *Singer*, 68 ff.
[25] Cf. R. A. Brower, 'The Heresy of Plot', *English Institute Essays, 1951* (New
York, 1952), 44 ff.
[26] Περὶ Ὕψους 33. 2.

account for the greatness of the *Iliad*, can their combination do so? Kirk declares (*Songs*, 82): '[Homer's] originality did not lie in the choice of specially appropriate epithets or phrases, but on the one hand in the whole conception and scale of the poem, on the other in the consistently fluid and adept handling of traditional phraseology. Not every singer of his time would be capable of systematic creation, of constructing such lines as his, of extruding clumsy locutions, as effectively as the main composer of the *Iliad* . . .' 'Constructing such lines as his' and 'systematic creation' seem to be all we have in definition of 'fluid and adept handling', and this seems to amount to keeping clumsy expressions out of the traditional diction. We have then a statement of the three criteria we discussed above: plan of the poem, adherence to the heroic style, and avoidance of awkwardness. But if no one of these criteria can take us far toward grasping the peculiar excellence of the *Iliad* it may be that the three of them together will not do so. So far we have a long poem with a good plot—though not exactly a tightly constructed one: Kirk is the first to admit that there are long developments of doubtful relevance and structural difficulties to boot—composed in the heroic style, and free *most of the time* from blunders. Does this make an *Iliad*? |

The answer of course is No. To say what does would require an extended work of rigorous criticism, and that is not possible here. But perhaps we can see the direction of such criticism. If we should try to define the *Iliad* in a general way, we might think of a long poem dealing critically with the heroic conception of life.[27] The values of a heroic society are everywhere presented to us in the *Iliad* and they are constantly presented to us in different aspects. Distinct figures—Agamemnon, Achilles, Odysseus, Paris, Hector—are dramatized for us in crucial situations, and their *attitudes* as they speak and act in conflict with each other constitute the real force of the poem.[28] For these conflicts to be expressed, there must be a plot

---

[27] Cf. W. Sale, 'Literary Values and the Homeric Question', *Arion* (Autumn 1963), 86–100. The values of the warrior as such are not the only values in the *Iliad*: there are hedonistic values (Paris), patriotic values (Hector), domestic values (Andromache), and so on. But they are all seen in relation to heroic values, expressing themselves in a world where heroic values are dominant.

[28] 'Attitudes' as in I. A. Richards, *Principles of Literary Criticism*.

to occasion them. The fierce and immediate resentment of Achilles in Book 1,[29] his meditated disillusionment in 9, his reconciliation with his own society, with the gods and with his enemy Priam in 24—these are some of the things that make up the *Iliad*, and they require the plot of the poem to work themselves out. But these key attitudes and these great dramatic moments involving the chief characters of the poem could not have their value if we did not see them in the perspective of the whole heroic universe. Book 22 would be an admirable piece of poetry if it were all we had. But the encounter between Achilles and Hector gains immeasurably from other encounters elsewhere in the poem, for example from the chivalrous encounter of Glaucus and Diomedes in 6, which presents a different and strongly contrasting view of the heroic mode and heroic feelings. And this scene itself has its significance partly determined by the ἀριστεία of Diomedes which im|mediately precedes it.[30] Sarpedon's sense of the obligations of the warrior-prince in 12 enlarges our understanding of Achilles' rejection of the embassy earlier in 9. And again, the scenes on Olympus—and Ida—define the lot of man by contrast as they do that of the gods directly.[31] There are very few 'incidental' themes in the *Iliad*. The term implies an ornamental view of poetry, an inadequate understanding of the interrelation of the parts of the dramatic structure.

[29] Some good remarks on the moods of Achilles in *Songs*, 353.

[30] The differences between Diomedes' temporary exaltation in battle with the gods (*aut ope Palladis / Tydiden superis parem*) in 5 and his humanity in the encounter with Glaucus in 6 have been taken by some as symptomatic of multiple authorship. It is true that if we are looking for large sections of the *Iliad* which could be removed with the least disturbance to the whole, 5 would (in my opinion) be as good a candidate as 10. But that does not mean that its function in the whole structure is unclear. The extravagant heightening of Diomedes' powers in 5, with Athene by his side, until the encounter with Apollo (which anticipates 16), serves to delimit human potentiality. And the almost puppetish godlikeness of Diomedes in 5 contrasts with Achilles' deliberate and self-sufficient sense of his powers and limitations in battle later, e.g. 19. 420 ff., 21. 99 ff., 22. 15 ff.)

Diomedes has his moment of brittle greatness when the gods are with him (or, if you will, he outdoes himself as he is caught up, like a *berserker*, in the fury of battle); once the gods (or his battle mania) leave him, he becomes very human again, even a little ordinary. 6. 141 ff., where he states this himself, far from being out of harmony with 5, is a clear reference to it. And the return to humanity which the Glaucus–Diomedes scene represents is a transition from the battle to the latter domestic scenes of 6.

[31] The similarities and the differences between the quarrels of men and of gods in 1 is an obvious example.

To offer a succession of scenes so comprehensively evaluat-
ing the human situation, to present them in a dramatic
trajectory which reaches a climax in 22 and ends in the
resolution of 24, requires an artistic construct of the highest
order. It will seem evident to many (Kirk and Lord on the
whole among them)[32] that such a construct could only be the
design of a single mind. But that point need not be made here.
The fact is | that the construct exists, and to describe it is to
describe what the *Iliad* is.

This construct is in the first place a certain proper order of
scenes. The scenes must be of the right length, and they must
follow each other in the right sequence. To use Homer's own
critical term here, the poet must sing κατὰ κόσμον.[33] But a
scene is not a given block of material, as some critics seem to
assume. It is itself made up of parts, and will be more or less
effective, and more or less able to fulfil its function in the
economy of the whole, as those parts are more or less well
arranged. The κόσμος of Homeric poetry, then, involves the
quality of language as well as the construction of the plot.

Of this language there are two essential things to be said.
First, we must throughout the poem have persons (and the
poet himself) speaking the same language and expressing the
same assumptions. For the unity lies, as we said, in a constant
critical presentation of heroic values. The homogeneous formulary language is necessary to accomplish this. Everyone,
from Zeus to Thersites—who echoes, as we know,[34] the speech
and attitudes of Achilles—must in some way talk the same.
And the Homeric descriptions and similes, which are also
dramatic scenes, must continually evoke and further define this
heroic world. Herein lies the 'golden light' which Homer sheds
over the world, and herein is the nobility and strangeness of the
language, which must be everywhere 'holy and sweet and
wonderful'.[35] The heroic diction, which the poet uses because he
is in a tradition that knows no other, forces the concentration

[32] The carefully qualified discussion in *Songs*, chs. 10 and 11, 'Structural Anomalies
in the *Iliad*', and 'Structural Anomalies in the *Odyssey*', gives us a 'main composer' for
each poem.
[33] 8. 489. The art of the bard resembles that of the general.
[34] e.g. Whitman, *Homer and the heroic Tradition* (Cambridge, Mass., 1958), 161.
[35] M. Parry, 'The Traditional Metaphor in Homer', *CP* 28 (1933), 30–43.

on heroic values which is the ultimate subject of the poem.

But, and this is the second point, this language must vary with his speakers. We must be aware, from everything that they say, and that Homer says about them, that Achilles and | Odysseus are in the same world, but we must also be aware that they occupy distinct and different positions in that world. The irreducible character and attitude of each man must show through. The analysis of formulary diction shows us that there can be no or very little individual vocabulary and individual combination of single words.[36] Therefore the individuality which is so obviously there, and so much a part of the poem's greatness, must lie in the *juxtaposition* of formulae. Achilles and Odysseus must use the same phrases: but they combine them into speech in separate ways.[37] This process can work between one half-line and the next, or even between smaller parts of lines, but it is easiest for us to see it in the case of whole lines. For example: Achilles' words in 1. 155–7 are not compelled by grammar or connection of formulae. The way they move from indignation ('The Trojans never wronged *me*!') to the nostalgic sense of his distance from home, culminating in the overpowering line 157, is peculiar to Achilles. No one else in the poem speaks in this way.[38]|

[36] The idea that *all* of Homer is formulary, commonly attributed to M. Parry, was already held and stated by A. Meillet (see *Ep. Trad.* 9–10). Parry himself was rather cautious about this, although as early as his University of California M.A. thesis he clearly felt that virtually all of Homer was somehow *traditional*, and in his later articles (e.g., 'About Winged Words', *CP* 32 (1937), 59–63) he all but stated that formulae filled the whole of the text. Recently debate has centred on the exact definition of 'formula', e.g., Hainsworth in *CQ* 58 (1964), 155–64, answered by J. A. Russo in *YClS* 20 (1966), 221 ff. But the points to bear in mind are that (1) a very large amount of Homer is demonstrably formulary, (2) if we had more ancient Greek heroic poetry, this amount would increase, and (3) the attempt to distinguish parts of Homer which are not formulary is bound to fail, because we do not have enough evidence. What we are beginning to be able to do is to see how the poet combined his formulae in effective ways. This will be one important future direction of Homeric criticism.

[37] See A. Parry, 'The Language of Achilles', *TAPhA* 87 (1956), 1–7, now reprinted in G. S. Kirk, *Language and Background of Homer* (Cambridge, 1964) [and as ch. 1 of the present volume].

[38] 157 is not *evidently* formulary: we do not find it elsewhere (or its parts) in the *Iliad* or the *Odyssey*. But to argue that its effectiveness derives from the poet's having transcended the tradition to invent a special phrase here is wrong: first, because elsewhere equally effective lines do turn out to be formulary, but more important, second, because a criticism is poor which could be undone by a chance discovery, such as that of another poem containing a regular use of the line.

No one else in our *Iliad* speaks in this way; but it is evident
that in the heroic tradition others could. It is easy to imagine an
epic poem (by Homer or someone else) in which Odysseus
quarrelled with Agamemnon and expressed regrets over his
distance from home in the same way.[39] But if Odysseus were
allowed such emotional and imaginative developments of his
thoughts in the *Iliad*, we should have lost something fundamental. No longer would these two characters be symbolic for
us of something like the passionate and practical attitudes to
life. The argument between the two men in 19, for example,
would lose much of its force. No one else in the *Iliad* questions
the heroic code—that is, the value of leaving one's home to
fight and win possessions—so profoundly as Achilles, although
in the legend, we note, it is Odysseus who is marked as the man
originally reluctant to leave his home, not Achilles. To express
this questioning, Achilles, here and throughout, has his own
use of the formulary language.[40]

The lines discussed are an essential part of one scene, the
third in the rapid succession of scenes that brings us into the
poem in 1. Now a reader—like Homer's audience—will not
normally isolate them for attention as we have done here. But
they will be for him none the less an essential part of that scene.
Granted, that if they were not in our text we should not
constate a lacuna.[41] But the scene would be poorer without |
them. Take them away or replace them by something else, and
we have a weaker conception of Achilles and hence a weaker
conception of the crucial conflict between him and Agamemnon.

This is a small example of how each scene, like the work as a
whole, must be 'systematically constructed', i.e. how the value

[39] Odysseus does quarrel with Agamemnon in 14. 82 ff., and the motive for his
anger has something to do with the necessity of staying away from home. Then there
is the statement in his rhetorically masterful speech in 2 (esp. 291 ff.) of the hardship of
separation from one's wife. The difference of the grim resignation of the speech in 14
and the sincere, but objective and policy-conscious, concession in 2 from anything
Achilles says or could be imagined as saying, is as good an illustration as any of my
point.

[40] See above, n. 37.

[41] The famous statement of Aristotle (*Poetics*, 1451[a]31 ff.) on what constitutes the
wholeness of a work of art has been interpreted too literally to mean that all the parts
of a work of art are essential in the same way as the moving parts of a machine. But
art is not like that.

of each scene depends on a particular economy of treatment within it. Now here is a larger example. The τειχοσκοπία and with it our introduction to Helen ends with the famous lines about Helen's brothers, for whom she has looked in vain in the Achaean army:

> So she spoke; but the fruitful earth already possessed them
> Back in Lacedaemon, in their own native land.

Then the poem turns abruptly to the battlefield and the preparations for the duel: κήρυκες δ' ἀνὰ ἄστυ . . . To the whole scene, marked by the surprising affection between Priam and Helen and Helen's rueful memories of her homeland and those she knew there, the poignant brevity of these lines forms a perfect close. A popular modern translator, however, replaces them by the following: 'The truth was that both her brothers were long since dead and buried in their native country, after they had ambushed a pair of rival twins from Messene: Idas and Lynceus. None of the four survived that fight.'

The addition is altogether out of harmony with the scene. The legendary information about the death of Helen's brothers, so like what we actually find elsewhere in the *Iliad*,[42] here destroys the effect of the closing lines of the scene. It is essential that we think not of how or why Castor and Pollux died, but only of the fact that they are dead, and dead back in Lacedaemon, and that Helen is so far from the life she bitterly regrets (εἴ ποτ' ἔην γε) that she does not know it. Certainly the scene is *recognizable* if we read it in the translation with the translator's appendage; but the effect of the ending is no longer | of the *Iliad*. If a number of similar changes were made in the body of the scene, it would indeed no longer be the scene we know in the *Iliad* at all.

We must consider further that the ostensible purpose (or so it has been said) of the scene is to introduce us to the captains of the Greek army. But it fulfils in fact the much more important purpose, here in the third book, of introducing us to Helen herself. This second purpose is woven into the first, after the

---

[42] I mean the filling-in of genealogical or other background, usually by means of *Ringkomposition*—(see *Lesky*, 84), e.g., 9. 555 ff., where it helps to create Phoenix's particular type of garrulity.

dialogue between Priam and Helen begins, by a succession of carefully placed touches throughout the scene. Helen does not merely declare her feelings at the beginning of the dialogue (172–6). She intersperses her description of the Greek leaders with delicate references to herself.[43] Thus in 230–3, she varies the pattern of the dialogue by singling out Idomeneus on her own, instead of answering Priam's questions as she has done before. The strongly formulary responses[44] οὗτός γ' Ἀτρεΐδης (178), οὗτος δ' αὖ Λαερτιάδης (200), οὗτος δ' Αἴας ἐστί (229) here give way to the more abrupt Ἰδομενεὺς δ' ἑτέρωθεν (230). This shows us the poet's skill in varying a pattern at the right moment, and it indicates as well that Idomeneus is a less outstanding leader (physically and otherwise) than Agamemnon, Odysseus, and Ajax. But more important, the spontaneity of the response makes it more natural for Helen to touch on her own past:

> πολλάκι μὲν ξείνισσεν ἀρηΐφιλος Μενέλαος
> οἴκῳ ἐν ἡμετέρῳ ὁπότε Κρήτηθεν ἵκοιτο.   (232–3)

The four lines about Idomeneus could have been left out without disturbing the 'fluidity and adeptness' of the formulary | flow of the narrative. Once he is introduced in 230–1, 232 is not compelled, and again 232 could stand without 233. But the whole scene would have been less what it is if any of these lines were not there. On the other hand, it would surely not have been improved, quite the reverse, if the theme of Idomeneus had been further developed. A complex and delicate reference would have become a digression.

Another self-revelation in the midst of Helen's description is line 180, again uncompelled by grammar or sense, after her identification of Agamemnon:

> δαὴρ αὖτ' ἐμὸς ἔσκε κυνώπιδος, εἴ ποτ' ἔην γε.

---

[43] Cf. K. J. Reckford, 'Helen in the *Iliad*', *GRBS* (Spring 1964), 13.

[44] The formulary language of the *Iliad* and the *Odyssey* gives a kind of ritual quality to all the poetry, although this quality obviously does not inhibit spontaneity and dramatic force. But within the world created by this language, some passages are more ritual, emphasize more the recurrent patterns of life, than others; e.g., arming-scenes, sacrifices, or the passage in 1 (432 ff.) describing the arrival of the ship in Chryse, on which Russo comments in *YCIS* 20 (1966), 228 f.

The poignancy of the last phrase needs no comment; but it is revealing to consider that almost every reader of Homer has been struck by it, whereas few who were not thumbing the pages of the text or of Schmidt's Parallel-Homer will recall the other occasions the identical formula is used in the identical part of the line, and to express similar regret (*Il.* 24. 426, *Od.* 19. 315, 24. 289). Its peculiar effectiveness here comes about partly from the context, because we have already caught Helen's character and mood, so that we know exactly how to intone the formula. But it also helps to create the context, deriving much of its emotion from the juxtaposition with κυνώπιδος. This formulary word occurs always in this grammatical case, always in this position in the line. Sometimes it joins the preceding group of words, sometimes the following. The closest parallel here is 18. 396 ff., Hephaestus to Thetis, who has come to ask him to forge new armour for her son:

> μητρὸς ἐμῆς ἰότητι κυνώπιδος, ἥ μ' ἐθέλησε
> κρύψαι χωλὸν ἐόντα . . .

Hephaestus is a character with whom our emotions are little involved. The complaint about his mother is scarcely more than amusing. The juxtaposition of formulae, in accordance with the simple and uncharged tone of the speech, is straightforward: 'Hera is a *bitch because she wanted* to hide me and my lameness | away.' But the juxtaposition in the τειχοσκοπία has, even apart from its context, far more edge. The remoteness of the past expressed by εἴ ποτ' ἔην γε does not follow in any simple way from the self-castigation of κυνώπιδος. The thought is 'If I were not a bitch, then I should not have to say that Agamemnon *was* my brother-in-law; I would not have failed to appreciate so great a family, and all this would not seem so beautiful and so far away.' It can be observed that the juxtapositions of εἴ ποτ' ἔην γε, in the other instances where this formula occurs in its fixed position after the bucolic diaresis, are likewise less delicate and complexly significant than here in 3. 180.[45]

[45] The student of M. Parry may object that it is not easy to find a formula filling the last two feet of 3. 180 and beginning with a vowel which will at the same time complete the sense in any satisfactory manner. (ἀλβιοδαίμων, as in 182, only in the nominative, jars, and ἄλγε' ἐχούσης [cf. *Od.* 17. 142] seems odd too. Such expressions

This analysis of what creates in the τειχοσκοπία the qualities a good reader can recognize in it could be extended. But the fundamental point is, I believe, already clear: its particular qualities depend on an exact economy of the formulae within it. Change the order of statements, add a bit here, subtract a bit there, develop the thoughts in some other manner, and we no longer have the great scene we know. But the *Iliad* is made up of | a succession of such scenes, arranged in a fairly exact order. Some of course are more moving, or more highly charged, than others. But they must all be such as to create the impressions they now create, or we shall not have our *Iliad*. We should at best have a poem with discernible similarities to our own. And to be such as to create the impressions we receive now, they can endure very little change. The margin of flexibility of language and structure in our version of the *Iliad* is, I submit, rather small, certainly much smaller than one would have to assume in order to sustain the theory of an orally transmitted *Iliad*. For the minimum of change that we can calculate from the only evidence we possess, the observed tradition of Yugoslav poetry (as opposed to the inferred tradition of ancient Greek heroic poetry), will imply changes that would destroy the exact texture and therewith the essence of our poem.

It follows that the name 'Homer', if by this we mean the author of our poem, must be reserved for the poet who composed the *Iliad* at the time when it was put into writing. The poets who preceded him, even if we imagine them as

are obviously flat; we do not even know whether they are possible at all. ὀκρυοέσσης, cf. 6. 344, is of course not possible, because it almost certainly is a MS distortion of κρυοέσσης—see Chantraine, 1. 45.) Was not this particular juxtaposition arrived at by an intuitive process of elimination?

Such an observation has force, and one should never ignore it in assessing the style of the *Iliad*. It is in fact very difficult, I believe not possible at all, to find in our texts a formula which could replace εἴ ποτ᾽ ἔην γε and give an adequate sense. But just though the observation is, it does not constitute an objection. We know what the sense of the line is, because we have it in our text. Any different development—say a new sentence beginning with something like ἀλλά με δαίμων—brings us to the realm of the completely conjectural. If anything plausible could be constructed, it would give us a different sense. We must define the sense of what we have and measure its exact relevance to the effect of the passage. No one could seriously argue that the self-revelation of Helen as we have it in Book 3 is an automatic creation of the formulary system.

singing poems the length of the *Iliad* and dealing with the same theme or group of themes, cannot have been responsible for the essential quality of what we possess. We must conclude that Kirk's belief that 'There is no compelling reason from the point of view of transmission why the *Iliad* and the *Odyssey*, once they gained wide repute, as they presumably did in the lifetime of their monumental composers, should not have been handed down from singer to singer with comparatively minor deviations' is, as he says of Lord's theory of the necessity of dictation, 'fallacious and must be absolutely rejected as it stands'. It is hard to imagine why or under what conditions singers would hand down to each other a poem of that length; more important, if they did so, it would not be the *Iliad*.

Having argued (incorrectly, I think) that Yugoslav poetry itself reveals the possibility of sufficiently accurate oral repro-duction, Kirk (*Poetry*, 277 ff.) puts forward arguments in favour of such reproduction | which insist on the difference between Slavic and Homeric poetry. (*a*) The 'powers of memory of the Greek ἀοιδός can have been no less than those of his modern Yugoslav equivalent.' No doubt; but memory in our sense is not in question here, since bards in general do not memorize. (*b*) 'But his formulary equipment and his dramatic and imaginative capacities must have been far superior.' Certainly, given the superiority of the *Iliad* and the *Odyssey* to anything which has appeared in the Yugoslav tradition. But do greater poetic resources make for closer imitation, or the reverse? Kirk's 'monumental composer' was presumably very far from an imitator. (*c*) The formulary structure of the *Iliad* and the *Odyssey* is more 'rigid'. The Homeric line is evidently more formulary than the Yugoslav. The style as such was more developed, and the Homeric poet dealt more in blocks of words, so that there is in Homer less play of individual words. It does seem true, as Kirk well sums up the matter, 'that the traditional language of the Greek oral poets was much more highly organized, as it was much richer, than that of any modern oral poet of whom we know'. In other words, Yugoslav heroic poetry, though it is far more formulary than any literary poetry could well be, still cannot

be analysed into the same systems of formulae as M. Parry was
able to analyse in the *Iliad* and the *Odyssey*. All this is true; but
it does not follow that Homer and his successors were better
verbatim transmitters than the singers of modern Yugoslavia.
Homer himself, by Kirk's own account, was presumably not.
But there is no reason why his immediate successors at least
should have been, either. A rich and complex tradition imposes
restrictions and at the same time provides greater possibilities
of expression. We have seen that in the τειχοσκοπία the exact
tradition of heroic diction made possible juxtapositions of
formulae which we meet nowhere else in the *Iliad* and the
Odyssey, and which create a unique texture of meaning for this
scene. Other scenes can be shown to possess the same creative
individuality within the clear framework of the tradition.
Parry was the first (as far as I know) to adduce the analogy of
rhyme in modern poetry as | an element directing the thought
of the poet.[46] Of the rich stock of formulary language in
Homer, it could be said, as Proust remarks of modern poetry:
'comme les bons poètes que la tyrannie de la rime force à
trouver leurs plus grandes beautés.' The *tyrannie de la formule*
can be observed in Homer. But neither its purpose nor its
effect, when the tradition was in its vigour, would have been to
make of the ἀοιδός an accurate imitator.

Kirk's final point (*d*) to show that Homer's contemporaries
and successors would have copied a poem in performance
more accurately than their modern Yugoslav counterparts
relates not to the diction of the poems, but to their thematic
material. 'As for thematic changes ... It may be conjectured
that there too the Greek oral tradition, at any rate by the time
of Homer, was more highly organized than any modern
equivalent.' By 'more highly organized', Kirk seems to refer to
a narrower variation in choice of theme on any single subject
than was allowed in Greek heroic poetry. This is very doubtful. The number of epic themes alluded to in the *Iliad* and the
Odyssey themselves[47] is an impressive indication of how careful
and particular a selection the author or authors of these poems
made from the thematic material available to them.[48] Some of

---

[46] *L'Épithète*, 166 ff.   [47] *Lesky*, 37 ff.
[48] So Aristotle, *Poetics*, 1451ᵃ 16 ff.

these themes are so close to those of the *Iliad* in particular that a number of theories of the old Analytic school have been formed to show that the poet or poets of the *Iliad* drew their material directly from them.[49] If we add the other epic tales we know from the Cyclic Epics, and then consider how many themes will be simply unknown to us, it becomes questionable whether the plots of the *Iliad* and the *Odyssey* were fixed in comparison with the Yugoslav material, as Kirk appears to suggest.

An argument of a different sort, one which is not so much concerned with the question of transmission as with our evaluation of the Yugoslav analogy in general, is Kirk's theory of *creative* and *reproductive* stages in traditions of epic poetry. This | theory is obviously related to his theory of transmission. A reproductive phase, if such existed, would help make accurate oral transmission possible. We shall only note here that arguments (*a*) to (*d*) can all be used at least as well to show that the Homeric poet was more creative than his Yugoslav counterpart, as to show that he was a better reproducer. Kirk's implied argument appears to be that the Greek poet was simply better, therefore better at whatever he put his hand (or voice) to, whether creation or reproduction.

   The argument of distinct stages in the tradition of unlettered heroic song has, however, its own interest. It is set forth most clearly in *The Songs of Homer*, 95 ff., where the author imaginatively describes the 'life cycle of an oral tradition'. There are four stages: *originative*, *creative*, *reproductive*, and *degenerate*. The first represents the origins of narrative poetry. In the case of Homer it might have begun with Mycenaean poetry, perhaps slightly later.[50] No one would want to question the existence of this phase, though some might describe it differently from others. The second phase is that of Homer. He would be a special type within this category: the *monumental* creative singer. By *creative*, Kirk means possessing the qualities which Parry and Lord ascribed to unlettered singers in general: the ability to absorb the formulary and thematic tradition and turn it to account by improvising new and distinct versions within

---

[49] *Lesky*, 38.   [50] See Kirk, *Dark Age and Oral Poet*, *passim*.

the received framework. As long as no clear demarcation is set between these first two stages (for creativity must have existed as soon as poetry did), few would want to dispute the value of these first two categories.

It is with the third stage that Kirk's theory becomes controversial. The *reproductive* singer is still a bard.[51] He still works within the formulary tradition. But he cannot really do anything new. At most he *varies* or *contaminates* (see *Poetry*, 279). He uses the tradition for 'memorization and to facilitate the | transposition, often though not always unintentional, of language or minor episodes from one acquired song to another' (*Songs*, 97). This is the stage in which we find all the bards of modern Yugoslavia. But it existed in Greece too. 'Such reproductive singers must have existed for a time in Greece—particularly, one would conjecture, in the mid-seventh century BC; but we have no direct knowledge of them' (ibid.).

Economy of hypothesis is perhaps the point of view from which we should regard this proposed third stage of development. We know that there was something that could well be called a *creative* stage: for we have the *Iliad* and the *Odyssey*. We know also that there was a rhapsodic period when the *Iliad* and the *Odyssey*, known as such and ascribed to Homer, were recited by professional reciters, like Plato's Ion, who carried a staff and were called *rhapsodes*. This is Kirk's fourth or *degenerate* stage.[52] Exactly how and when this change took place is

---

[51] Though *Dark Age and Oral Poet* states otherwise. Kirk later abandoned this view, which would make the Yugoslav bards into rhapsodes.

[52] When the rhapsodic stage begins, to what extent it spelled the end of improvising song, and how much it coincided with the replacement of the lyre by the ῥάβδος, are questions not easily answered. The extremes are given us by the descriptions in the *Odyssey* of Phemius and Demodocus on the one hand, and by Plato's Ion on the other. Between these two points it is very difficult to trace the development. Schadewaldt argues that Homer is already *archaisierend*, and that though he describes singers accompanying themselves on the lyre, he himself sang (or *spoke*, for Schadewaldt denies any clear demarcation here) with the staff. Wade-Gery too wants Homer to be a rhapsode in this sense. Webster holds the now more conventional view that the shift from the lyre to the staff was the shift from creative improvisation to recitation from written texts. Stressing the fact that Hesiod portrays himself as receiving a ῥάβδος from the Muses, he sees the crucial change as taking place between Homer and Hesiod. Notopoulos and others argue that Hesiod is as much oral poetry as Homer, which would confirm Schadewalt's point that the staff did not necessarily mean the replacement of creative singing by the virtuoso recitation of an Ion.

impossible to say. But the introduction of an intermediate stage about which nothing can be known hardly brings us closer to a solution. Nor does this conjectural stage help to explain the transmission of the *Iliad* and the *Odyssey* on Kirk's hypothesis. For, in the first place, the inferior poetic capacities of Kirk's reproducing | bards would not seem necessarily to be an advantage in the matter of accurate reproduction. And even if we agree with Kirk that they were somehow such an advantage, still by his own account such inferior improvising bards existed 'particularly . . . in the mid-seventh century BC'. We still have to get from a monumental composer whose acme was at the latest 725 BC to our reproducers of the mid-seventh century whose feebleness of poetic gift could guarantee that we have the song that was originally sung. If, however, he were to revise this conjecture, so that the reproductive stage begins immediately after Homer (as he seems to say, *Poetry*, 278: 'the Homeric poems came at the end of the true oral tradition'), then we shall still have to assume, as I argue above, that these reproducers approached medieval scribes in accuracy, which is against all evidence, and we shall moreover have the difficulty of explaining why the most creative of all singers in the Greek heroic tradition was straightway followed by the least creative and most reproductive.[53]

The conjectural third stage of reproductive bards does not, when you follow out its implications, help Kirk's hypothesis of transmission. It will be of still less help to those who are disinclined to adopt the hypothesis in the first place. Moreover it is, by the definition of its creator, a stage about which virtually nothing can be known. Surely those inferior, contaminating, | pointlessly varying, mechanically reproducing

[53] Actually, the complete sentence in Kirk's article, of which the beginning is quoted above, is yet more confusing: 'The third factor [in accurate word-of-mouth transmission] is that the Homeric poems came at the end of the true oral tradition, so that their oral transmission depended for much of its course not on singers but on reciters or rhapsodes.' The statement here is hedged; but it suggests that the creative singer *par excellence*, Homer, was almost immediately followed, not by inferior ἀοιδοί of the reproductive phase (whom Kirk himself cannot seem to fit in here) but by rhapsodes of the degenerate stage, men noted for their clumsy additions to the text (*Songs*, 97) (though there are also 'non-decadent aoidic expansions', 322), but still responsible for keeping the delicate unwritten text in its essential shape from generation to generation over the centuries.

bards should be forthwith put out of their misery by the rapid demolition of the stage of development invented to house them.

Two further arguments for the 'reproductive stage', not directly connected with the problem of transmission, have their own interest. The first is that such a stage would help to explain some of the difficulties of our texts.

Kirk distinguishes (*Songs*, 316 ff.) three main kinds of difficulties. The first consists of structural anomalies (*Songs*, 211 ff.). As he is aware, no reader or scholar will precisely agree with another on what is a structural anomaly, or on exactly how they should be explained once they are recognized as such. But most scholars (Page is the notable exception) will now be ready to agree with Kirk's main judgement that such anomalies are fewer and less anomalous than they appeared to old-line Analysts, and that in any case they are an inevitable aspect of the conditions of traditional improvising song.

The second and third kinds of difficulties defined by Kirk consist of places where the usually clear flow of Homeric narrative gives way to expressions that appear to us awkward or obscure or both (316 ff.).

One of these two kinds of difficulty involves really bizarre expressions such as ἠέ ποθι πτολέμοιο μέγα στόμα πευκεδανοῖο (10. 8: p. 205). These Kirk would assign to his *degenerate* stage. Since they do not therefore involve the 'reproductive stage', we shall only point out here that mere reproducers (whether bards or rhapsodes) are unlikely candidates, on the face of it, for the authorship of such bold, though puzzling, expressions as this. Would a reciter be likely to invent, or introduce, the powerful *hapax legomenon* πευκεδανοῖο?

The other kind of difficulty Kirk can assign to his reproductive stage. It seems to consist (320 ff.) of passages in the *Iliad* which sound like (the judgement here is inevitably subjective) passages from the *Odyssey*, such as the scene between Priam and Hermes in 24; and passages which remind us of Hesiod, | like 12. 20–2, the list of Trojan rivers. In Kirk's view, such passages are too good to be the work of the thoroughly degenerate rhapsodes, but at the same time jar too much with the tenor of the *Iliad* (passages from the *Odyssey* do not seem to come into play here) to be the original work of the monumental composer. They are in between. So there is a stage in between.

It helps us to be aware of such passages in the *Iliad*—though, as I said, and Kirk is admirably willing to concede, not all will agree what they are: the whole of the apocalyptic beginning of 12, as it stands, for example, strikes me as one of the most powerful passages of the *Iliad*. But to assign them to a stage later than Homer, even more to assign them to a particular stage for which there is no clear evidence, requires a knowledge of the exact state of the tradition when Homer sang the *Iliad* such as we can never possess. There is no sure sign that the *Odyssey* is much later than the *Iliad*. If so, we cannot say that 'Odyssean' passages—what one has in mind in reference to the Hermes–Priam scene is mostly its rambling tone—were not known to Homer, or that he had not sung such himself, or that he did not choose to introduce such a scene in Book 24.

Or again, the *Theogony* is the great catalogue poem of all time. The *Iliad* and the *Odyssey* do not obviously depend much on the catalogue technique.[54] Therefore scholars tend to regard catalogues, when they do appear in Homer, as 'Hesiodic'. The 'Boeotian School' theory owes its origin to such general considerations. But there is, after all, no reason to believe that lists of names were not a common element of poetry when Homer sang. There is indeed reason to believe that they were, because the catalogues in Homer reveal a strict adherence to the formulary structure defined by M. Parry.[55] Homer used them when they seemed to him useful and effective, and there is | nothing specifically 'Hesiodic' or 'Boeotian' or 'late' or 'reproductive' about them at all.[56]

Such passages are the first of Kirk's two illuminating,

[54] The (unpublished) Harvard diss. of C. R. Beye (1958) and the (unpublished) University of California (Berkeley) diss. of J. H. N. Austin (1965) both argue for the importance of the catalogue style as an informing principle in the *Iliad* and the *Odyssey*.

[55] Parry, *Formules et métrique*, esp. 23 ff.

[56] This does not mean, of course, that we *know* that, e.g., the κατάλογος was composed by the probably single composer of the *Iliad*. It bears signs of earliness (see Page, *History*, with Parry and Samuel's comments, *CJ* 56 (1960), 85). But we do not finally know if it is Homeric in this sense or not.

though inadequate, reasons for the proposition of a 'reproductive stage'. The second is yet more interesting, and brings
us back to the heart of the question. Kirk is, as we have seen, in
general reluctant to accept the analogy so confidently drawn
by Lord between the Homeric poems and the modern Yugoslav tradition. Yet he must concede that both Homer and the
Yugoslav poetry are unmistakably the products of a true oral
tradition of heroic song. But the difference in quality, admitted
by Lord but otherwise passed over by him, seems to Kirk
crucial. So he invents a stage where poetry is still clearly the
product of the old tradition, but by definition vastly inferior to
the poetry of an earlier, Homeric stage, or to that of one
equivalent in value.

If we take Kirk's biological (almost Spenglerian) metaphor
of a 'life-cycle' literally, it would seem to follow that at some
point in the Yugoslav tradition poems were being sung that
were not much inferior to the *Iliad* and *Odyssey*, or at least
were far closer to them in value than anything we now find in
that tradition. 'The creative stage in Yugoslavia', he says,
'ended at some time in the past: probably quite recently, in the
last century' (*Songs*, 96). This is his concession to the notion
implicit in Lord, and to some extent in Parry, that the
difference between an unlettered and a literary tradition of
poetry is so much more important than any difference between
one unlettered tradition and another, that the latter can be for
practical purposes ignored. Ćor Huso, then, coming at the end
of the true creative period, may well have had qualities
comparable to those of Homer, though his work perhaps never
equaled the 'monumental' compositions of the *Iliad* and the |
Odyssey. But it is certain that the quality of the recorded
Yugoslav material is mediocre by Homeric standards. Therefore it must represent a declining stage in the tradition.

Now Kirk's judgement here may be fundamentally correct.
The solution he proposes to the problem, however, may be
quite unnecessary. For there is a simpler solution, which Kirk's
respect for the convictions of Parry and Lord may have kept
from him. This is that the Yugoslav tradition, so far as we can
know anything about it, was never more than immeasurably
inferior to the Homeric. It may once have been more flourishing

than it is now; doubtless it was more widespread. But there is no reason to assume biological stages to explain its present lack of greatness, and then to import these stages into the Greek tradition for symmetry.

Artistic traditions can never be wholly explained in positivistic terms. No definition of the economic conditions and intellectual history of Florence in the thirteenth century would explain the *Divine Comedy*, though now that we have it, we can see much of Dante's debt to his contemporaries and predecessors. He is a fairly isolated figure, in a rich tradition, to be sure, but one which never produced anything to equal his work; and possibly Homer was that. But similarly nothing in the economic conditions and intellectual history of nineteenth-century Russia could have guaranteed the existence of the tradition of the Russian novel, represented by at least four major figures. Perhaps Homer was one in such a constellation. We need not grow mystical about the emergence of the great artist to see that while he works in a tradition, the quality of his work cannot be either predicted or explained; and we need not be mystical about art in general to see that while we can trace the development of traditions, their height and intensity can be neither predicted nor explained.

These and like considerations can remind us not to hope to explain Homer by the historical and cultural conditions in which he lived. But at the same time the existence of a rich | tradition can make the emergence of a great artist easier, or at least easier for us to understand. And we can consider with profit some of the conditions which favour or discourage the growth of a great poetic tradition.

M. Parry's work was at first designed to describe the *tradition* of Greek heroic song. His original antithesis was not between the oral and the lettered poet, but between the poet of a *traditional*, and the poet of an *individual*, style. So much was he concerned to establish the tradition, and Homer's participation in it, that he consistently underplayed the uniqueness of the creator (or creators) of the *Iliad* and the *Odyssey*. The tradition had never been understood; it was Parry's purpose to make it understood, and few scholars have realized their aims so well.

He is not to be blamed for not having stressed, or only in rare moments, the distinctiveness of Homer.[57] To do so, he clearly felt, would have run the danger of confusing Homer with the other kind of poet, the poet of individual style. But this feeling should not lead us, his successors, to confuse Homer with the splendid tradition which he represented. Such a confusion of Homer and the tradition is possibly an error from which Homeric studies are only now beginning to emerge.

In Parry's later work, the traditional–individual antithesis was replaced by the antithesis of the oral and the literary poet. The occasion for this change in terminology was Parry's | discovery of the analogy between Serbian and Greek heroic poetry. This analogy too, I believe, is only now beginning to be understood. For the tendency has been, on the part of some, to dismiss it altogether, and on the part of others, Parry's followers, to accept it to the point of making another error: the assumption that different traditions of unlettered song are similar in value and complexity. Here again, Parry himself led the way. In his desire to show that Yugoslav poetry has more light, in many ways, to shed on the *Iliad* and the *Odyssey* than does Virgil or Milton, he underplayed the evident difference between the Yugoslav and the Greek traditions themselves. Because the style of Homer was an oral style, and part of an oral tradition, and because in Yugoslavia Parry found and heard bards singing songs in an oral style which was part of an oral tradition, the difference between those traditions seemed to him of far less account than the immensely exciting similarity.

Again, it is up to us, Parry's successors, not to stop where he

[57] e.g. the conclusion to his first long article in *HSCP* 41 (1930), 147, where he comments on the 'wondrously forceful' line

κεῖτο μέγας μεγαλωστὶ λελασμένος ἱπποσυνάων

(16. 776 = 24. 40)

and shows how it 'is made up of verse-parts found in other parts of the poems'. Even here, of course, his discussion leaves entirely open the possibility that the whole line might have been traditional—or the two lines, because the previous line in both cases is virtually the same. One might go on to observe that the line is indeed 'wondrously forceful' in 16, where it forms a handsome conclusion to a paragraph of agitated description of battle, beginning with a violent simile; but that it is not especially effective, let alone 'wondrously forceful', in 24. 40. What has Achilles' horsemanship to do with this scene? The distinctiveness of Homer, I suggest, is what makes the line, traditional or not, so masterful a stroke in 16.

stopped. Kirk is one of the severest and subtlest critics of the Yugoslav analogy, but even he, as we have seen, seems at moments to assume that one oral tradition must be much like another, and his stages of development seem partly occasioned by this assumption. And yet he is fully aware that the tradition of Greek heroic poetry, in the beauty and complexity of its process of expression, was something far beyond any other tradition of heroic poetry we know, Norse poetry being possibly the nearest thing to an exception. Parry and his direct successors, Lord and Notopoulos, are of course also aware of this, but their awareness rarely becomes operative in their criticism.

An external difference in the traditions of ancient Greek and Yugoslav poetry has been oddly overlooked. It is that unlettered culture in Yugoslavia has been a rural, one might almost say, backwoods phenomenon, existing alongside a literary urban culture. This means, first, that much of the best poetic talent would be lost to it; and, second, that there was an entirely different relation between literacy and the power of unlettered | song from any that could have existed in Ancient Greece at the end of the eighth century. One of the few scholars to have seen this clearly is T. B. L. Webster, who says in his review of Kirk's book:

Unlike the Greek poems, the Yugoslav poems were not continually brought up to date, perhaps because the social status of the audience (unlike the Greek) continually declined; literacy killed the Yugoslav poets because it brought them into touch with a higher culture; there is no reason why it should have had the same effect on Greek oral poets, and the transition may have been much more gradual.[58]

When a Yugoslav poet learns how to write, a whole literary culture, the culture of the cities of his own country and of what we call the civilized world, becomes accessible to him. It is a culture of books and newspapers. If he abandons the traditional formulae which have enabled him to improvise his heroic narratives, this is not necessarily, or even probably, because he is corrupted by a new technique. It is because he has become part of a different world, a world with new values and new habits of thought.

[58] *JHS* 83 (1963), 157.

An example in Lord's article on dictation shows, I believe, the disadvantage of overlooking this point. He says:

There are in Yugoslavia a number of oral poets who can write. Their first attempts at writing were mere recordings of the songs which they knew. When they go beyond this and begin to break the formula patterns in which they have thought poetically all their lives, the results are not felicitous. They abandon such imaginative introductions as 'Once in the days of old, when Sulejman held empire', for prosaic beginnings like, 'In the bloody year of 1914, on the sixth day of the month of August, Austria and all Germany were greatly worried'. They become wordy and stilted to the point of being unconsciously mock heroic. The natural dignity of the traditional expressions is lost and what remains is a caricature.[59]

Lord speaks as if the process of writing itself had created the 'prosaic beginning' which he | cited. But the poet who sang that was obviously trying to copy the style of the newspaper, or possibly the school textbook. The implication that if Homer or a contemporary learned to write, this is what would have happened to them, is wrong.

We do not know exactly how the advent of writing affected the Ionian singer of the end of the eighth century BC. No modern analogy will take us far, because the conditions of that time have no modern analogue. We know from Homer principally, and in the second place from archaeology, that there was in Greece a civilization highly developed both economically and artistically. We know that, before the introduction of the Phoenician alphabet, this civilization was innocent of the art of writing,[60] and that it had developed without any pervasive influence from surrounding cultures which did know writing. One could in short be an extremely civilized person, living in the heart of an advanced culture, without having any notion of what writing is. That is not possible today, nor was it in the 1930s, or long before that.

When a man of that society learned writing, after the introduction of the Phoenician characters, he was thereby

---

[59] *TAPA* 84 (1953), 129.
[60] Need I say that the use of Linear B centuries earlier (probably, in my opinion, by Minoan scribes with a limited knowledge of Greek: so Kirk, *Songs*. 26 f.) does not constitute an exception?

initiated into no organized literary culture. There is to my knowledge virtually no evidence for Phoenician influence on Greek society apart from the alphabet. When an Ionian of 725 BC, singer or other, learned to write, he learned only that: how to put down marks which would afterwards remind him of something said or sung. These marks entailed no new style of saying or singing whatever.

These conditions cannot be found in the modern world. We can find societies where the idea of writing is unknown, like parts of New Guinea. Some of these societies have perhaps some song, but writing can be introduced to them only by persons of vastly superior, or at any rate vastly more organized, culture. Or we can find societies, like that of Yugoslavia, where unlettered song, of a fairly high order, has existed for centuries in country | districts but where much of the business of life is carried on in cities where life depends on literacy. In both cases, the introduction of writing involves the introduction of new ways of thought.

In the absence of any valid analogy, we are driven back to conjecture. It seems to me a reasonable guess that the Ionian singer of 725 BC, trained in the use of a formulary technique far more subtle and elaborate than any other we know, would, if he had learned how to manipulate the magical σήματα which had come to him from the Phoenicians, not be inclined to change his thoughts or modes of expression at all. Why should he? The epic style, such as we know it from the Homeric poems, would at first have remained unchanged. Only in later generations, when the use of writing had made possible the growth of new indigenous ways of thought, would the style of heroic poetry decline and give way to new forms of expression.

All this is conjecture, of course, but some conjecture is more reasonable than other. What we can conclude with some certainty is that the analogy which Lord assumes between conditions in twentieth-century Yugoslavia and eighth-century BC Ionia is very shaky, and that his statements about the effect of writing on improvising poets cannot be applied in any simple way to the composer of the *Iliad*. The corollary to this is that the notion explored by Lesky, Whitman, Wade-Gery, and

Bowra,[61] that Homer himself knew the art of writing, is in fact
| not ruled out by modern research into the processes of oral
poetry, as Lord and Dow would have it.

The two principles which Lord has articulated concerning
composition and transmission of poetry in the improvising
style are, we remember, that (1) an orally composed poem
cannot be handed on by the tradition of oral song without
fundamental change and (2) 'the [oral] poet's powers are
destroyed if he learns to read and write'. Kirk has rightly felt
that it is all too easy for such principles, supported by the
dubious analogy of Yugoslav and ancient Greek poetry, to
become standard belief. But Kirk's attack on the first principle
has not been successful. The second principle, which Kirk
accepts, seems in fact the weak point of Lord's argument, for it
rests on the weakest part of the Yugoslav–Homeric analogy.

If Lord's first principle is correct, as I believe it is, the *Iliad*
will somehow have been put into writing at the time of its
composition. Lord has insisted on dictation as the only way this
could have been done because of his (as I believe) mistaken
notion of the impossibility of a bard who can write. If the man
who, on this hypothesis, put the poems into writing was more
an amanuensis than a recording scholar in the manner of Parry
and Lord, then perhaps the difference between this sort of
dictation and actually writing by hand would not be enor-
mous.

In either case, we have the striking coincidence that in the
Iliad and the *Odyssey* we have poems far longer than improvised heroic poems are likely to be, longer than the usual
conditions of improvised singing (as we learn from the *Iliad*

---

[61] (1) Bowra, *Heroic Poetry* (London, 1952), 240–1. Bowra in his later Andrew
Lang lecture (*Forerunners*), 9 ff., written after Lord's proposition of the dictation
argument, 1953, came out more for dictation than for knowledge of writing on the
part of the poet himself. (2) Wade-Gery, in his *Poet of the Iliad* (Cambridge, 1952),
argues not only that the poet could write, but also that the alphabet was adopted for
the recording of hexameter poetry. (3) Whitman, like Bowra, leans more to the
dictating than to the writing poet, and argues that the availability of writing was
partly responsible for the creation of the large epic: 'the monumental purpose of the
large epic is profoundly served by anything which bestows fixity of form.' Rather
than Whitman's abstract notion of 'permanence' as the value of writing in
composition, I would suggest the usefulness of writing in enabling the poet to
compose a long but coherent work without immediate dependence on the vagaries of
his audience.

and the *Odyssey* as well as from comparative studies) would suggest or allow; and that in this very same period, the use of writing becomes available. It seems difficult not to see in the use of writing both the means and the occasion for the composition, in the improvising style, of poems which must have transcended their own tradition in profundity as well as length, just as that tradition itself surpassed all subsequent traditions of heroic song.

I I

Classical Philology and Literary Criticism

IF we ask the question, what chiefly distinguishes literary criticism today from what it was fifty or one hundred years ago, our answer would likely contain some reference to such things as New Criticism and Close Reading. New Criticism is no longer new; the attitude to literature of which it was one manifestation is no longer an innovation in academic and intellectual life. But this attitude, now so firmly established, is something new in a broader historical sense, if we compare the humane study of literature now with what it was in 1920 and earlier. What it involves essentially is a concern with the way words are used to produce the specifically literary experience, and it embodies the observation that artistic writers of prose and poetry use words in ways other than merely to convey information or merely to delight; that the pattern of words in a work of literary art contains a density of meaning which is lost in any paraphrase; and that this density of meaning is often dependent on the multiple significance of single words.

Has Classics, the study of Greek and Roman literature, the most traditional of humanistic studies, followed this develop-ment of literary criticism? Do we in the better American universities teach the ancient authors with a technique compar-able in sophistication to what we see in departments of English and modern foreign literatures? The answer I think is Yes. But we must here note an important qualification. The ancient languages are difficult. They are difficult partly because of the accident of transmission: our manuscripts are exceedingly imperfect, we often simply do not know what the ancient authors wrote. Hence an important, traditionally, one might

Ventures (magazine of the Yale Graduate School) (Spring 1967), 30–4.

say, the most important, aspect of classical philology has been the attempt to discover exactly what the ancient authors did write. This study is known as textual criticism.

But even if we had autograph manuscripts of the ancient authors, their language would still be difficult to understand: indeed, much of the charm and the interest of the study of Greek and Latin derives from our involvement in ways of thought and modes of expression which are not our own. Modern western languages, by comparison, seem in their essential syntactic and idiomatic structure, to be simply translations of one another. |

The difficulty of Greek and Latin for a modern reader has always enforced a closer attention to the text than is needed in the study of, say, an English or a French author. The classicist has to be concerned with the exact way in which words are used in order to find out what those words mean even on the simplest, most denotative level. So that, in a sense, the classicist has always been a Close Reader.

But the close and exact examination of ancient texts with a view to their interpretation, carried out by means of syntactical and morphological analysis and the adduction and verification of parallels, the basic and traditional philological method, has suffered from a grave defect. The interpretation which is the end product of the philologist's labour means a plausible translation. Only when the scholar can put a Greek or Latin expression into English words which can be somehow shown to have the same meaning as the Greek or Latin words can the process of interpretation be said to be complete. But translation is of course inadequate. If the translation had entirely the same meaning as the original, there would be no further need to deal with the original text. But this fact of the essential inadequacy of any translation of a Greek or Latin expression is often lost sight of by the philologist. What we often find in learned commentaries on ancient texts is, first, the isolation of a group of words the meaning of which is not immediately clear. Then the commentator will cite a number of possible interpretations, that is translations, with some account of the arguments by which these interpretations were reached. The commentator then either chooses one of the existing translations or offers one of his own, giving at the same time arguments to show that this is the right one. The implication is usually that there is only one right one, that the alternatives offered by himself and his predecessors are mutually exclusive, and that the right translation more or less exhausts the meaning of the original. This in effect implies a theory of univocal meaning. Even though it is obvious to the most self-assured scholar that there can be no one-to-one correspondence between the individual words of the Greek text and the individual words of an English translation, he acts as if there were a kind of one-to-one correspondence between the English expressions by which he translates Greek expressions and the Greek expressions themselves.

It has been one of the chief insights of modern criticism, however, that individual words and expressions in poetry and in artistic prose often bear a multiplicity of meanings, and that the significance of a whole passage or even of a whole work often depends upon this multiplicity. It can be analysed in terms of ambiguity, or of denotation and of connotation, or, as the case may be, by the associations which are created by the author for individual words and expressions from other parts of the work. Once we grant this, we see that mere choice between alternative translations is no longer adequate. We can define and discuss the range of meanings of words in an ancient text, but we can never replace that text by any translation. Interpretation ceases to be translation, as it has too often been in the past, and becomes an indirect discursive approach to the author's meaning. It is in this kind of interpretation that | the student of Greek and Latin has too often fallen behind his colleague in English and modern foreign literatures.

We offer a single example. In chapter 65 of Book 2 of his history of the Peloponnesian War, Thucydides sums up in a series of epigrammatic sentences the character and achievements of Pericles, the man whom he considers the greatest statesman in Greek history. In the eighth sentence of that chapter he says, to give a rough translation, that 'Pericles was able by his reputation, his intelligence, and his evident incorruptibility to control the populace freely, and he was not so much led by them as he himself led them'. A difficult phrase in this sentence is the one of which I gave a literal and provisional translation, 'to control the populace freely'. The difficulty lies in the word *freely*. A typical commentary will consider such possibilities as 'i.e. without hesitation' or 'as a free man should' or 'in such a way as to be consistent with the freedom of individuals and the dignity of well-born men'; and the commentator will choose one of these interpretations.[1] But did Thucydides, by the word translated as *freely*, mean any single one of these things, or even merely a combination of them? We must consider the meanings which are elsewhere attached not only by Greek authors in general but also by Thucydides himself, within this work, to the Greek words ἐλεύθερος, 'free', and δοῦλος, 'slave', and to their derivatives and compounds.

Slave and *free* refer in the first instance to the distinction between a free man and a slave, and they are often so used in the text of Thucydides in passages where no discussion of ambiguity would have any point. But in Thucydides and other fifth-century authors, they had acquired another more political range of meaning, whereby *free* meant 'not in any way under the domination of a foreign power'. Thus the Persians were traditionally represented in the fifth century and later as having attempted to *enslave* the Greeks even though there was no question of the Persians' actually wanting to reduce individual Greeks to the condition of slavery. By the time of the Peloponnesian War (431 BC) these terms were freely used in such a sense by Greek powers to each other: the Athenians, in maintaining by force an empire in which other states had to remain in alliance with Athens and pay tribute, but were otherwise free, were, as Thucydides tells us, frequently accused of *enslaving* Greece. The Spartans, by the same token, presented themselves as the *liberators* of Greece. Thucydides presents these accusations and these claims with a certain ironic detachment, but makes it clear that they were largely accepted throughout the Greek world. In 432 BC, on the eve of the Peloponnesian War, Thucydides gives to Pericles a speech in which he turns this political meaning of the word *slavery* against the Spartans. They had presented to Athens an ultimatum, demanding that the Athenians rescind a decree against

[1] See e.g. Gomme, *A Historical Commentary on Thucydides*, vol. 2 (Oxford, Clarendon Press, 1956), ad loc.

Megara, a hostile neighbour. Pericles urges that the Athenians not yield to this ultimatum, saying that although it involves a small concession, that concession would be fatal to Athens. When of two equal powers one orders the other to do something, be | it great or small, without submitting to arbitration, that order amounts to *enslavement*.

Elsewhere, Pericles presents the Spartan demands as an attack on Athenian *intelligence*. The word which I translate here as intelligence—γνώμη—means also 'political decision' or 'policy'. The test of that *policy* or *intelligence*, Pericles says, is the Athenians' ability to maintain it intact in the face of Spartan threats. Elsewhere in this and other speeches, Pericles is made to identify with this *intelligence* a whole complex of features of Athenian political life: their ability to use discussion to formulate the best plans; their possession of walls which render them invulnerable to attacks by land; their possession of capital, which enables them to make long-range plans which can be maintained through varying circumstances; their control of the sea, and their naval skill. All these aspects of *intelligence* guarantee to the Athenians a kind of *freedom* which embraces not only personal and political autonomy, but also the Athenians' entire control over their political environment.

The Spartans represent a threat to this supremacy of Athenian *intelligence*, but a threat which that *intelligence* (remember that the word γνώμη also means 'decision' or 'plan') can effectively deal with. We are told this by Thucydides himself as well as by the words he attributes to Pericles. A danger which cannot be so easily dealt with is something like the great Plague of Athens in the year 430, which Thucydides describes in Book 2 with a vivid power which is as much poetic as it is scientific. Thucydides emphasizes the suddenness and the unpredictability of the plague. Pericles calls it a παράλογον, something which happens contrary to λόγος or 'rational planning'. But λόγος is very frequently identified in Thucydides' terminological system with γνώμη or 'intelligence'. It is fully in accordance with the new meaning which Pericles has added to the Greek words we translate as 'slave' and 'free' that he says in his last speech, referring to the Plague, 'whatever is sudden and unexpected and happens with the greatest παράλογον enslaves the mind'. *Freedom*, we learn by an examination of the way in which Thucydides uses his Greek words, means the ability of the mind to control political circumstances.

What about Pericles' relationship to his own fellow Athenians? The γνώμη, or 'intelligence', of the Athenians, as Thucydides presents it, is something vested in Pericles himself, and not in the Athenians as a whole, except to the degree that they follow his policy. Pericles' first speech begins with words which are, literally translated, 'I am, Athenians, of the same γνώμη as before'; and in his last speech, after the Plague and Spartan invasion have undermined Athenian morale, he speaks of the Athenians' *weakness* of γνώμη. *Freedom* for the Athenians in Pericles' view consists in their maintaining the policy which he has persuaded them to adopt and which enables them to maintain dominion over political circumstances both concrete and psychological. They are *free men* only in so far as they control political affairs, and they are *enslaved* when adversity leads them to alter their *policy* or *intelligence*. But by the same token, Pericles himself, as a political leader, | is only *free* as long as he can secure the agreement of the populace in his political planning, which he must do by constitutional means, for Athens remains a democracy. Thucydides does not expressly bring up the possibility that Pericles might control the populace by other means; but if he did the Athenians would of course have lost their *freedom* in another sense.

If we keep all these things in mind, and each of them can be demonstrated by clear associations of words in the text of Thucydides' history, we begin to see how complex the meaning of the simple phrase 'he controlled the populace freely' is. Of course it means that he controlled them without making himself a tyrant. And of course it also means that he was not forced, as were other political leaders, with whom Thucydides compares Pericles in this same chapter of Book 2, to alter his views to please the people. The freedom belongs to Pericles in relation to the people, and to the people in relation to Pericles. But it also has a further meaning which embraces both Pericles and the people who elected him. It means that he was able to lead them so as to maintain the sovereignty of Athenian intelligence over all political circumstances, both predictable ones, such as Spartan hostility, and unpredictable ones, such as the almost supernatural visitation of the Plague. This sovereignty, for Thucydides, is the final meaning of *freedom*.

The method of exploration of an ancient text of which I have here given a single compressed example involves the possibility of multiple meaning. It is not new, even in Classics, and may seem obvious to some. But it is a method which has, in fact, been practised much more in criticism of modern than in that of ancient authors. And it is an important one because it can, as I hope we can see here, reveal much of the larger patterns of an author's thought. If we accept this, we can draw the sobering but finally cheerful conclusion that both the techniques of the classical scholar and our understanding of the great authors of antiquity can be brought to a much higher stage of perfection.

12

Herodotus and Thucydides

TRANSLATION of Herodotus brings up the question of *archaism*. No one can know how a native of Halicarnassus in the mid-fifth century would have felt on hearing Herodotus recite his prose. But we do know something of how an Athenian would have felt. To an Athenian it must have sounded a little exotic, a little old-fashioned in diction, a little Homeric. But rendering these qualities in English is another matter. If we have learned anything about translation in the last fifty years, it is (let us hope) that there is no one 'archaic' style. Shakespeare used forms of language no longer common in his day; so did Milton. We do not feel that Shakespeare is 'archaic' at all; Milton we feel is, in a particular way. Among deliberate archaizers there can be a great difference: Howard Pyle, basing his style on the Bible and Malory, conveys a quality different from Robert Louis Stevenson, who archaizes largely from ballads. Among the archaizing translators of the late Victorian age who set a whole style, there are differences. Sandys's Pindar is unreadable; Jebb's Sophocles, however difficult it is to read through a whole play, still astonishes with its accuracy and felicity.

The question is whether Herodotus should be read in an archaizing version at all. J. Enoch Powell thought so, but most readers find his translation as unpalatable as his politics. The mild archaizing of George Rawlinson, however, is something else. His version (Modern Library, Everyman) is still the most widely read in English. In the Penguin Classics, de Sélincourt, who rejects archaism almost altogether, has not displaced him in popularity.

Arion (Autumn 1968), 409–16. Review article on Herodotus, *The Histories*, trans. and intro. by A. de Sélincourt (Penguin, 1954; rev. edn. 1965) and Thucydides, *The Peloponnesian War*, trans. and intro. by R. Warner (Penguin, 1954).

Candaules' queen, whose husband has shown her naked to Gyges, has her revenge:

'Gyges,' she said, as soon as he presented himself, 'there are two courses open to you, and you may take your choice between them. Kill Candaules and seize the throne, with me as your wife; or die yourself on the | spot, so that never again may your blind obedience to the king tempt you to see what you have no right to see. One of you must die: either my husband, the author of this wicked plot: or you, who have outraged propriety by seeing me naked.'

For a time Gyges was too much astonished to speak. At last he found words and begged the queen not to force him to make so difficult a choice. But it was no good; he soon saw that he really was faced with the alternatives, either of murdering his master, or of being murdered himself. He made his choice—to live.

'Tell me,' he said, 'since you drive me against my will to kill the king, how shall we set on him?' (de Sélincourt, p. 17)

Compare:

Then she addressed these words to him, 'Take your choice, Gyges, of two courses which are open to you. Slay Candaules, and thereby become my lord, and obtain the Lydian throne, or die this moment in his room. So you will not again, obeying all behests of your master, behold what is not lawful for you. It must needs be, that either he perish by whose counsel this thing was done, or you, who saw me naked, and so did break our usages.' At these words Gyges stood awhile in mute astonishment; recovering after a time, he earnestly besought the queen that she would not compel him to so hard a choice. But finding he implored in vain, and that necessity was indeed laid on him to kill or to be killed, he made choice of life for himself, and replied by this inquiry, 'If it must be so, and you compel me against my will to put my lord to death, come, let me hear how you will have me set on him.' (Rawlinson, pp. 8–9)

Though de Sélincourt on the whole is a bit briefer than Rawlinson, here he uses more words: the score is 148 for de Sélincourt against 139 for Rawlinson. Herodotus, always briefer than either, uses 110 words.

De Sélincourt is certainly cooler and neater than the nineteenth-century translator. He has some genuine elegance, and this is one of the better Penguin translations. Occasion|ally his neatness and sharpness make him clearly superior to Rawlinson, as for example in the last sentence in the paragraph above, where the detached rendering of Gyges' self-preservation comes out with a distinct edge, whereas it is almost lost in Rawlinson, and thus does some justice to the effect Herodotus achieves by the brief sentence and the asyndeton. (The dash, however, overplays it a bit: 'He chose to live' might have been better.)

None the less, Rawlinson's translation remains the more satisfactory substitute for Herodotus' sinuous and dramatic Greek. For one thing, it is most of the time more accurate. Thus Gyges in the Penguin version 'saw that he really was faced with the alternatives'—modern idiom, but with no special edge or force. Rawlinson's phrasing catches some of the metaphorical hint and the Homeric tone of ὥρα ἀναγκαίην ἀληθέως προκειμένην. In de Sélincourt the husband is 'the author of this wicked plot': hardly modern, but also not in the Greek, which has only 'he who planned these things'. De Sélincourt's queen complains that Gyges has 'outraged propriety'; Rawlinson's, that he 'did break our usages', which is further from the Edwardian drawing room, and closer to ποιήσαντα οὐ νομιζόμενα, the first expression in the *Histories* of the νόμος, or rule of custom, theme. We may here look back to Gyges' reaction when Candaules first suggests he see the queen naked. 'Master, what an improper suggestion!' says de Sélincourt, as opposed to 'What most unwise speech is this, master, which you have uttered?' Rawlinson's wordiness (six words in Herodotus, δέσποτα, τίνα λέγεις λόγον οὐκ ὑγιέα) is still truer to the Greek than de Sélincourt's Trollopean neatness.

Moreover, there is more passion in the older translation. The queen's 'So you will not again, obeying all behests of your master' has a regal indignation and a consciousness of power not quite attained by 'so that never again may your blind obedience . . .'.

A last comparison. Herodotus, at the end of 1. 5, leaving behind the delicately derisive account of the mythological past, turns to his proper theme. Here he rises, in that imperceptible way of his, to a statement of some eloquence: he will deal with the small cities of the past as well as the great, for those that are now great were once small, and those that were once great

have lost their greatness. τὴν | ἀνθρωπηίην ὦν ἐπιστάμενος
εὐδαιμονίην οὐδαμὰ ἐν τωὐτῷ μένουσαν ἐπιμνήσομαι ἀμφο-
τέρων ὁμοίως. Of course he gets the effect by transparent
means, the long-drawn words at the beginning, the separation
of ἀνθρωπηίην from εὐδαιμονίην, of οὐδαμά from μένουσαν;
but the effect is there: it is a grand and elegiac statement.

Rawlinson renders some of that: 'I shall therefore discourse
equally of both, convinced that human happiness never continues long in one stay' (p. 6). The last three words are archaic
English rightly used. De Sélincourt gives us: 'It makes no odds
whether the cities I shall write of are big or little—for in this
world nobody remains prosperous for long' (p. 15). The tone is
dry and a little mean. His Herodotus gives a kind of salty
wisdom, but loses the magnanimity that gives the wisdom its
perspective.

More teachers and students than mere innocent readers are
going to buy these books, and we might here bring up a few
practical matters. De Sélincourt is happily free from the
footnotes which Rawlinson could not resist, footnotes which
occasionally explain something we want to know, but more
often annoy us by telling us that Herodotus must be all wrong
on some date or military calculation, where we didn't want to
know. Like other Penguin translators of historians, de Sélincourt does, but only now and then, indulge in the insane
practice of putting part of the *text itself* into a footnote. This
must have been editorial practice; other translations like
Warner's Thucydides and Grant's Tacitus do it much more
frequently. 'If Herodotus were alive now, he would have put it
into a footnote.' If my aunt had balls . . . Rawlinson has, on the
other hand, one great practical superiority: each paragraph is
numbered, as in the Greek texts we use. The advantage in ease
of reference is immediately apparent, if you try to look up a
specific passage in Herodotus from some secondary historical
work. The editors may have felt that these numbers would be
unaesthetic; more likely they felt it would remind us that, after
all, Herodotus is not alive now.

Rex Warner, the Penguin Thucydides translator, had two
notable predecessors, Thomas Hobbes, and the Victorian

translator Richard Crawley. The Hobbes translation[1] passes |
for a classic, but more for its date, one suspects, and the
illustrious name of its author than for its intrinsic worth.
Occasionally it stamps out a fine, vigorous phrase, but most of
the time it is wordy and slack and, moreover, quite inaccurate.
Mr Warner's intense admiration for the Hobbes version is
puzzling.

Crawley's style is Victorian English parliamentarian. He
misses the abruptness, the brilliant hyperbata, the continual
unexpectedness, the poetic imagination, the amazing freedom
with word-order, the 'création perpetuelle' of Thucydides.
But he has a real style of his own, a dignity with some force,
and, above all, he draws easily on a living rhetorical tradition.
Hence, though toned down and domesticated, Thucydides'
speakers come through with antitheses that work, with
rhythms that impose themselves as we read, with the genuine
ring of political passion. In this he far surpasses Hobbes.
Warner falls short of both.

The Corinthians (in Crawley) berate the Lacedaemonians in
432 BC:

Time after time was our voice raised to warn you of the blows about
to be dealt us by Athens, and time after time, instead of taking the
trouble to ascertain the worth of our communications, you contented yourselves with suspecting the speakers of being inspired by
private interest. And so, instead of calling these allies together before
the blow fell, you have delayed to do so till we are smarting under it;
allies among whom we have not the worst title to speak, as having
the greatest complaints to make, complaints of Athenian outrage and
Lacedaemonian neglect.   (p. 38)

We get something here of the continual Thucydidean play on
contrasts, of his power to focus on the single word, of the
unexpected transition whereby a sentence that seemed to be
ending takes on a new direction and new life. The anaphora of
'time after time' is Crawley's, but it renders the whole balance
of the Greek sentence. We contrast, as we read the sentence
aloud, 'the worth of our communications' with 'the speakers',
just as Thucydides wanted. 'Before the blow fell' and 'till we

---

[1] The quotations below are from the edition by David Grene in 2 vols. (Ann
Arbor, 1959).

are smarting under it' bring in a stock metaphor not in the Greek, but the complementary rhythm of the two | phrases reinforces the matching sense. 'Allies among whom' keeps up the passionate impetus of the statement, so does the staccato opposition of 'worst title to speak' and 'greatest complaints to make'. 'Complaints' immediately following carries the sentence farther without relaxation of intensity, and the rhythmical balance of 'outrage and neglect' maintains that of the Greek verbs ὑβριζόμενοι and ἀμελούμενοι. Art here still works as rhetorical urgency. There are no loose words floating about not caught up in the compulsive current of the Corinthians' protest.

Hobbes:

For although we have oftentimes foretold you that the Athenians would do us a mischief, yet from time to time when we told it you, you never would take information of it but have suspected rather that what we spake hath proceeded from our own private differences. And you have therefore called hither these confederates not before we had suffered but now when the evil is already upon us. Before whom our speech must be so much the longer by how much our objections are the greater in that we have both by the Athenians been injured and by you neglected.   (i. 38)

Warner:

Many times before now we have told you what we were likely to suffer from Athens, and on each occasion, instead of taking to heart what we were telling you, you chose instead to suspect our motives and to consider that we were speaking only about our own grievances. The result has been that you did not call together this meeting of our allies before the damage was done; you waited until now, when we are actually suffering from it. And of all these allies, we have perhaps the best right to speak now, since we have the most serious complaints to make. We have to complain of Athens for her insolent aggression and of Sparta for her neglect of our advice.

(p. 49)

Hobbes, for all his Jacobean credentials, is dull and | Warner is duller. They have much the same fault. Both fail to perceive the emphases, the rhetorical form, of Thucydides' Greek. Neither therefore can render the rapid contrasts with any adequacy, or maintain anything like the intensity of the original. And the clearest signs of the comparative failure of both translators as against Crawley are the junky and pointless words, words we don't know what to do with, as we read the passage aloud. Hobbes's 'from time to time' is only a little worse than Warner's 'on each occasion'. Thucydides' ἑκάστοτε both reaffirms the force of the initial πολλάκις, and offers the characteristic Thucydidean variation, his broken symmetry. Hobbes is weak and inaccurate, Warner here just weak. Crawley loses the variation, but sustains the drive of the sentence.

Hobbes's last sentence is wrong (he takes οὐχ ἥκιστα with εἰπεῖν, in the sense of μακρότατα), but has some balance and power despite its wordiness. Warner's two sentences manage to destroy any balance or force. For example, 'now' ingeniously placed so as to undo the natural opposition of 'speak' and 'make'.

Crawley's superiority to both his predecessor and his follower is evident in almost every passage of the work, narrative as well as speech. Consider the brilliant simplicity of the last sentence in the account of the battle at the Assinarus river (8. 84).

Hobbes: 'And suddenly the water was corrupted; nevertheless they drunk it, foul as it was with blood and mire; and many also fought for it' (ii. 497).

Crawley: '. . . the water, which was thus immediately spoiled, but which they went on drinking just the same, mud and all, bloody as it was, most even fighting to have it' (p. 451)

Warner: 'The water immediately became foul, but nevertheless they went on drinking it, all muddy as it was and stained with blood; indeed, most of them were fighting among themselves to have it' (pp. 486–7).

The power of the sentence is almost beyond words, what the author of *On the Sublime* put into the category of sublimity deriving from sheer greatness of thought. Crawley succeeds in remaining closest to that greatness of thought. The succession of horrors, 'immediately spoiled . . . mud . . . bloody', ending with the terrible detail 'most even fighting | to have it', is direct and controlled. Each element stands out with its proper stress, as in the Greek. Hobbes is queer, with the queerness of one who doesn't quite get the Greek. Why 'suddenly' for εὐθύς? Why the quality of afterthought in the last clause? Warner has the schoolboy's literalness. He remembers the Greek imperfect can often be translated by the English past progressive. So the simple καὶ περιμάχητον ἦν τοῖς πολλοῖς becomes the cumbersome clause which disturbs, in the tense of the verb, the narrative sequence.

Warner here, incidentally, is possibly derivative: his 'foul' may have come from Hobbes, his 'muddy as it was' from Crawley's 'bloody as it was'; his 'fighting ... to *have* it' also from Crawley. But he has neither the perception of Thucydides' Greek nor the living English style of his closest predecessor.

13

The Language of Thucydides' Description of the Plague

THE historical work of Thucydides is to a striking degree informed by a number of general and theoretical conceptions. Those conceptions involve such matters as the nature of man, the inevitability and the predictability of the historical process, the relations between passion and judgement, and between words and action. The degree to which such conceptions can be discovered in the structure, and even in the style, of Thucydides' work, is one of the features which distinguish him as an historian. Although the conceptions themselves, and the part they play in Thucydides' record of events, have not yet been adequately defined, their presence has been recognized, and a number of scholars have attempted to locate the sources from which the historian could have drawn them.[1] Many have looked to the Sophists. In the matter of style at least, this is a connection which dates back to antiquity. Of recent studies, the careful and imaginative discussion of the historian's intellectual background in John Finley's *Thucydides* (1942)[2] still gives us what is to date the most thorough argument for Thucydides' debt both to the Sophists and to other contemporary representatives of the Greek Enlightenment, notably Euripides.[3]

[1] A good selective bibliography, including the most important works dealing with all these subjects, can be found in H.-P. Stahl, *Thukydides*, Zetemata 40 (Munich 1966), 172 f. My doctoral dissertation, '*Λόγος* and *Ἔργον* in Thucydides' (Harvard, 1957) argues that the opposition between words and action, related to, if not deriving from, the theories of Gorgias of Leontini, pervades the History of Thucydides, and largely controls its diction and syntax.

[2] Cambridge, Mass., Harvard University Press.

[3] More recently, G. Ludwig, *Thukydides als sophistischer Denker* (diss., University of Frankfurt, 1952); A. Parry, op. cit.

Bulletin of the Institute of Classical Studies 16 (1969), 106–18. Reprinted by permission of the Institute of Classical Studies.

In recent years a number of attempts have been made to find the sources of Thucydides' thought in Greek medical theory. C. N. Cochrane, in a book entitled *Thucydides and the Science of History* (1929),[4] argued that Thucydides was obviously a scientist in vision and temperament, and that the only true scientific thinkers of fifth-century Greece were the medical writers. Assuming the veracity of the traditional biography of Hippocrates, and that Hippocrates was in fact the author of most of the works in what we now usually call the Medical Corpus, Cochrane imagined a meeting in Thrace between the Father of Medicine and the Father of Scientific History;[5] and although his discussion of both Thucydides' ideas and the ideas of what he assumes to be the Hippocratic treatises is disorganized and unspecific, he none the less manages to find a number of connections of thought between the historian and the doctor, on the basis of which he hails Thucydides as the intellectual child of the fifth-century Greek school of medicine.

A single example of Cochrane's method may suffice. He says (p. 27) the following:

In his account of the plague, Thucydides follows precisely the Hippocratic procedure. After the general introduction (ii. 47–8), in which he describes the outbreak and its gravity, he begins (49) by what in Hippocratic terminology is a κατάστασις—a general description of the conditions, climatic and otherwise, prevailing during the summer in which the plague broke out. |

Cochrane's argument, followed by some later scholars, is that although Thucydides does not actually use the word κατάστασις, he adheres to the concept by prefixing to his account a description of the meteorological conditions prevalent when the plague broke out. The author of *Epidemics* 1 and 3 begins with his first κατάστασις in these words:

In Thasos, during the autumn, about the time of the equinox and close to the setting of the Pleiades, there was much rain, continuous and falling softly, with winds from the south. The winter was also marked by southerly winds, only slight north winds; and droughts. Altogether it was like spring into the winter. The spring had winds

---

[4] London, OUP.   [5] pp. 15 f.

from the south and was cold. The summer was mostly cloudy. No rain. Etesian winds, few and light, blew irregularly.[6]

Thucydides (2. 49) in fact begins his account of the Plague in this way:

It was agreed that of all years, this one was particularly free from sickness of every other kind. And if anyone was ill with anything else, it was all resolved into this. Others, from no discernible cause, suddenly in the midst of health, were seized by, first, powerful fevers of the head, and reddening of the eyes and inflammation . . .

The swift and dreadful symptoms continue.

This is what Cochrane refers to as Thucydides' κατάστασις. The sentence in fact contains not a single reference to the weather, or to any conditions other than the good health that people generally enjoyed when the Plague struck them. The medical writer's κατάστασις was clearly designed to explain, on meteorological grounds, the particular shape a disease took, or to provide the data for such an explanation. Thucydides' statement that the Plague arrived in a year which had been exceptionally healthy tells us of course nothing about why this disease took the shape it did. It explains rather why, when it did come, it was such a shock. It is the uncanny suddenness of the disease—brought out by the word order: ἐξαίφνης ὑγιεῖς ὄντας—that interests the historian and the psychological effect of this. There is, in short, no κατάστασις in Thucydides, and no analogy at all, in this respect, between him and the author of the *Epidemics*.

A more recent monograph attempts to arrive at a similar conclusion by a more philological route. This is *Thukydides und die hippokratischen Schriften* by Klaus Weidauer (1954).[7] Unlike Cochrane, Weidauer realizes that the treatises of the Medical Corpus differ from each other in style and thought, presumably also in authorship and date. He nonetheless believes that the particular usage of certain key words in Thucydides—πρόφα-σις, εἶδος, and φύσις—can be traced to the usage of one or another extant medical work, and that the ideas which these

---

[6] My translation of the text of W. H. S. Jones, Hippocrates, vol. 1 (Loeb Library, London, 1928). *Epidemics* 1. 1.

[7] Heidelberg, Carl Winter.

words imply, especially the word φύσις, likewise show Thucydides to be the pupil of a medical school.

Weidauer's work is more useful than that of Cochrane, but it is doubtful that he has proved his point. That Thucydides' use and conception of φύσις as the unchanging constitution of the human animal has some relation to what we find in some medical writings, is true. But other usages, in the Sophists and in the dramatists, make it appear that these conceptions were not unique to the doctors and to Thucydides. Weidauer's unnecessarily narrow approach does not demonstrate the exclusive connection he wants. And the underlying romantic and biographical tendency of his work appears when we find that he too subscribes to what we might call the Fathers' Club theory of ancient literary history. Finding, for a single, slightly anomalous usage of the word εἶδος in Thucydides 3. 62, a parallel in the medical treatise *Epidemics* 1 and 3, and noting that the sicknesses so scrupulously observed in that medical treatise are mostly said to take place in Thasos, he theorizes that when Thucydides was stationed on that island in 424 BC, or afterwards when he may have remained during his subsequent exile on the nearby mainland, he and the medical author may have met: as if this fairy tale could tell us anything about how Thucydides uses words.[8]

Scholarly views of Thucydides' purpose in composing his History may be divided roughly into the optimistic and the pessimistic. The optimistic view holds that Thucydides wished to | formulate general laws of human behaviour which were to enable a future statesman to do his job better. The pessimistic holds that while Thucydides was interested in revealing permanent aspects of the human condition, he did not feel that the suffering and destruction in which his *History* so largely consists—and the extent of which, in 1. 23, is made to justify his contention that this was the greatest of all wars—could be avoided in the future. They are a sickness for which, as far as he could see, there is no cure.[9] Of these two poles of criticism,

---

[8] εἶδος, 29 f.; meeting between Thucydides and the author of *Epidemics* 1 and 3, 72 f.

[9] Among strong representatives of the optimistic view see e.g. J. Finley, *Thucydides,* esp. 50, 83; for a strong statement of the pessimistic view see H.-P. Stahl, op. cit., *passim.*

Weidauer belongs to the optimistic: he is so positive that he becomes almost a champion of the optimistic view. It may be useful to look at one or two of his arguments to see just how vulnerable that view is.

First, φύσις. The most striking use of this word to refer to a constant structure in humanity, to an unchanging 'human nature', the sense which Weidauer believes to be derived from medical writers, is in 3. 82. 2. Here Thucydides, after describing the horrors attendant on the total breakdown of society in the Corcyrean revolution of 427 BC, makes this sad and general comment:

Many terrible things befell the cities of Greece in the course of the War, things that happen and always will happen, as long as the nature of men remains what it is (ἕως ἂν ἡ αὐτὴ φύσις ἀνθρώπων ᾖ) although their occurrences will be more or less violent and they will differ in their forms (τοῖς εἴδεσι διηλλαγμένα) according to the changes of circumstance that prevail at any moment.

The words φύσις and εἶδος as they are used in this passage are not,. despite Weidauer, purely medical; but they are what we we might call *scientific*; and in fact Thucydides has a distinct way of using scientific language—for example, words of quantitative measurement—at moments when his narrative is such as to affect us emotionally most deeply.[10] Weidauer argues, however, that the concept of a permanent human nature, as it is conceived in this passage and others, proves that Thucydides believed that one could predict human behaviour and plan effective action on the basis of such prediction.[11] The effective action, informed by the truth of Thucydides' History, would be that of the statesman. But what, we may ask Weidauer and his fellow utilitarians, ought a statesman to do? Surely the first thing would be to prevent the kind of disintegration of society which Thucydides has been depicting in Corcyra. But the sentence cited for the view of human nature which is to be the foundation of the statesman's instruction states that things like this disintegration will always occur precisely because of the constancy of human nature. In short, the sentence states the direct contrary of the thesis which it is being called up to support.

---

[10] e.g. 3. 113. 6; 7. 29. 5 and 30. 3; 7. 87. 5.   [11] Op. cit. 34, 70.

Another frail reed for the optimistic view is the phrase τὸ σαφὲς σκοπεῖν. This occurs in the famous statement of Thucydides' methods and aims in 1. 22. The historian there says that he will be satisfied if his work 'is judged useful by those who want a clear picture (ὅσοι βουλήσονται τὸ σαφὲς σκοπεῖν) of what happened in the past and of what will, in accordance with the state of human things, happen again in a similar or analogous fashion'. Volumes have been written, in comment on this famous sentence, concerning Thucydides' alleged cyclical view of history and the like. But the first thing for us to observe is that while it appears to offer a kind of predictability of human affairs—'this is what happened: like things will happen in the future'—it is a kind of predictability which expressly rules out any possibility of so improving the human condition that similar things will not again occur. Like the sentence about the Corcyrean revolution, it offers no hope at all of any cure, based upon Thucydides' own work or upon anything else.

In this sentence, the word 'useful'—Thucydides hopes that men will find his work 'useful' in getting a clear picture of what happened—has proved a particular stumbling-block for modern scholars imbued with positivism and utilitarianism. It was widely interpreted, and the sentence frequently mistranslated so as to support the interpretation, to mean that Thucydides intended his work to be of practical use. Not until Kapp's review of Schadewaldt's book on Thucydides | was this ghost laid to rest. Kapp established what should have been obvious, that the 'usefulness' was limited to the reader's acquisition of a clear picture.[12]

Weidauer, however, wishes to bring in practical value from the words τὸ σαφὲς σκοπεῖν themselves. To get a *clear picture*—τὸ σαφὲς σκοπεῖν—or *clear knowledge*—σαφῶς εἰδέναι—of something must mean to gain an understanding of the facts not obviously on the surface; and the only purpose of this, he argues, somewhat begging the question, is to plan one's actions on the basis of such knowledge. An example: in 415 BC the general Nicias writes home from Sicily (Thucydides 7. 14). He gives a dark picture of the military situation in Sicily and then

---

[12] *Gnomon* 6 (1930), 76–100, esp. 92–4.

concludes: 'I could have written a different and more pleasing letter, but nothing more useful than what I have written, if you are to make your plans on a basis of clear knowledge of how things are here'—Nicias, that is, assumes that the clear knowledge, the unvarnished picture, he is communicating to the Athenians will lead to good planning; so, Weidauer argues, must Thucydides himself have done.[13] But Nicias in Thucydides' account is a prime example of a leader as incapable of good planning himself as he is of inducing it in others. In Book 6 his good advice to the Athenians not to sail against Syracuse led them only to redouble their armament; at the point when he writes this letter the only proper plan is to sail back home. Nicias' 'clear knowledge' does not suggest this course to him, nor does this knowledge, communicated in his letter to the Athenians, lead them to suggest such a course. No more ironically self-defeating argument than Nicias' letter could possibly be thought up to urge the case that Thucydides believed the knowledge he was communicating to posterity would be a key to salutary political action.

Shortly after the beginning of the description of the Plague, Thucydides gives us a statement of his purpose and methods in describing it. He says:

Each man, physician or layman, can say what he understands concerning this sickness, and can give the causes which he believes to have had the power to effect so drastic a change. I will tell what it was like, and will set forth those features of it from which, if ever it should attack again, a man will be best able to know beforehand what it is and not to fail to recognize it.

Thucydides suggests that the Plague may at least occur again, in the lifetime of one of his readers, and that if it does, it is likely to have the same features as the one he himself experienced. An accurate description therefore will give his readers a *clear knowledge*—ἀφ' ὧν ἄν τις σκοπῶν ... προειδὼς μὴ ἀγνοεῖν—of his Plague and any recurrence of it.

Since the sentence is clearly analogous to Thucydides' general statement of his methods and aims in 1. 22, it is not surprising to find it interpreted by those who believe in

[13] Op. cit. 68–9.

Thucydides' practical intentions in much the same way. The knowledge, Weidauer argues, must be to provide an eventual cure.[14] In the whole sentence he finds a great similarity to the early medical writers, in particular the author of *Epidemics* 1 and 3. That writer also refrains from giving us any *causes* for the sicknesses he describes—although he does of course offer us his κατάστασις, absent, as we have seen, in Thucydides. Like Thucydides, the medical author describes with great objectivity and accuracy the symptoms of the sickness—'what it was like'—and in one passage at least, speaks of being able to prognosticate the future course of the sickness—a possibility we may assume to be implicit in Thucydides' knowledge of the sickness so as not to fail to recognize it—προειδὼς μὴ ἀγνοεῖν.

The author of *Epidemics* 1 and 3 makes only occasional references to treatment; but from the fact that this is a medical treatise in the first place, and from the explicit accounts of other early medical treatises, we know that the doctor would offer treatment, treatment made appropriate to the course of the disease as the doctor has carefully described it. Thucydides in this sentence makes no mention of cure at all; but, by analogy with the medical treatises, Weidauer argues, again begging the question, Thucydides must have intended his account to be a guide to future treatment of the disease.[15]

The argument, characteristically combining the optimistic and the medical interpretations of Thucydides, is invalid on its own grounds, for it substitutes mere assertion for genuine inference. But it is also expressly contradicted by what Thucydides says of the Plague, which is that no treatment of any kind had the slightest effect on it. 'They died,' he says, 'some in neglect, and others receiving every kind of treatment': οἱ μὲν ἀμελείᾳ, οἱ δὲ καὶ πάνυ θεραπευόμενοι. 'There was absolutely and utterly no cure at all'—my English adverbs are a clumsy attempt to render the stress of the unique Greek syntax: ἕν τε οὐδὲ ἕν κατέστη ἴαμα.

Thucydides' insistence on the uselessness of any kind of remedy, medical or otherwise, for the Plague is indeed an essential part of his description, because, as we shall shortly see

---

[14] Op. cit. 60.   [15] Op. cit. 65 f.

in more detail, he wants to present the sickness as an inhuman
or even superhuman visitation, a demonic enemy against
which no human weapon could avail. But if this is so, it both
invalidates the connection, drawn by Weidauer and others,
between the doctors and Thucydides' general purposes in this
passage, and at the same time is just one more nail in the
coffin—to use an appropriate image—of the optimistic interpretation of Thucydides' History.

Now it is quite possible that although Thucydides' purposes
and the general scheme of his description are at clear variance
with the medical writers, he is none the less their pupil in the
detail of his description and in the spirit in which that
description is set forth. This is the argument, widely accepted
today, of those who see in Thucydides' description of the
Plague primarily a brilliant piece of scientific observation. This
view in turn is likely to be connected to at least a limited
optimistic interpretation of the History. Thus the intelligent
and versatile Budé editor, Jacqueline de Romilly, takes the line
that the whole passage is inspired by a fierce scientific passion.

Thucydides here shows, she says, 'a proud confidence in the
scientific value of analysis'—'une fière confiance dans la valeur
scientifique de l'analyse'. When Thucydides says (2. 47) that
doctors could do nothing against the disease, he reveals, she
claims, his regret at their lack of experience, 'which if they had
had it, would have enabled them to treat it successfully'—'ce
qui leur eût permis de le soigner'—a regret which nowhere
appears in Thucydides' text. And finally, generalizing at will,
she says of 48. 3, where Thucydides states the purposes of his
description, that he there speaks 'like a scientist well acquainted
with the value of rigorous observation', and that 'he surely
expects that this description with all its exactitude will be of
practical utility for men of the future'—'il compte bien que
cette description si exacte présentera une utilité pratique pour
les hommes de l'avenir'.[16]

Of this expectation, again, there is no sign in Thucydides'
text. Mme de Romilly's reasoning is that the actual description
of the Plague is *scientific*, in a modern sense; and that therefore
Thucydides must have had in mind, whether he says so or not,

[16] *Thucydide, Livre II* (Société d'Edition 'Les Belles Lettres'; Paris, 1962), p. xxxi.

the purposes which modern scientists are at least popularly supposed to have. We can safely leave this reasoning behind and consider instead its major premise, the scientific nature of the actual description of the Plague.

Thucydides, in his description of the Plague and elsewhere, is extraordinarily observant and precise, and with the notion *per se* that he possesses many of the virtues that we associate with the exact sciences, there is little reason to argue. The difficulty is that those who insist on Thucydides the Scientist are likely to do so, as Cornford long ago observed, on the basis of a dichotomy of *science* and *art*. Thus the fine Victorian translator Richard Crawley took the description of the Plague to be an outstanding example of Thucydides' historiographical excellence. 'In five short pages', Crawley says in his preface, 'Thucydides has set forth the symptoms of the disorder with a precision which a physician might envy, and the suffering and moral anarchy which it produced with a vividness which may teach the lover of picturesque description how much force there lies in truth and simplicity.'

We note that Crawley here opposes truth to 'the picturesque'—how, we may wonder, does | the sheer dramatic power of Thucydides' description fit into this scheme? And we may wonder also how the translator could feel that 'simplicity' was an adequate term for the remarkable and expressive word order which distinguishes Thucydides' description. But Crawley at least did not, like more recent commentators, under the influence of theories of the Hippocratic origin of Thucydides' thought, assert that the historian was here writing a sort of modern case-history. He did not, like John Finley for instance, speak of the historian's 'cool, impersonal tone',[17] or like J. de Romilly later, following Denys Page, who is again followed by as sceptical a critic as Arnold Gomme, state that 'the vocabulary [of the description of the Plague] is ... entirely technical.'[18]

In an entertaining as well as influential article in the *Classical Quarterly* of 1953, Denys Page appeared to do what his predecessors in this line of enquiry had failed to do, viz., to give a specific demonstration of both the technicality of

[17] Op. cit. 70.   [18] Op. cit. p. xxx.

Thucydides' vocabulary in his account of the Plague and of his specific indebtedness to the attested usage of the medical writers. This article has had an odd history. It consists of two parts. The author first attempts to establish the coincidence of Thucydides' method and vocabulary, especially in chapter 49, the chapter in which the physical effects of the disease are described, with the method and vocabulary of ancient medical writers. Having to his satisfaction established this point, Page goes on in the second part of his article to conjecture the name of the disease itself. It turns out to be measles. This second part of the article, containing the identification of the disease, was answered in the following year by a doctor well-versed in the Classics. His name was MacArthur, and his article appeared in the *Classical Quarterly* in 1954. MacArthur's arguments have effectively convinced most readers since that the fatal sickness was not measles, but typhus. It is unnecessary here to enter into any of the details of that debate.

Thus Page saw the second part of his study, for which the first part was designed only to clear the way, refuted. But the philological arguments of the first part seem only to have been embraced with greater warmth by subsequent scholars. It is now taken for granted, e.g. by Gomme in his *Commentary*, that the primarily technical nature of Thucydides' vocabulary in the description of the Plague is securely established.[19]

I believe that the technicality of Thucydides' vocabulary here is far from certain. Before examining Page's arguments in detail, we must make clear what is meant by *technical*. Greek medical writing of the fifth and fourth centuries BC did not on the whole contain what we usually mean by *technical* terms. That phrase for us usually means what is illustrated by C. P. Snow's complaint that when he asked the company at a cocktail party of humanists what a *power tool* was, no one could tell him. That is, a *technical term* is a word incomprehensible to anyone not familiar with the technical facts. There are few such words in the early Greek medical writers. Thus—to choose one of Page's stronger examples—the word αἱματώδης, which appears in Thucydides 2. 49. 2 and in some medical

---

[19] A. W. Gomme, *A Historical Commentary on Thucydides* (Oxford 1956), vol. ii, *ad* 2. 48. 3.

treatises, may possibly have been a professional medical term chosen in preference to words like αἱματόεις and αἱματ-ηρός in order to avoid poetic association and to stress the physical appearance of blood; but it would have been instantly comprehensible to any Greek reader whether he was acquainted with medical treatises or not. A word like *hematoid* in English is technical in the common modern sense: it is likely to be meaningless to anyone not a medical doctor—or a classicist.

There are, in early Greek medical prose, a *few* words which *are* 'technical' in this strict sense: e.g. the word ἀπόστασις, which W. H. S. Jones translates as *abscession*, will be incomprehensible in its medical context if one does not know the medical definition. No such words appear in Thucydides' description of the Plague.

If we look closely at the way in which Page phrases his philological arguments, we see that he does not directly claim that Thucydides, in describing the disease, used a technical | vocabulary. He does imply this, and later commentators assume that this is what he meant; but he does not say it. What he sets out explicitly to demonstrate is the much more modest point that Thucydides does not use in this passage a language greatly at variance with common medical usage. This point appears to be true. It is with the widely accepted implications of Page's arguments, rather than with their declared assertions, that one must take issue.

Page's method is to list the words Thucydides uses to describe the sickness, and then to find the same words used in the same sense by one or more medical writers. In most cases, he succeeds in doing this, and where he does, he is likely to add: 'a standard medical term'. The words in question are indeed 'standard medical terms', if we understand this phrase to apply to words which are at the same time 'standard *non*-medical terms'. Thus the word βήξ (Thucydides 2. 49. 3) is used regularly by the doctors to mean 'cough'. Page does not add, and for his strict purposes need not add, that everybody else in Greece also used βήξ when he meant 'cough'.

Again, Page tells us, and I quote: 'πταρμός, "sneezing", and σπασμός, "convulsion", are standard medical terms. φλύκταινα

is the standard term for an exanthem of the blister type . . .'. He
then gives two examples of φλύκταινα in the medical writers.[20]
If we do not stop a moment and think—and many of the
readers of Page's article seem not to have done so—we may
conclude that these are in some reasonable sense technical
terms. But the Greeks in fact had, as we do, one word for
'sneeze', the verb πτάρνυσθαι, appearing e.g. in the *Odyssey*;
and the regularly corresponding noun πταρμός, in Aristophanes and Plato as well as, of course, the doctors.[21] σπασμός,
'convulsion', which can correspond in all senses with the
common verb σπᾶν, 'to draw or pull', is the common noun
referring to a more violent and less usual physical phenomenon
than sneezing. It therefore appears especially often in the
doctors, but we find it also in Herodotus, Aristophanes,
Sophocles. Sophocles can in fact use it in a poetic and
metaphorical mode of expression which would have been
impossible if the word were strictly technical, i.e. if its basic
meaning were not immediately transparent: *Trachiniae* 1082:
'the wrench of *atê* burned him'—ἔθαλψεν ἄτης σπασμός. Of
the poetic possibilities of this sort of language, more in a
moment. φλύκταινα finally, although LSJ contrives to give the
word a technical cast in Thucydides here by translating it as
'pustule', just means 'blister' and so appears often enough in
Aristophanes. ἐγὼ δὲ φλυκταίνας γ' ἔχω, cries poor Dionysus as
he rows across the lake and tries to keep up with the Frogs. No
one has bothered to suggest that Aristophanes here is using 'a
standard medical term'.

What is true of these terms is true of the great majority of
words listed and discussed in Page's article, and in so far as this
is the case, his argument, as subsequent scholars have interpreted it, loses all its point. The argument in fact receives
support only from those very few words in the Plague passage
which among early Greek authors appear only in Thucydides
and the medical writers, or which appear only in Thucydides
and the medical writers in the sense which they bear in the
Plague passage. These words are so few that we can rapidly
survey them here: of αἱματώδης, 'bloody', I have spoken;

[20] D. L. Page, op. cit. 101.
[21] For this and the following Greek words, see the references in LSJ.

ἕλκωσις, 'wound' or, in more technical translation, 'lesion', Page says: '[is] not found elsewhere in fifth-century Greek, [but] is common in the doctors'. That is true, but ἕλκωσις is a normal nominal formation from the verb ἑλκοῦν, which appears in Euripides. στηρίζειν Thucydides uses with ἐς and accusative to refer to a disease 'settling' in a part of the body; so at least once, in a late treatise, does the medical corpus. It seems to be the case that the verb in the sense of 'settle' does not appear elsewhere with ἐς + the accusative in early Greek. But with the dative in this sense it appears often enough: for example in Homer and Euripides. ἐξανθεῖν, to 'flower' or 'burst forth', Page claims for this category, but himself points out that it appears in poetry;[22] thus at the end of Aeschylus' *Persians*: ὕβρις γὰρ ἐξανθοῦσ᾽ ἐκάρπωσεν στάχυν ἄτης: 'Violence has flowered; its harvest is ruin.'[23] ἀποκριθῆναι is, as Page says 'a standard medical term in the doctors, especially signifying the secession of an element from a compound ...' The word appears in Thucydides' | description, but in a quite different sense, one that does not appear in the doctors, of diseases other than the Plague being *resolved* into the sickness. And finally, what may be the *unique proper member* of this set, ἐπεσήμαινεν (Thucydides 2. 49. 7), used intransitively of a symptom which appears and reveals the disease: 'for those who did recover from the Plague, the *seizure* (ἀντίληψις, a word that does *not* appear in the doctors) of their extremities (i.e. apparently the loss of their extremities) marked (ἐπεσήμαινεν) the previous course of the disease.' ἐπισημαίνειν does seem to be almost technical and it does occur only here and in the doctors, although it should be pointed out that σημαίνειν in a similar sense is found in Euripides.[24]

Against this category of truly technical terms, which turns out on inspection to be a category with only one dubious member, must be set a fair number of words of physical reference in the description which either do not appear in the medical corpus at all, or appear there in a different sense: e.g. φλόγωσις, 'inflammation', not attested in the doctors, who use

---

[22] ἕλκωσις, Page, op. cit. 101; στηρίζειν, 106; ἐξανθεῖν, 107.
[23] *Persians* 821–2.
[24] ἀποκριθῆναι, Page, op. cit. 107; ἐπισημαίνειν, 106.

φλεγμονή; ἀντίληψις, as I mentioned above; or ἀσθένεια in the sense of 'disease', always used by the doctors to mean 'physical weakness accompanying a disease'.[25]

I hope enough has been said to show that the vocabulary of the description of the Plague is not entirely, is not even largely, technical.[26] I should like to suggest a directly contrary conclusion, that Thucydides, like Plato, had something of an abhorrence, or an aristocratic disdain, for technical terminology, either of his own or of others' making. The evidence for such a conclusion is that Thucydides succeeds in giving us so physically precise a description without using the quasi-technical vocabulary which we in fact find in the early medical treatises, e.g. ἔπαρμα, 'swelling'. But another sign, or shall we say symptom, of this tendency in Thucydides appears in the one sentence where he actually speaks of the medical writers and shows that he had read them (49. 3): 'vomitings of bile ensued, all the kinds of vomiting which the doctors have named' — ἀποκαθάρσεις χολῆς πᾶσαι ὅσαι ὑπὸ ἰατρῶν ὠνομασμέναι εἰσὶν ἐπῆσαν. The rapid and devastating account of the physical manifestations of the sickness occupies little over a single page of the Oxford text. To have dwelt on the medical classification of the forms of bile vomited by its victims would have slowed down the account in a way contrary to Thucydides' real purpose, which is to present the onslaught of the pest in as dramatic a form as possible. It would also have introduced the very sort of clinical and technical language he chose to avoid.

The beginning of a typical case from *Epidemics* 1 and 3 may show what I mean: 'Silenus lived in the flat land near the quarter of Eualcidas. After fatigue, drinking, and unseasonable

[25] φλόγωσις, see Page, op. cit. 101; ἀντίληψις, see ibid., 108; ἀσθένεια, 109.

[26] Mr S. W. Scott has kindly pointed out to me the analogy of the scholarly argument concerning the language of the Gospel According to St Luke and the Acts of the Apostles. Attempts to show that the author of these works was a physician on the grounds that they contain technical medical terms were effectively refuted by Henry J. Cadbury in his monograph, *The Style and Literary Method of Luke* (Harvard Theological Studies VI; Cambridge, Mass., 1920), esp. pp. 39–51. In an editorial note in the same volume describing the history of the controversy (pp. 51–4), G. F. Moore points out that the Greek medical writers themselves did not use a language in our sense technical. Moore also cites (p. 54) an impressive list of scholars who argued that the fact that the gospel-writer used the word κραιπάλη proved him 'to be versed in Greek medical literature'. If only it were true that one needs to be a medical savant in order to be aware of having a hangover.

exercise, he got a fever. The pains began in his loins. Then heaviness of the head and stiffness of the neck. From his abdomen on the first day, bilious matter, unmixed, frothy, saturated. Urine black, with black sediment; thirsty; dry tongue; at night did not sleep. Second day: sharp fever, evacuations more copious, lighter; urine black; difficulty at night; became slightly delirious. Third day: general exacerbation; oblong rigidity of the hypochondrium from both sides towards the navel, without hardness beneath; evacuations light, blackish . . .'; and so on until 'Eleventh day: died'.[27]

The language is clipped, factual, a kind of scientist's shorthand. There are few verbs, and those there are receive little stress. There is a good deal of mildly technical language, mostly in the nouns and adjectives: ὑποχόνδριον, 'hypochondrium'; κατακορέα, 'saturated in colour'; ὑπόστασις, 'sediment'; ὑπολάπαρος, 'without swelling or hardness beneath'. The Loeb editor, W. H. S. Jones (introduction, p. 149) expresses his admiration for the style of this treatise: 'Pretensions to literary form it has none, yet no Greek writer, with the possible exception of Thucydides, has used language with better effect. Often ungrammatical, sometimes a series of disconnected words, the narrative is always to the point.'

It is possible that Thucydides felt this bare, detailed, and undramatic medical style to be | a kind of ideal, just as Stendhal claimed he would have liked to write like the *Code Napoléon*. But if he did, he did not let this feeling show in his own description of the Plague. The style of that description is observant and exact, but it shows what Wade-Gery, speaking of the historian's style generally, called 'a poet's precision'.[28] It is grammatical, but it stretches the limits of Greek grammar. It is dramatic and imaginative, controlled throughout by the writer's determination to show the awful and overwhelming power of the sickness. The sentence-construction is various, often containing powerful and unexpected verbs in emphatic positions, or after a climactic catalogue, resolving itself into an epigrammatic summation.

A few examples: the very beginning of the description, 47. 3: 'The Lacedaemonians had not been many days in Attica

27 W. H. S. Jones, op. cit. case 2, p. 187.   28 *OCD*, p. 904.

when the Plague first began to afflict the Athenians. It was said to have struck previously more than once in Lemnos and in other places, but never had there been so great a pestilence, never had thus destruction of humanity been recorded to occur.' The verb 'struck' is ἐγκατασκῆψαι—λεγόμενον καὶ πρότερον πολλαχόσε ἐγκατασκῆψαι. This is the first appearance of it in a prose writer, and it is not in the medical corpus. Sophocles uses it, and before him, Aeschylus: 'The ills wherewith god smote the Persians'—κακῶν δ' ἃ Πέρσαις ἐγκατέσκηψεν θεός.[29] Thucydides frequently puts his metaphors into verbs, and the suppressed image here, of a thunderbolt, is the same as what appears in Sophocles' *Oedipus Tyrannus*, a work which may have been inspired by this same Plague:[30]

> ... The fiery god
> Hath struck: a loathsome plague, he drives our town.
>
> ... ἐν δ' ὁ πυρφόρος θεός
> σκήψας ἐλαύνει λοιμὸς ἔχθιστος πόλιν.[31]

It is easier to talk about Thucydides' poetic vocabulary than about his sentence structure, but the latter is no less part of the extraordinary rhetoric of his description. The information he is giving in the opening sentence is that although the Plague was said to have appeared often elsewhere, *it had never appeared on so large a scale*. The Greek is οὐ μέντοι τοσοῦτός γε λοιμὸς οὐδὲ φθορὰ οὕτως ἀνθρώπων οὐδαμοῦ ἐμνημονεύετο γενέσθαι. The most drastic adversative particle μέντοι; the quantitative word τοσοῦτος thrown into relief by γε; two new words for the Plague in swift succession, λοιμός and φθορά; the negative thrice repeated, οὐ ... οὐδέ ... οὐδαμοῦ, in crescendo, and the adverb which conforms to the negatives in sound, οὕτως, displaced from its verb, so that grammatically it goes with the verb ἐμνημονεύετο, but in meaning with the noun φθορά; and the whole crescendo uncoils itself like a whip with the crack on the longest negative οὐδαμοῦ.

He continues with negatives: 'For neither were doctors of any avail—at first treating it in ignorance; rather they them-

[29] *Trachiniae* 1087; *Persians* 514.
[30] Cf. B. M. W. Knox, 'The Date of the *Oedipus Tyrannus* of Sophocles', *AJPh* 77 (1956), 133–47.
[31] *OT* 27–8.

selves most of all died inasmuch as they most of all came in contact with it; nor did avail any other human art whatsoever'—οὔτε ἄλλη ἀνθρωπεία τέχνη οὐδεμία. The beginning of the sentence is perfectly ambiguous; the doctors treated the sickness at first *in ignorance*, because they had had no experience. Later we find that experience did not enable them to treat it any better. Or, they tried to treat it at first—present participle θεραπεύοντες—*because* they were ignorant of its true nature.[32] 'And finally', Thucydides says, 'they all gave up any attempt to fight, by the evil overcome'—τελευτῶντές τε αὐτῶν ἀπέστησαν ὑπὸ τοῦ κακοῦ νικώμενοι.

The final overwhelming force of the disease is expressed by the long passive participle at the end of the sentence—ὑπὸ τοῦ κακοῦ νικώμενοι. This device appears throughout the description. Later we have: 'Since at the end even their families grew too weary even to lament those who were dying, by the greatness of the evil overcome'—ὑπὸ τοῦ πολλοῦ κακοῦ νικώμενοι. And in the description of the physical effects of the sickness: 'So burning was the sickness ... that what seemed most pleasant to them was to hurl themselves into cold water: *and many of those who were not looked after actually did it*—*into wells*,'—καὶ πολλοὶ τοῦτο | τῶν ἠμελημένων καὶ ἔδρασαν (note the violence of the verb) ἐς φρέατα—'by the thirst unceasing fast held'—τῇ δίψῃ ἀπαύστῳ ξυνεχόμενοι.

Since treatment is a futile cover for the disease, the verb 'to treat' falls into the same pattern: 'They died, some in neglect, others being elaborately treated'—ἔθνῃσκον δὲ οἱ μὲν ἀμελείᾳ, οἱ δὲ καὶ πάνυ θεραπευόμενοι. In order to put the participle at the end of that lapidary sentence, the author has put the poetic verb ἔθνῃσκον (instead of ἀπέθνῃσκον) first. But most often the verbs describing the victims of the pest occur at the sentence end: νικώμενοι, πάσχοντας, ξυνεχόμενοι, διεφθείροντο, θεραπευόμενοι, ἔθνῃσκον, νικώμενοι, διαφθαρῆναι, ἐπανοθνῃσκόντων, διαφθαρήσεται. These verbs all appear at the end of sentences in the seven chapters devoted to the Plague.

The Plague itself is likely to appear in active verbs at the

---

[32] The latter interpretation was suggested by E. Kapp, op. cit. 92. It is denied by Weidauer, op. cit. 82–3, on the bizarre grounds that if this was the meaning, Thucydides ought to have written προειδὼς μὴ θεραπεύειν in 2. 48. 3.

beginning of sentences: 'the malady descended'—ἐπικατιόντος τοῦ νοσήματος; 'it penetrated through the entire body'—διεξῄει γὰρ διὰ παντὸς τοῦ σώματος; 'overpowering men's rational faculties, the sickness . . .'—γενόμενον γὰρ κρεῖσσον λόγου τὸ εἶδος τοῦ νοσήματος and so forth. Both devices are used at once in dramatic juxtaposition in a sentence about the impossibility of burial: 'The sanctuaries in which they were camping out were full of dead bodies, as the men in the same place died; so overwhelming was the evil. . .'—τά τε ἱερὰ ἐν οἷς ἐσκήνηντο νεκρῶν πλέα ἦν, αὐτοῦ ἐναποθνῃσκόντων· ὑπερβιαζο-μένου γὰρ τοῦ κακοῦ. The verb ὑπερβιάζεσθαι, it might be added, is in effect a ἅπαξ λεγόμενον; it does not appear again in Greek until Josephus uses the identical phrase—ὑπερβιαζομένου γὰρ τοῦ κακοῦ; and then in the Thucydidean imitator Proco-pius.[33] Thucydides fashions his own vocabulary to render the appalling power of the sickness.

To show with any degree of thoroughness what kind of language Thucydides uses in his description of the Plague, in chapters 47–54 of Book 2, would require a commentary of a sort which the prevailing belief in the passage as a case-history, adopting the methods and imitating the style of the medical writers, has so far made impossible. I hope, however, that I have succeeded in giving the general idea.

The question remains, Why?—because the dramatic texture of the passage is not after all a virtuoso display designed to ensure that Thucydides would be imitated not only by later historians, but also by poets like Lucretius and Virgil.

Two passages tell us pretty clearly. First, 2. 54, summing up the effects of the sickness. 'Such a disaster caught up the Athenians and crushed them'—τοιούτῳ μὲν πάθει οἱ Ἀθηναῖοι περιπεσόντες ἐπιέζοντο—'as the men died within and the land without was wasted'—ἀνθρώπων τ' ἔνδον θνῃσκόντων καὶ γῆς ἔξω δῃουμένης. The Plague is a πάθος and it is equated with the War. The second half of the sentence, with its characteristic word order—θνῃσκόντων and δῃουμένης ending the two matched clauses—and with its funereal succession of long syllables, is a kind of Thucydidean counterpart to Achilles' words in *Iliad* 1: 'Agamemnon, we'll soon be struggling home

---

[33] LSJ s.v.

if we escape death at all, *if War and Plague alike are to whelm the Achaeans*':

εἰ δὴ ὁμοῦ πόλεμός τε δαμᾷ καὶ λοιμὸς Ἀχαιούς.

The Plague is a πάθος, like war, and in fact, it is a partner of war. War, Thucydides tells us clearly in 1. 23, consists of πάθη. It is in fact to be measured by suffering and destruction. It is precisely in this context that Thucydides first speaks of the Plague: 'This war was great in its duration, and disasters— παθήματα—occurred in it to Greece such as no others in a like space of time may be compared to them.' He lists these disasters, which prove his contention that this was the greatest of all wars: cities captured and laid waste; exile; slaughter; then earthquakes, eclipses of the sun; drought; famine; and, *not least pernicious and in part utterly destructive, the pestilential Plague*. As Thucydides rises to this, the greatest | disaster of all, his language assumes that unique and almost apocalyptic poetic power which we observed in the description of the Plague itself: καὶ ἡ οὐχ ἥκιστα βλάψασα καὶ μέρος τι φθείρασα ἡ λοιμώδης νόσος, where the nine words that intervene between the first article ἡ and its noun probably set a syntactical record.

The πάθη that accompany the War and become a part of it, in Thucydides' presentation, are not those we should expect from the thoroughly enlightened, thoroughly scientific, historian. For he adds the disasters of the cosmos to what we think of as purely human misfortunes. Of these, the Plague is the greatest and most destructive. Now in Thucydides' scheme, war in any form, even the smallest battle, is a παράλογον. That is, it is incalculable; it defies human reason. Human reason, in the History, appears in its most powerful and comprehensive form in the planning and authoritative mind of Pericles. For the War proper, Pericles makes the fullest possible provisions, although, as Hans-Peter Stahl has recently shown with great cogency, even the course of the war itself, that is, the Spartan attack and its effect on the Athenians in Book 2, does not proceed in perfect conformity with Periclean foresight.[34]

The Plague is a παράλογον beyond all others, and is essentially part of the war. It represents the most violent incursion of

[34] Op. cit. 76 f.

the superhuman and incalculable into the plans and constructions of men. Immediately after the Funeral Speech, the
strongest assertion of the power of the mind to control the
world, the Spartans come again. They instantly give way to
this new and more horrible enemy. Much of the language of
the description of the Plague, in fact, suggests that it comes as a
military attack: verbs like ἐπιπίπτειν, ἐσπίπτειν, νικᾶν, ξυναιρ-
εῖν.[35]

It is in short the most sudden, most irrational, most incalculable, and most demonic aspect of war in Thucydides' view of
history. This is precisely how Pericles is made to speak of it, as
a δαιμόνιον.[36] As such, the Plague offers the most violent
challenge to the Periclean attempt to exert some kind of
rational control over the historical process. Thucydides ultimately leaves it undecided whether Periclean will and fore-
thought—γνώμη—in fact is able to meet this challenge. The
History here becomes a kind of dialectic: the statesman's heroic
last speech (of course he must have given others, but the
historian leaves them out) restores his control and ensures the
partial victory of the Archidamian War. But he himself dies of
the Plague, and the City is governed by all those men who, in
the historian's words, 'were more on a level with one another',
and who eventually destroyed Athens. Only the memory of
the City and the Empire is left victorious; physically, the war is
lost.[37] In this dramatic process, so profoundly observed and so
artistically patterned by Thucydides, the Plague, as the superhuman enemy, has a vital role. The language in which the
Plague is described is determined by that sense of its meaning.
The attempt on the part of the Optimists to see the description
of the Plague as a modern scientific treatise, and to persuade us
that Thucydides saw it as a thing subject to rational human
control, obscures for us not only the compassionate poetry
with which the Plague is in fact described, but also the whole
meaning of the History.

[35] This point is made with abundant evidence in the excellent study of the
vocabulary of the description of the Plague by G. M. Parassoglou, Honours thesis
(Yale University, 1968).

[36] 2. 64. 2.

[37] The individual soldier's triumph in the Funeral Speech is one of δόξα and γνώμη
rather than of fact—ἔργον: 2. 43. 2–6; and so is that of the whole City in Pericles' last
speech: 2. 64. 3 f.

14

Thucydides' Use of Abstract Language

IN the following essay I shall consider it my purpose not so
much to make a point about Thucydides, as to suggest a
method of study which I think is a useful one for him and for
many other Greek writers as well. This method is hardly a new
one, but as far as I can see it has not been adequately exploited
in Classical studies: I mean making a study of *the means of
expression which a given writer had at his disposal* as the starting
point for the elucidation of his thought; in other words to view
a writer's creation first in terms of the raw materials he used.
Certainly the student of the plastic arts is careful to do that: we
are always told what sort of marble, or bronze, or clay a
sculptor or architect chose to shape to his purposes; and often
we can see how the final shape he created depends on the
material chosen. Sometimes we are shown the phenomenon of
a certain incongruence between material and form: cases where
an artist uses in one medium devices which were evolved in,
and are more appropriate to, the use of another. Clay vases
with handles which are too thin because the potter is copying
the shape of metal vases, and similar phenomena. I suggest that
the literary artist is no less dependent on his medium: he deals
with language in a particular state of evolution: what he says,
the final purport of his work, will always be intimately
associated with the language which is his raw material. Occasionally he will be seen to struggle against certain qualities in
his literary medium: he may, like Pindar, use abstract nouns as
if they were tangible entities; and in doing so he will be like the
sculptor who tries to give to marble some of the qualities of
bronze. A difference between the literary and the plastic artist
is that the material of the former is always immediately bound

Language of Action (Yale French Studies), 45 (1970), 3–20. Reprinted by permission of
Yale University Press.

up with *thought*: the language which a poet or an historian adopts itself contains a way | of looking at the world; and in making his individual statement, a writer will have to begin from the assumptions implicit in his raw materials.

Such an approach may I feel be particularly valuable for Greek writers before the age of philosophy, for the reason that their language differs much from ours, and that its special qualities have accordingly been little understood. We are likely to consider the poets as isolated phenomena, as if every poet not only created his poems, but also fashioned as it were from nothing the style and the mode of language they are written in. And yet it should be obvious that if we are to understand say, Pindar, we must first understand the lyric style which existed before he wrote: we must find out what is common to him and to Simonides and to Bacchylides, and then look for the variations from this common style in order to see what he was really trying to do. Even in the case of a—for us—isolated representative of a tradition, this is so. The genius of Homer (assuming only for the purposes of argument that he is the author of both the *Iliad* and the *Odyssey*) will only be understood by seeing clearly the particular use he made of a pre-existing formulary tradition. In the case of prose-writers on the other hand, our knowledge is obscured by a worse failing: the tradition of ancient rhetoric has been so strong as to dominate most modern presentations: so that we look at the history of Greek prose from the point of view suggested by Aristotle in his *Rhetoric*,[1] and see, or think we see, how from primitive beginnings the fully developed rhetorical period slowly but inevitably evolved.

The truth is that we do not have in Greek literature on the one hand a group of poets, each unique and a law unto himself, and on the other, a development of prose moving from infantile simplicities toward an Aristotelian final good of developed rhetorical device. What we have instead is a constant development of language from Homer to Hellenistic literature *which embraces all forms of writing*. The chief characteristic of this development is the formation and the | gradual ascendancy of abstract modes of expression. We can watch the

[1] 3. 9.

progress of this development, which occurs with especial clarity in the Greek world, much greater clarity than is shown by the analogous development of the European languages, since they from their earliest stages have been troubled by influences from without. We can see how at a distinct point of time—around the middle of the fifth century—prose begins to displace poetry as the most serious vehicle of thought, largely as a consequence of this very development. And, most important, we can gain new insight into every ancient writer along the way, by seeing how he reacted to the stage of the development which was contemporary with him.

Let us consider the specific problem of Thucydides' style. I was led to a study of this a good many years ago by the discovery that, while I admired the style of Thucydides immensely, more indeed than that of any other Greek prose-author, not everyone shared my unqualified enthusiasm. It may have begun one day with a professor who handed me back a prose piece I had done for him with the comment that it sounded like one of Thucydides' speeches. That this was by no means a compliment struck me as a very interesting judgement on Thucydides as well as myself. About the same time I discovered that already in antiquity the style of Thucydides had met with criticism, and notably that Dionysius of Halicarnassus had written at some length to demonstrate the vices of the historian's Greek style.[2] Dionysius, writing from a strict school-rhetorician's point of view, and condemning Thucydides accordingly, is rather like Bentley criticizing Milton from the point of view of a stricter, and tamer, standard of English than Milton's own. And he is valuable as a critic in the same way as Bentley: what he is least capable of understanding is likely to be most characteristically Thucydidean.

We are faced with the paradox that the Greek writer who above all others, even Plato and Aristotle, values *intellectual clarity and certitude*, is often so extremely difficult and involved that we must struggle to figure out what he is saying, and not rarely fail in the | attempt. Who has read through the text of Thucydides without at dozens of places exclaiming: 'Here he is

---

[2] *De Thucydide ad Aelium Tuberonem*, esp. ch. 24. Dionysius mixes some admiration of the historian's style with his censure.

really not giving us a chance!' Was this the last joke on the world of the exiled Athenian military leader, or of the disappointed Periclean imperialist? Surely not, because anyone who admires Thucydides must feel that the power of his language is inevitably bound up with his obscurity. Often we feel that it is precisely his desire for accuracy, for a precision that will not relax into cliché at any point, that makes his sentences so complex and so elliptical. But it is not enough to say that he sought precision, or even, as Wade-Gery so finely put it, a 'poet's precision'.[3] Albin Lesky is nearer the mark when he speaks of the remarkable antinomy in Thucydides, whereby 'beneath his serene detachment and lucid objectivity there is the agitation of a passionate and troubled spirit'. This, says Lesky, is what accounts for his density and variation in expression, qualities which, as Lesky sees with great insight, appear most clearly when his style is compared with that of his older contemporary Gorgias.[4] This brief statement of Lesky brings us closer, I think, than any previous analysis to understanding Thucydides; but to grasp its full implications we must define this passionate and yet scientific style more closely and consider it in the large context of the development of abstract expression in the Greek language.

One obvious peculiarity of Thucydides' style is a fondness for antithesis, for balancing one sort of thing against another. An equally obvious peculiarity is variation, what the Greek rhetoricians called μεταβολή.[5]

These two features are persistent and obvious. Aristotle (*Rhet.* 3. 9) distinguished the λέξις εἰρομένη, the *running*, *paratactic*, or *strung-out* style, from the λέξις κατεστραμμένη, the *bound* or *periodic* style. Modern historians of Greek prose have added the *antithetical style*, represented by Thucydides and the orators Antiphon and Gorgias as an intermediate step between Herodotus, master of the strung-out, and Isocrates and Demosthenes, masters of the periodic, | styles.[6] We thus get a

---

[3] *OCD*, 904.

[4] *Geschichte der griechischen Literatur* (Berne/Munich, 1957–8), 524.

[5] See [Longinus] *De Sublimitate* 23. 1 and Russell's note ad loc.

[6] e.g. A. Croiset, *Histoire de la Littérature grecque*, vol. 4 (Paris, 1898), 629. By κατεστραμμένη, Aristotle in fact probably had something like the 'antithetical' style in mind, and would have included Thucydides. See G. Kennedy, 'Aristotle on the Period', *HSCPh* 63 (1958), 283–8.

development which subserves the idea of progress, with the
infantile simplicities of Herodotus developing through the
energetic adolescence of Thucydides to the rounded maturity
of Isocrates. This is one way of presenting the history of Greek
prose; but if something may have been gained in smoothness
and fullness of expression between Thucydides and Isocrates,
between the masters of the antithetic and the periodic styles, so
much may have been lost as well.

Antitheses are the most prominent feature of Thucydides'
style. Only a little less prominent is variation. The effect of this
is partly to counteract the effect of antithesis. That is, you first
balance off one thing against another, then you introduce an
imbalance by phrasing the two corresponding parts differently.
Thucydides does this constantly, and there is a huge book by a
Dutch scholar[7] which lists example after example of this sort of
variation, and tries to classify them.

Let me give a few simple examples. The simplest antithesis
rapidly sets up one word against another. The Corinthians
complain of Lacedaemonian sluggishness in the first congress
in Lacedaemon in Book 1. 'You, Lacedaemonians, are alone
inactive, and defend yourselves, not by *doing* anything, but by
looking as if you *would* do *something*.' That is Crawley's
translation, and it is very ingenious, but the contrast is much
more focused on single words in the Greek: οὐ τῇ δυναμει τινά,
ἀλλα τῇ μελλήσει ἀμυνόμενοι: defending yourselves not by
power or *action*, but by *delay*. An abstract noun in the dative
singular (δυνάμει) is balanced against another abstract noun in
the dative singular (μελλήσει). They have the same endings (-
ει-) and the same number of syllables, although the first is
rapid, being an anapaest (∪∪—), the second, appropriately
slow, a molossus (— — —). The symmetry is virtually complete, and there is almost no *variatio*. But we can observe these
points: that the effect of the swift staccato | style is to
concentrate our attention on particular words; and that these
words are likely to be abstract nouns.

Here is an example from Book 2. The Peloponnesians are
attacking the town of Stratus in Acarnania: '[They] advanced

[7] J. Ros, *Die METABOΛH (Variatio) als Stilprinzip des Thukydides* (Paderborn,
1938).

upon Stratus in three divisions, with the intention of encamp-
ing near it and attempting the wall *by force* if they failed to
succeed *by negotiation*.' Here the English does get the contrast of
word against word, but the Greek again is a little different. 'So
that they could encamp near it, ὅπως ἐγγὺς στρατοπεδευσάμε-
νοι, and then, if they could not persuade *by words*, *by action* they
could attempt the wall: εἰ μὴ λόγοις πείθοιεν, ἔργῳ πειρῶντο
τοῦ τείχους. The words that are made to leap out at us in the
sentence are the dative plural λόγοις and the dative singular
ἔργῳ, and these fundamental Thucydidean terms are what
Crawley translated by *negotiation* and *force*. Moreover there is a
subtle *variatio* in this sentence: the one word is plural, the other
is singular. The one goes with a verb used absolutely: πείθοιεν,
'persuade'; the other goes with a verb that has an object:
πειρῶντο τοῦ τείχους, 'make an attempt on the wall'. Furthermore, λόγοις is a real instrumental dative: they will try to
persuade *by words*; whereas ἔργῳ is adverbial and in a sense
unnecessary: they will make an attempt on the wall, and this
will be an example of *action*. ἔργῳ in fact is not needed for the
immediate meaning of the sentence, and Thucydides puts it in
because he wants to present this simple bit of narrative too as a
manifestation of the fundamental opposition he is always
seeing between *words*, or *conception*, and *deeds*, or *actuality*. But
at the same time that he establishes, with a touch of violence,
this symmetry, he breaks it a little bit, by *variatio*, making the
corresponding words fulfil slightly different functions in the
sentence.

A third example of antithesis and variation is the most
famous sentence in the whole *History*: Pericles says of the
Athenians: φιλοκαλοῦμέν τε γὰρ μετ᾽ εὐτελείας καὶ φιλοσοφοῦμεν ἄνευ μαλακίας. This is a highly Gorgian sentence, and
rhythm plays an obvious part. 'We love things of beauty with
economy, with restraint, and we love things of the mind
without softness.' Or, less literally, in Crawley's | translation:
'Cultivating refinement without extravagance and knowledge ·
without effeminacy . . .'

φιλοκαλοῦμεν, 'we love beauty', matches φιλοσοφοῦμεν, we
love wisdom, *or* knowledge', perfectly, in rhythm, grammar,
and sense. The verb φιλοσοφεῖν first appears in Herodotus, a

generation before Thucydides, who probably took the word from his predecessor; φιλοκαλοῦμεν is Thucydides' own coinage, made by him to match Herodotus' word.[8] μετ' εὐτελείας, '*with* economy', does not quite match ἄνευ μαλακίας, *without* softness'. You could describe this sort of variation in two ways: either by saying that it establishes a symmetry and then breaks it up; or by saying that it forces into a symmetrical pattern phrases and thoughts that are not entirely commensurate with each other. And notice again, first, that antithesis and variation turn on abstract words; and second, that here too, though not so obviously, Thucydides has seen things in terms of his fundamental opposition between thought and actuality. The first half of the epigram sounds ethereal in translation, but in fact refers to a specific external reality, the Periclean building programme, responsible for the Parthenon among other things, so fiercely attacked by his political opponents who complained particularly about its high cost. The second half contrasts with the first, moving into the sphere of *mind*; only the term μαλακία, 'softness', brings us back to the external world, being primarily a military term: Pericles is saying that the Athenians can be intellectuals, and still defeat the Spartans.

Another example, with greater syntactical variation. The Thebans, allies of the Spartans, make a speech in Book 3 (chs. 61–7) urging the Spartans to exterminate the male population of Plataea, a small town on the border between Theban-dominated Boeotia and the territory of Athens. The Plataeans, the Thebans say, are war-criminals: their crime has been to help the Athenians in their attempt to conquer all of Greece. The Plataeans, on trial, make an eloquent plea for their lives, employing poetic and archaic language and reminding the Spartans of their old alliance with Plataea at the time of the Persian wars: | the Plataeans then gave their lives for Greek liberty. The decisive battle against the Persian invaders was fought on their soil. The graves of the Spartans who died in that battle have been cared for ever since by the Plataeans; and Sparta in particular has an obligation to maintain her old sacred alliance with Plataea, rather than destroy all her citizens to

[8] I am rejecting without argument the possibility that Thucydides in this sentence is actually repeating words spoken by the historical Pericles.

please the ruthless Thebans. The Thebans in reply reject what they regard as a sentimental appeal to irrelevant past history. The Plataeans are guilty *now* because they have become the lackeys of Athens *now*. 'Do not', they say in their peroration to the Spartans, 'let *us* be disregarded in your decision because of the Plataeans' *words*; make here a *demonstration* to all of Greece that you will insist on contests not of *words* but of *acts*; if the *acts* are good ones, the *report* of them can be brief [the Thebans had earlier complained of the length of the Plataeans' speech] but of *wrong-doing men, words*, adorned with *poetic forms*, are a *mask* of hypocrisy.'

The Thebans' rejection of history as a standard of judgement in favour of immediate political pressures, and their reduction of a complex situation to a simplistic morality reflecting immediate political interest (all pro-Spartans are good; all pro-Athenians are evil) are expressed by a series of antitheses. '*us*' — in Greek the first person plural of the verb περιωσθῶμεν, 'be neglected *or* pushed aside' — is contrasted with words, λόγοις, the dative plural of the noun. παράδειγμα, 'example', object of a verb, is set against λόγων, 'words', in the genitive plural, which is itself instantly contrasted with ἔργων, 'acts', again in the genitive plural. Another genitive plural noun ἀγαθῶν, 'of good men', is contrasted, external reality with words, with the noun 'report' on which it depends; then the genitive plural of a participle ἁμαρτανομένων, 'of wrong-doing men', is contrasted in sense with ἀγαθῶν, but as reality versus words with λόγοι, 'words' or 'speeches', which 'become' a 'deceptive reality', προκαλύμματα. The speakers ring the changes on the relation of words to reality as they see it in this rapid dialectic. The antitheses are syncopated (*a* being the first member of an opposition with *b*, which then becomes the first member of a new opposition with *c*) and varied in grammatical form, verbs balancing nouns, etc. The basic device of antithesis is varied and refracted to a | degree beyond what we could find in any other Greek author. And yet the intensity of intellectual analysis and the elaborateness of the syntactical structure are not frivolous: they deepen our dramatic sense of the Thebans' brutal mode of self-justification as they brilliantly and systematically annihilate all the emotional claims which the Plataeans had presented in their self-defence. Thucydides' syntactical modulations operate in his history like the sophistic conceits of a Euripidean Jason or Menelaus. Here, within the spectrum of Thucydides' opposition of inner and outer reality, the Thebans represent one extreme, rejecting all moral standards, intellectual criteria, and psychological motives in favour of the immediate pressure of political interest.

This distinction between thought and actuality, between λόγος and ἔργον or more or less obvious equivalents, is the real idiosyncracy in Thucydides' style. It comes up again and again—I have counted some 420 examples altogether in the eight books of the *History*, and it comes up where we least expect it, in moments of great pathos for example, as in the long sentence where Pericles describes the death of the Athenian soldiers in the first year's fighting, where ἔργον appears twice, λόγος once, and various equivalents appear ten times.[9]

This is the one feature of Thucydides' style which critics have most singled out for attack, and even his admirers have been put out by it. Thomas Arnold, commenting on the sentence I just referred to, laments that such fine pathos is adulterated by a frigid rhetorical device, and A. Croiset chooses another example of it[10] to demonstrate that Thucydides allowed a mannerism to stand in the way of what he was saying. More recently, J. D. Denniston, in an attempt to set up Herodotus as the standard of Greek prose-writing, has stated that while Gorgias was misled by a craze for verbal antithesis, Thucydides was misled by a craze for logical antithesis. 'In particular,' he says, 'he drags in the λόγος/ἔργον contrast in season and out of season. It spoils one of his noblest utterances in the Funeral Oration... etc.'[11] | But this cannot be right. Thucydides depends far too much on these antitheses for us to brush them off as inconvenient mannerisms. Antitheses of this sort are in fact to Thucydides what poetic ambiguity is to Shakespeare; and Arnold's impatience with the Greek historian is remarkably like Samuel Johnson's impatience with the idiosyncracies of the English poet. Johnson, in a famous

[9] 2. 42. 4.
[10] 1. 70. 6; in his edition of Thucydides 1 and 2.
[11] *Greek Prose Style* (Oxford, 1952), 13.

remark, says of Shakespeare: 'A quibble is to Shakespeare what luminous vapours are to the traveller: he follows it at all adventures: it is sure to lead him out of his way, and sure to engulf him in the mire. It has some malignant power over his mind, and its fascinations are irresistible ... A quibble was to him the fatal *Cleopatra*, for which he lost the world, and was content to lose it.' In recent times scholars of English literature have done much to show just how indispensable a part quibbles, or ambiguities, do play in Shakespeare's verse. Consider the analyses of William Empson. And similarly, our purpose in understanding Thucydides should be to explain what the effect is of *his* fatal Cleopatra, the abstract antithesis with variation, turning on a contrast of thought and reality.

To do this, I think we have to go back and look, however briefly, at the entire history of the development of abstract language in Greek literature. The Greeks in fact were the first people to develop an abstract vocabulary. T. B. L. Webster has described very well a number of features of this development in an article called 'From Primitive to Modern Thought in Ancient Greece'.[12] I should like here to offer a *schema*. I propose five stages of abstraction. Thucydides and Gorgias before him will fall neatly into the third. They are: the concrete abstraction, the proverbial abstraction, the social abstraction, the dogmatic abstraction, and the tentative abstraction.

The concrete abstraction is where we start, and it is of course a contradiction in terms. There are in fact virtually no abstractions in Homer. In the Homeric world everything is felt as a substantial thing, a concrete for our abstract, and this is what gives us that wonderful sense of solidity and reality when we read Homer. Courage, for example, is nothing one could ever argue about, or try to define, | in the manner of a Platonic dialogue. ἀνδρεία, the classical noun meaning 'courage', does not occur in Homer, and a word like μένος resists definition. It is something *lived*, rather than *thought*. A man has it, and when he dies, he loses it: ἀπὸ γὰρ μένος εἵλετο χαλκός; 'the bronze took his μένος away'.

The proverbial abstraction is developed by Hesiod, and reaches its height in the archaic writers, Pindar and Aeschylus

<hr>

[12] *Acta Congressus Madvigiani* ii. 29–46.

and Herodotus. Here we get real abstract words, like εὐβουλία, 'good counsel', πενία, 'poverty', φιλοφροσύνη, 'benevolence', and the like.[13] δικαιοσύνη, replacing the old substantive δίκη, seems to appear first in Herodotus.[14] These abstract words exist in their own right, but they are largely limited to proverbial expressions, or else are immediately bound up with some dramatic context. A brief example from Hesiod: after pounding in upon his brother the value of economy, he says:

$$εὐθημοσύνη \ γὰρ \ ἀρίστη$$
$$θνητοῖς \ ἀνθρώποις \ κακοθημοσύνη \ δὲ \ κακίστη.$$

'εὐθημοσύνη is best for men and κακοθημοσύνη is worst.' These words stand alone as true abstractions, but they really sum up a long concrete argument which precedes them; and they are locked, so to speak, in a proverb. εὐθημοσύνη means literally *good disposition*, i.e. household economy and orderliness; κακοθημοσύνη is its contrary. Both words appear to have been invented by the poet, to make possible the next proverbial jingle we find in these lines. Unlike other abstracts ending with -σύνη and -ία coined in this period, these words did not survive in the living language. The reason for Hesiod's frequent clumsiness is that he is such a great innovator in language.

The third stage, which I will come back to, is the social abstraction. Abstract words now appear completely free. They are independent entities, and they can dominate whole passages of writing. But they are *social*, because they always imply a clear human state or | a clear mode of behaviour: they have not lost dramatic and human reference.

The fourth stage is Aristotle, and I call it dogmatic. Abstract words are entirely independent: they need not refer to any human state or behaviour. And they are the true realities. When Aristotle has reduced something to a set of terms, which is the procedure of all his treatises, he feels that he has really explained and fixed that thing forever. ὕλη and οὐσία—matter and essence—mean nothing in a human context, but they are what the world is made of. Aristotle's universe is constructed

---

[13] Such words, including φιλοφροσύνη, do occur in Homer, in a proverbial use, e.g. *Iliad* 9. 256. But this is exceptional.

[14] See now E. A. Havelock, 'Dikaiosune, an Essay in Greek Intellectual History', *Phoenix* 23 (1969), 49–70.

not of spheres, but of pinpointed and hypostasized abstract words.

The fifth stage is the one we are in, and I mean, by calling it *tentative*, that it comes when the confidence of early philosophy has been shaken, so that we no longer can reduce everything to a sure terminology; on the other hand, abstract language, developed originally by philosophers, is so widespread that we can't get along without it. A return to a Homeric or even a Gorgian naïveté is impossible: and so we use complicated abstractions, but without feeling that they accurately describe things for once and for all. When we read this kind of sentence—to choose an example—it is obvious that the writer is not claiming for his intellectual abstractions the kind of lasting accuracy that Aristotle, or Descartes, claimed for his:

Sade's fictional counterworld was articulated in stages; each of them marks a further advance in the profoundly dramatic emergence of an ideology that is at the same time an affective tonality.

I have chosen here, partly for amusement, an example from English, because, as I indicated, I believe the scheme works for modern languages as well as for Greek, although the stages are less clearly articulated in English. A more serious example of the fifth stage in Greek literature might have been taken from the treatise *On the Sublime*.

Let us return to the third stage, the social abstraction, which is the one I said Thucydides and Gorgias represent. Abstract words abound; at times they appear the staple of the argument; but they always have a human and dramatic reference. Here is a paragraph of English prose which represents this stage to perfection. It is from | the end of Jane Austen's *Persuasion*:

In his preceding attempts to attach himself to Louisa Musgrove (the attempts of angry pride), he protested that he had for ever felt it to be impossible: that he had not cared, could not care for Louisa; though, till that day, till the leisure for reflexion which followed it, he had not understood the perfect excellence of a mind with which Louisa's could so ill bear comparison; or the perfect, unrivalled hold it possessed over his own. There, he had learnt to distinguish between the steadiness of principle and the obstinacy of self-will, between the darings of heedlessness and the resolution of a collected mind. There he had seen everything to exalt in his estimation the woman he had lost, and there begun to deplore the pride, the folly, the madness of resentment, which had kept him from trying to regain her when thrown in his way.

Captain Wentworth makes his final choice in the novel by distinguishing between social abstractions, with a precision that would have won the admiration of Prodicus.

This third stage marks the first real triumph of abstract language, and in Greek literature it coincides with the development of prose as the major vehicle of man's understanding of the world. We find at this stage an excitement, even an exhilaration, with prosaic language which does not recur later. The novel sense of the power of abstract prose, not only as a means of persuasion, but even more as a way of seeing and controlling the world, is what we find in Gorgias and Thucydides, and no fourth-century writer quite has it. By the fourth century already, men had grown familiar with abstractions that could be juggled by the Sophist and the Scientist.

The social abstraction begins in the poets. We find it in the writers of elegy, in Pindar—in a special form—and in Aeschylus; and in Aeschylus, I might add, most clearly in a play which some heretics have considered not genuine on the grounds that it reveals 'sophistic influence': that is, the *Prometheus Bound*. But by its nature it demands prose, and this is because it demands antithesis for its full expression. It is interesting, in view of Thucydides' own view of history, to note that Themistocles is the first man we can cite as having used it as the chief material of discourse. In Book 8, | chapter 83, Herodotus describes a speech of Themistocles which may well have been actually delivered by him, and Herodotus' description makes it fairly clear that this speech must have turned on abstract antitheses. We can further note as an amusing example of the power of abstractions, Themistocles' attempt to hold up the Andrians (for their want of service to the Greek cause) after the battle of Salamis: We have with us two great gods, he said, Persuasion and Necessity; and the Andrians reply that their land is already possessed by two contrary divinities, Poverty and Resourcelessness. But if Themistocles, as a fifth-century 'First Modern Man', was the first practitioner of the social abstraction in political speech, we must look to Gorgias as the earliest writer we have who shows this mode of language in developed form.

Gorgias' *Defence of Helen* is the most important document we possess in this respect. It seems to have been written in Attic, and it is tempting to suppose that Gorgias read it when he visited Athens as an ambassador from Leontini in 427, and that Thucydides heard it then. But we cannot know; and in any case Finley has presented many cogent arguments to show that the sort of style it represents was known in Athens earlier than 427.[15] If we consider Thucydides himself, this seems reasonable. Most of us develop what style we have by our early twenties, and very likely most of our ideas. Thucydides was apparently in his twenties by the beginning of the Peloponnesian War in 431: so it is unlikely that he adopted a new and revolutionary style in 427 as a result of Gorgias' visit. The *Defence of Helen*, or things like it, was probably known in Athens earlier in any case: according to Philostratus, Gorgias was already growing old in 427.

The style of this remarkable speech represents the triumph of antithetical language. Consider the first sentence:

κόσμος πόλει μὲν εὐανδρία, σώματι δὲ κάλλος, ψυχῇ δὲ σοφία, πράγματι δὲ ἀρετή, λόγῳ δὲ ἀλήθεια· τὰ δ' ἐναντία τούτων ἀκοσμία.

The glory of a city is εὐανδρία ['goodmanness']; of the body, beauty; of the mind, wisdom; of a thing, excellence; of speech, truth; the contrary of | these is ingloriousness.

Here is a writer—the first—who exploits brilliantly the potentialities of the Greek language, such as the stem-shift from κόσμος to ἀκοσμία, to put forth general and abstract nouns as isolated and dynamic entities. He operates almost entirely with antitheses, presenting abstractions as a contrasting series of staccato imperatives. No style, if only we could take it wholly seriously, contrives more to give a sense of the mastery of intellectual language over the world. His words give the impression that all the essential elements of a situation have been selected and placed in their proper categories. To this end

---

[15] 'The Origins of Thucydides' Style', now in *Three Essays on Thucydides* (Cambridge, Mass., 1967).

he employs a repetitive but impressive array of rhetorical device: assonance and end-rhymes, paronomasia and isocolon, hypnotic iambic rhythms, come forth with bewildering neatness and rapidity. And notice, in contrast with Thucydides, the perfect symmetry, the absence of the *variatio* that makes Thucydides so complicated. Gorgias' style implies an unqualified confidence that the intellect can divide and distil the essentials of a situation into balanced antitheses.

This is the method which Gorgias then employs to exonerate Helen. With six abstract phrases, he presents six possibilities for the cause of Helen's desertion of her husband: it was

either the will of fortune, or the counsels of the Gods, or the decrees of Necessity; or it was because she was seized by force, or persuaded by words, or captured by Love.

ἢ γὰρ τύχης βουλήμασι καὶ θεῶν βουλεύμασι καὶ ἀνάγκης ψηφίσμασιν ἔπραξεν ἃ ἔπραξεν, ἢ βίᾳ ἁρπασθεῖσα, ἢ λόγοις πεισθεῖσα, [ἢ ἔρωτι ἁλοῦσα].

The accuracy of diction and the placing of words succeed in giving the impression that if these six possibilities have been isolated and contrasted, then *there are no others*. All Gorgias has then to do is to show that for each possibility, Helen herself is not responsible, and she is entirely exculpated.

Now to be sure the *Defence of Helen* is a joke. We are not meant to feel that either Gorgias or his audience care about Helen one way or the other. But it is not enough to dismiss the speech as 'a brilliant display of rhetoric'. Gorgias' method is an intellectual—a dialectic—one, and the rhetoric subserves the action of the analytic mind. | And after all, is this not largely the method of all philosophy? The philosopher tells us, You've got to accept view A or B or C; and I will disprove them all; or else I will disprove A and B and leave you inescapably saddled with C. When we are not wholly convinced by a philosopher, and most of us rarely are, it is usually not because we disagree with his logic step by step, but because we don't accept his terms to begin with. We say to ourselves, consciously or unconsciously, I am not going to let you reduce my world to that particular set of abstract categories. The prose of Gorgias contains in naked form the essence of philosophic method.

But there is one passage in the *Defence of Helen* which in content as well as in method is not a joke, but is meant by Gorgias in dead earnest. He has six categories, things that might have motivated Helen; and the fifth of these is significantly, λόγος. When he comes to λόγος, he enters into a striking disquisition which lasts for 35 lines of Diels's text. λόγος, he says, is a great wielder of power: λόγος δυνάστης μέγας ἐστίν; its body is small, even invisible, but it can accomplish godlike deeds: θειότατα ἔργα ἀποτελεῖ. It can quell fear, and take away pain, and arouse joy and increase pity: δύναται γὰρ καὶ φόβον παῦσαι καὶ λύπην ἀφελεῖν καὶ χαρὰν ἐνεργάσασθαι καὶ ἔλεον ἐπαυξῆσαι ... Now Gorgias in this remarkable statement is obviously not talking about the words with which Paris may have persuaded Helen, but about his own style. Playing on the traditional opposition of λόγος and ἔργον, where λόγος had meant 'mere words' and ἔργον had meant 'reality'. Gorgias is saying, My style, my way of dividing the world into abstract concepts, is not mere words. It is in fact superior to what we call reality because it can *create* this reality. No modern semanticist, telling us we live in a world not of things but of words, has gone further than this. Gorgias has invested his brand of λόγος, the first dominantly abstract speech in the Western World, with a mythical being and a demonic capability.

Thucydides' style too is characterized by abstraction and antithesis, if not so exclusively so as that of Gorgias. And the influence of Gorgias, and to a lesser extent that of Prodicus, has been evident to | his readers from antiquity onwards. But does he make anything new of the Gorgian style? I suggest this answer.

The sort of confidence that Gorgias expressed, that the free human intellect, by a sophisticated and analytic language, could dominate the outside world was transformed by Thucydides into the *central problem of history*. Thucydides felt that man was constantly close to a situation where *things*, outside reality, pure force and pure chance, were his masters. The beginning of the 'Archaeology',[16] Thucydides' sketch of early Greek history, presents such a situation: men moving from one

[16] Esp. 1. 2. 2.

place to another, no cities, no financial resources, no ships or commerce, the ever-present possibility that there may come along all of a sudden *another* stronger than you, and dispossess you of the little you have. Here is a world where words and thought count for nothing and power—δύναμις, the active manifestation of ἔργον—counts for everything. Civilization is a product of human conception. When it exists, words and thought can be important, whether the words be the traditional moral terms by which we live in time of peace, or whether they are the enunciated policy of a great statesman, like Pericles. But the tragic dialectic is that civilization itself cannot live in mere conception. It must be transformed into reality. And this reality must eventually appear in its most violent form: war. πόλεμος, war, is the ἔργον *par excellence*. (As in 1. 23. 1, the terms are often in fact used synonymously.) But when war comes, it overpowers men, and reduces them to their original state, or worse; worse, because the degeneration of civilization is even worse than the brutality and uncertainty of man's first state. Corcyra in revolution is even worse than Greece before the Greeks had a name. In a famous sentence in the description of that revolution (3. 82. 2), the historian says 'War is a hard master, and brings the moods of most men into harmony with their conditions'. ὁ δὲ πόλεμος ... βίαιος διδάσκαλος καὶ πρὸς τὰ παρόντα τὰς ὀργὰς τῶν πολλῶν ὁμοιοῖ. τὰ παρόντα—'immediate, going reality'— assumes control of everything, and all language, including moral and political terms, becomes meaningless. Parallels for such a situation in our own time are not far to seek. |

Writing after 404, Thucydides, who had seen his world destroyed, presents this dialectic as an inevitable process. None the less, the best that man can do, even if it is not to last, is to create a civilization. This means creating in reality a situation where the intellect—that is, words—is in control of things. But it is hard to do, and once done, never stable for long. In short, what Gorgias naïvely assumed to be easily done—and the assumption reveals his basic frivolity—Thucydides felt could be sometimes, for a period, accomplished by great energy and daring and genius. Civilization, he felt, is the result of such qualities.

History, then, is the story of man's attempt to impose his intellect on the world. The word that expresses 'intellect' is γνώμη, corresponding to the verb γιγνώσκειν, 'to come to know', and it is even more common in Pericles' speeches than λόγος. The central problem of history is, 'How, and when, can man impose his γνώμη on things outside himself?'

So he dramatized this problem also by his style. The style of Gorgias was bland assurance. That of Thucydides, using much the same forms of language, is *struggle*. Like Gorgias, Thucydides distils the world into abstractions. But those of Gorgias fall neatly into their boxes and match each other perfectly. Those of Thucydides, to use the mathematical metaphor, are never quite commensurate. They resist the intellect which wants to put them into order. Of the elements of Thucydides' world, as it is implied by his style, it might be said with Montaigne, 'Resemblance does not so much make one, as difference makes other'.[17]

In their unwillingness to submit to the intellect, the things of the world reveal the possibility of the παράλογος, the sudden incursion of reality which overthrows the best analyses of the greatest statesmen. The broken symmetry, the variation and the difficulty of Thucydides' style are always repeating his final message: that the most splendid vision of civilization ever recorded—Athens of the Funeral Speech—can be reduced to the survivors of the Sicilian Expedition in a rock-pit in Syracuse, with half a pint of water, and a pint of meal, each day.

[17] From *Sur l'expérience*.

15

The Making of Homeric Verse: An Introduction

Abbreviations for Works Reprinted in The Making of
Homeric Verse

CH   Ćor Huso: A Study of Southslavic Song. Extracts.

DE   'The Distinctive Character of Enjambement in Homeric
Verse', *TAPhA* 60 (1929), 200–20.

FM   *Les Formules et la métrique d'Homère* (Paris, 1928), translated.

HC   'The Historical Method in Literary Criticism', *Harvard
Alumni Bulletin* 38 (1936), 778–82.

HG   'The Homeric Gloss: A Study in Word-Sense', *TAPhA* 59
(1928), 233–47.

HH   'Homer and Huso: I. The Singer's Rests in Greek and
Southslavic Heroic Songs', *TAPhA* 66 (1935), p. xlvii (summary).

HL   'Studies in the Epic Technique of Oral Verse-Making: II. The
Homeric Language as the Language of an Oral Poetry',
HSCPh 43 (1932), 1–50.

HM   'The Homeric Metaphor as a Traditional Poetic Device',
TAPhA 62 (1931), p. xxiv (summary).

HPH   A. B. Lord, 'Homer, Parry, and Huso', *AJA* 52 (1948),
34–44.

HS   'Studies in the Epic Technique of Oral Verse-Making: I.
Homer and Homeric Style', *HSCPh* 41 (1930), 73–147.

MA   'A Comparative Study of Diction as One of the Elements of
Style in Early Greek Epic Poetry', M.A. thesis (University of
California, 1923).

TD   'The Traces of the Digamma in Ionic and Lesbian Greek',
Language 10 (1934), 130–44.

TE   *L'Épithète traditionnelle dans Homère: Essai sur un problème de
style homérique* (Paris, 1928), translated.

The Making of Homeric Verse: The Collected Papers of Milman Parry (Oxford, 1971;
reissued in paperback, OUP USA 1988), introduction, pp. ix–lxii. Reprinted by
permission of Oxford University Press.

TM   'The Traditional Metaphor in Homer', *CPh* 28 (1933), 30–43.
WA   'On Typical Scenes in Homer' (review of Walter Arend, *Die typischen Scenen bei Homer*), *CPh* 31 (1936), 357–60.
WF   'Whole Formulaic Verses in Greek and Southslavic Heroic Song', *TAPhA* 64 (1933), 179–97.
WW   'About Winged Words', *CPh* 32 (1937), 59–63.

References in the form 'MA 427' are to page numbers in *The Making of Homeric Verse*.

MILMAN Parry, who died at the age of 33 years on 3 December 1935, when he was Assistant Professor of Greek at Harvard University, is now generally considered one of the leading classical scholars of this century.[1] His published work was entirely concerned the epic tradition which is represented for us by the *Iliad* and the *Odyssey*. This published work, together with Parry's University of California Master of Arts thesis, selections from the notes he made in Yugoslavia during the winter of 1934–5 on Serbocroatian poetry and its relation to Homer, and an article descriptive of his field-work written by his student and assistant Albert Bates Lord, is here reprinted in its entirety.[2]

The first two works are the doctoral dissertations, or *thèses*, which Parry wrote to obtain the degree of Docteur-ès-Lettres at the University of Paris in 1928. This degree, the highest awarded by the French university system, is usually obtained by French scholars who have established themselves in university or *lycée* positions; most of those who get it do not do so do so until their mid thirties or later; it does not lead to academic position; it is designed to follow it, and it represents a kind of

[1] To the following people, who offered valuable suggestions to him in writing this introduction, the editor wishes to express his gratitude: E. A. Havelock, G. S. Kirk, Hugh Lloyd-Jones, J. H. Moore, J. A. Russo, and not least to his wife, Mrs Anne Amory Parry.

[2] The papers Parry left behind at his death included twenty-eight pages, double spaced, of typewritten notes for a course on Homer and Virgil to be given at Harvard. The misspellings and lacunae in these notes show them to have been typed by someone not familiar with Greek from imperfect recordings of dictation. The state of the text dissuaded me from reprinting them; but I have referred to them several times in the notes to this introduction, and have quoted many of the more interesting fragments there.

final initiation in the society of the learned. In 1923, Parry had spent four years of undergraduate and one of graduate study at the University of California in Berkeley, and had earned the degrees there of Bachelor of Arts and Master of Arts. In the following year, at the age of 23, with a wife and newly-born child and a most imperfect knowledge of French, he arrived in Paris. He spent his first year there in mastering the language, and only then devoted himself to his work for the doctorate. At the end of his four years in Paris, he had written the required major and minor *thèses* in French, and had had them published in book form, according to the requirements of that time. These books, now long out of print, have been translated into English for this volume by the editor. Parry then underwent the public *soutenance de thèse* with conspicuous success, and shortly afterwards, in the spring of 1928, returned to America, a young and | virtually unknown scholar, who had completed and published work which was to change the aspect of Homeric studies. He began his teaching career in the following autumn at Drake University in Iowa, and after a year moved to Harvard University, where he remained on the faculty until his accidental death six years later. While he was at Harvard, he published a series of articles in American classical journals, elaborating the argument of the French *thèses*. These articles are all collected here, in chronological order.

Between the years 1933 and 1935, under the auspices of the American Council of Learned Societies and of Harvard University, Parry made two trips to Yugoslavia, the first in the summer of 1933 and the second in the academic year 1934–5. His purpose was to check and confirm the conclusions he had drawn from close analysis of the Homeric texts by observing a living tradition of heroic poetry. Some of his later published articles reflect much that he learned in Yugoslavia. But the work he had undertaken there was to be carried on by his assistant, A. B. Lord, who accompanied him on the second and longer trip. The concrete results of his investigations in Yugoslavia were, first, the Milman Parry collection of records and transcriptions of Serbocroatian heroic poetry, now in Widener Library in Harvard University, a small part of which is in the public domain in the form of the published volumes of

Songs from Novi Pazar, edited by A. B. Lord,[3] and second, the volume of notes mentioned above, which was roughly arranged by Parry into book form and entitled *Ćor Huso: A Study in Serbocroatian Poetry* (here CH). Lord's article in volume 52 of the *American Journal of Archaeology*, 'Homer, Parry and Huso' (here HPH), reprinted at the end of this book, describes Parry's purposes and methods of work in Yugoslavia. The same article quotes the few introductory pages Parry had completed of a projected book on epic poetry.

The Homeric Question[4] is a modern phenomenon, although we can trace some of its roots, and discern adumbrations of some of its notions, | in ancient times. Ancient scholars and men of letters, that is to say, sometimes showed hints of an awareness that Homer was not like later authors, and that the Homeric poems had origins more mysterious and more complex than later poetic compositions; but these intimations and conjectures amounted to little. Throughout the duration of the Ancient World, and in a dimmer way through the Middle Ages, and on through the Renaissance, Homer remained the primordial great poet, the truest expression of the divine inspiration of poetry, one who (as the eighteenth century would put it) 'perused the book of Nature', and during most of

[3] These are the first two volumes of a projected series of *Serbocroatian Heroic Songs* by Lord, published by the Harvard University Press (Cambridge, Mass.) and the Serbian Academy of Sciences (Belgrade) in 1953 and 1954. The first volume contains English translations by Lord, some musical transcriptions by Béla Bartók, and prefaces by John H. Finley, Jr., and Roman Jakobson; the second, the Serbocroatian texts.

[4] The following pages do not attempt to give a comprehensive summary of the Homeric Question, but only to sketch some of the lines of thought which helped to determine the direction of Parry's own study. Good recent accounts of the history of the Homeric Question can be found in: M. P. Nilsson, *Homer and Mycenae* (London, 1933), 1–55 (the most thoughtful account); J. L. Myres, *Homer and his Critics*, edited by Dorothea Gray, with a continuation by the editor which contains good comments on Parry's own contribution (London, 1958); J. A. Davison, 'The Homeric Question' in Wace and Stubbings, *A Companion to Homer* (London, 1962), 234–65 (the most detailed account); A. Lesky, 'Die homerische Frage' in his *Geschichte der griechischen Literatur*, 2nd edn. (Berne and Munich, 1957–8), 49–58 (contains a good discussion of the importance of Parry's work); and now the fuller account in 'Homeros, II. Oral Poetry' and 'III. Mündlichkeit und Schriftlichkeit' in Lesky's new article for Pauly–Wissowa–Kroll, *Real-Enzyklopädie*, printed as a separate monograph (Stuttgart, 1967); see also H. L. Lorimer, 'Homer and the Art of Writing', *AJA* 52 (1948), 11–23.

this time the works chiefly associated with his name, the *Iliad* and the *Odyssey*, remained the most popular and the most exemplary works of literary art the world possessed. The analogy of the Bible—the *Iliad* and the *Odyssey* together as a secular Bible—is not inappropriate.[5] If the Homeric poems never had the binding theological authority the Bible once enjoyed in our culture, they were throughout antiquity read and known far better than the Bible is read and known now, or has been for some time. Lacking our historical sense, and possessing the *Iliad* and *Odyssey* too much as part of themselves, the ancients never envisaged, let alone accomplished, anything like a scientific investigation of the origins of Homeric poetry.

The two references most often made to ancient anticipation of the Homeric Question show how casual and isolated such speculation was. Cicero's account of a 'Pisistratean Recension'[6] does not imply a theory of the Homeric poems as an amalgam of traditional songs, but the contrary: it assumes the existence, previous to Pisistratus, of the established text of a literary creation. Josephus' suggestion[7] that Homer could not write was made revealingly by a Hebrew author arguing the superiority of Hebrew culture, and it led to no genuine theory of the composition of the *Iliad* and the *Odyssey*. A line of approach potentially more fruitful than either of these, because it derived from actual observation of Homeric diction, was that of the Alexandrian scholars who distinguished between significant adjectives and ornamental epithets;[8] but the implications of this observation for the origins of Homeric poetry were never guessed then, and can only be seen clearly now because we have Parry's work behind us. |

A vain and irascible Frenchman of the seventeenth century, the Abbé d'Aubignac, has the best claim to be the originator of the Homeric Question. Reacting against the reverence for Homer of his day, and drawing on criticisms of Homeric

---

[5] Cf. Myres, op. cit. (above, n. 4), 14 and 20.

[6] Cf. J. A. Davison, 'Pisistratus and Homer', *TAPhA* 86 (1955), 1–21; C. H. Whitman, *Homer and the Heroic Tradition* (Cambridge, Mass., 1958), 71 f.

[7] *Contra Apionem* 1. 12. See esp. Davison in Wace and Stubbings, op. cit. (above, n. 4), 246.

[8] See TE 148 f., and the references there.

poetry that had been uttered earlier by Erasmus, Scaliger, and others, he composed a polemic.[9] The poetry, the construction of plot, the characterization in Homer, he claimed, are poor, its morality and theology odious.[10] So far we have but an unimportant literary attack, without historical potentialities. But d'Aubignac went further: Homer, he argued, cannot be a standard for poetry, because there was in fact no man Homer, and the poems handed down to us in his name are no more than a collection of earlier rhapsodies. Individual perversity and awakening historical sense, the former in greater measure than the latter, mingle strangely in this first explorer of the question of what the Homeric poems are.

The same can be said of the freewheeling opinion of Richard Bentley: '[Homer] wrote a sequel of Songs and Rhapsodies, to be sung by himself for small earnings and good cheer, at Festivals and other days of Merriment; the Ilias he made for the men and the Odysseis for the other Sex. These loose songs were not connected together in the form of an epic poem till Pisistratus' time about 500 years after.'[11] Bentley is less extreme than d'Aubignac, in that he sees as author of the *Iliad* and

---

[9] François Hédelin, Abbé d'Aubignac et de Meimac (1604–76), *Conjectures académiques ou Dissertation sur l'Iliade*, written apparently shortly before 1670, but kept by friends of the author and not published till 1715 ('incertum amici an veterum amore' is Wolf's Tacitean comment [*Prolegomena*, n. 84; see below, n. 13]). Edited with a good introduction by V. Magnien (Paris 1925). Modern accounts of d'Aubignac's purposes and arguments vary curiously. See, in the works cited in n. 4 above, Lorimer 12, n. 6; Myres 47; Davison 243; Lesky 51. Miss Lorimer (who is specifically concerned with the use of writing, on which d'Aubignac has not much to say) dismisses him as of no importance, although he obviously anticipated much of Wolf's far more learned argument, and his work was known to Wolf. On the other hand, Myres speaks with absurd extravagance of 'd'Aubignac's scholarship and real sense of literary art'. Magnien shows conclusively that he did not read Homer in Greek, and a glance at almost any page shows that he had no understanding of Homeric art whatever. Lesky's description of the Abbé's work as a defence of Homer is hard to understand. Davison's account is reasonably accurate.

For earlier criticism of Homer used by d'Aubignac, see, e.g., pp. 19 and 81 of Magnien's edition.

[10] Parry comments in his lecture notes: 'It is significant that it was a contemporary of Corneille and Racine who was first shocked by the literary form of the Homeric poems. To a mind habituated to the classical conception of literature of the time with its rigid sense of form, its exclusion of all which was not strictly relevant, Homer when regarded frankly, must have been the most slovenly of poets.'

[11] See the excellent discussion of this famous remark by Lorimer, *AJA* 52 (1948), 11–12. Bentley's dates are 1662–1742, and the remark occurs in a treatise (*Remarks upon a Late Discourse of Free Thinking*) of 1713.

Odyssey a man named Homer; but that author lived far earlier than the formation of the epic poem ascribed to him, while the processes both of transmission and formation are left obscure. |

A third view of the question was suggested by the Italian philosopher Giambattista Vico (1668–1744).[12] He was with d'Aubignac on the matter of the one poet: there was no such man. But this assumption led him to a judgement very different from d'Aubignac's, a judgement at once more romantic and more deeply historical. He declared that the Homeric poems were the creation not of one man, but of a whole people, and that they owed their greatness to this origin. They are the true expression of the Greek genius in one age of its history.

It was another Englishman who, however, in this early and speculative period of the Homeric question, set forth in essence the view that even today appears to have the best claim on our acceptance. A diplomat, one of the great travellers, an archaeologist, a man with a sober historical sense and a true lover of Homer, Robert Wood (*c*.1717–71) set out to demonstrate the historical reality of the scenes and events in Homer. He had some success in this endeavour, travelling about and observing with a good eye the places we read of in the *Iliad* and *Odyssey*. This work, published in his *Essay on the Original Genius of Homer* in 1767, was to be confirmed by the excavations of Schliemann and Dörpfeld more than a century later. But Wood's sense of the poet Homer, as he considers the question in the last chapter but one of his *Essay*, is yet more interesting. He accepts, on historical grounds, the impossibility of a literate Homer. But this leads him to renounce neither the individuality of Homer nor his greatness. The problem forces him rather to a new concept. Homer was a different kind of poet from the later, literate masters. The mechanisms of literary craftsmanship were absent in him, so was the learning of a more refined civilization; but in their place was the power of unlettered memory. 'As to the difficulty of conceiving how Homer could acquire, retain, and communicate, all he knew,

---

[12] See B. Croce, *The Philosophy of Giambattista Vico*, translated by R. G. Collingwood (London, 1913), 183–96; and G. Perrotta, 'Le Teorie omeriche di Giambattista Vico', in *Italia e Grecia* (Florence, 1940).

without the aid of Letters; it is, I own, very striking', he says (p. 259 of the second edition, London, 1775), and goes on (pp. 259–60):

But the oral traditions of a learned and enlightened age will greatly mislead us, if from them we form our judgement on those of a period, when History had no other resource. What we observed at Palmyra puts this matter to a much fairer trial; nor can we, in this age of Dictionaries, and other technical aids to memory, judge, what her use and powers were, at a time when all a man could know, was all he could remember. To which we may add, that, in a rude and unlettered state of society the memory is loaded with nothing that is either useless or unintelligible; whereas modern education employs us chiefly in getting by heart, while we are young, what we forget before we are old.

Of course Wood gives us no clear picture of how an unlettered poet | operates. Yet even this he manages to suggest, though dimly. His Homer is above all the poet of Nature. Besides making him *veracious* (Wood exaggerates his historicity), this means that his knowledge is both more circumscribed and more distinct than that of later poets. As his knowledge was more distinct, so was his language, for 'the sense was catched from the sound' (281), and 'If his language had not yet acquired the refinements of a learned age, it was for that reason not only more intelligible and clear, but also less open to pedantry and affectation' (285–6). And 'this language [which] was sufficiently copious for his purposes ... had ... advantages more favourable to harmonious versification, than ever fell to the lot of any other Poet.' Wood then proceeds to speak of the usefulness of the free use of particles in hexameter verse in a way that anticipates Parry's own demonstration of the role of convenience of versification in Homeric diction.

For all its generality and its dependence on an unexamined concept of Nature, Wood's insight was in many ways the most valid conception until modern times of what sort of poet Homer was, and of how the *Iliad* and *Odyssey* came into being. Yet we can observe how this insight became obscured in the age of more exhaustive scholarship and more scientifically searching investigation that followed him. This was the age when men became conscious of the Homeric Question as such.

The nineteenth century, in so many fields of endeavour the laborious age of mankind, saw the full development of the Higher Criticism of Homer. The dominant movement of this period of scholarship was that of the Analysts, that is, of those who, in one way or another, saw our texts of the *Iliad* and *Odyssey* as combinations of earlier poems or fragments of poems. Their theories all rested on one assumption, an assumption which, because it was so fundamental, and in their eyes challenged by no alternative assumption, was never clearly stated by any of them. This was that there existed, previous to Homer, an 'original' text, or 'original' texts, of the Homeric epics, which either were written, or were possessed of the fixed form which only a written text can provide.

That this assumption could have so controlled, and (in the opinion of this writer) so vitiated, the work of so many men of learning and acumen appears all the more ironic when we consider the work which began their line of enquiry: the *Prolegomena* of Friedrich August Wolf (1759–1824), published in Halle in 1795.[13] Wolf, the first professor of *Philologie*, proved, or seemed to have proved, with a rigour and scholarly authority hitherto | unseen in the controversy, that Homer must have lived at a time when the alphabet was not yet in use. Homer therefore could not have been *read* by his audience, and so could not have composed, and would have had no occasion to compose, works of the length of the *Iliad* and *Odyssey*. This last point Wolf regarded as the keystone of his theory.[14] The texts, the original pieces of the poems, unwritten, composed around 950 BC orally for recitation by rhapsodes, were handed down by oral transmission until the 'Pisistratean Recension' in the sixth century BC. In the course of transmission they were

[13] 3rd edition by R. Peppmüller (Halle, 1884). The full title, exhibiting its author's *copia dicendi*, is *Prolegomena ad Homerum sive de Operum Homericorum Prisca et Genuina Forma Variisque Mutationibus et Probabili Ratione Emendandi*. For discussion, see, in addition to the works listed above (n. 4), Mark Pattison's biographical essay of 1865, in *Essays by the late Mark Pattison*, collected and arranged by Henry Nettleship (Oxford, 1889, 2 vols.), pp. 337–414 of vol. 1, esp. pp. 377–91.

[14] 112–13: 'Eodem pacto si Homero lectores deerant, plane non assequor, quid tandem eum impellere potuisset in consilium et cogitationem tam longorum et continuo partium nexu consertorum Carminum. Saepius eadem repeto: sed identidem repetendum est illud *posse*, cuius ex ipsa humana natura vis tanta est et firmamentum causae nostrae, et, nisi illud tollatur, nemo aliis difficultatibus, quibus ea fortasse laborat plurimis, angi et sollicitari debeat.'

much changed and probably expanded. The unity of the poems as we now have them is due not so much to Homer, the original creator of most of them, as to the later editors who fused them, not always successfully, into whole works.

Wolf saw, more clearly than his successors, that there were difficulties in this account, chiefly deriving from the coherent structure of our texts, and in particular, as he saw it, of the *Odyssey*.[15] And unlike his successors, he avoided precise conjecture regarding the shape of the original ingredients of the Homeric poems, or the manner of their formation into the poetic unities which he clearly saw.[16] The intent of the *Prolegomena* made it easier for him to avoid committing himself in this way. The work was to explain the critical principles which would guide his establishment of a text of Homer. The acquaintance with Alexandrian criticism which the recent publication (by Villoison in 1788) of the scholia of Venetus A afforded him, had persuaded him that the textual problems of Homer were fundamentally different from those of other authors, and that there was no possibility of approaching, in his case, a hypothetical original manuscript. The *Prolegomena* sought to explain this state of things. The theory that nothing of Homer was written down until the time of Solon or Pisistratus, and that there already existed at that time a large number of variants, provided the explanation. There was no necessity to conjecture in detail what preceded the creation of the written text, although Wolf's general comments on this matter were what made his treatise so important an intellectual document. |

Like his successors, Wolf lacked any clear concept of what an oral tradition is like. He does not distinguish between the rhapsode, like Plato's Ion, who memorizes, and the bard or

[15] 114: 'Difficultates illas, quae mirifica forma et descriptio horum ἐπῶν partiumque dispositio obiicit'; 117–18: '. . . de Odyssea maxime, cuius admirabilis summa et compages pro praeclarissimo monumento Graeci ingenii habenda est.' Elsewhere he speaks of the remarkable unity of style in the Homeric poems: 138: 'Quippe *in universum* idem sonus est omnibus libris, idem habitus sententiarum, orationis, numerorum.'

[16] He does, however, occasionally anticipate the later tendency to seize on certain passages as *late* or *inferior* or *unHomeric*; e.g. about Book 4 of the *Odyssey* from l. 620 on, he says (133): '. . . neque hic Homerum canentem audimus.' Cf. 137–8 on the last six books of the *Iliad*.

minstrel, like Phemius and Demodocus in the *Odyssey*, who, as Parry was to show, improvises from a poetic store of formulae, themes, and tales. He argues in general terms that the original poems would not have changed completely in the course of oral transmission, but that they would have received some modifications, additions, and subtractions.[17] He cannot imagine the actual fluidity of an oral tradition of song, which makes it inconceivable that a passage of poetry sung in 950 could have been preserved without the use of writing until the sixth century. While arguing vehemently against the use of writing by Homer, Wolf has to assume the kind of fixity of form which is only possible when writing exists. Yet his very uncertainty on this matter reveals an intuition of the inherent difficulties which was lost in the generations of work that followed his.[18] In the course of the nineteenth century, the unlettered poet was largely accepted. But his work, apart from the accidents which for better or for worse befell it between first composition and the final formation of our texts, was conceived as not fundamentally different from the pen product of a later poet. Lachmann (1793–1851) reverted somewhat to the Viconian concept of a *Volkspoesie* by suggesting that the *Iliad* was an amalgam of popular *Lieder*, leaving little place for a dominantly creative Homer. But this theory, it was reasonably held, failed to account for the actual unity of the poems,[19] and on the whole, nineteenth-century Homeric speculation,

[17] e.g. 104: 'Haec autem reputanti mihi vehementer errare videntur ii, qui putant litteris non usum Homerum statim totum immutari et sui dissimilem reddi necesse fuisse.' Then a few lines below (104): 'In primis vero recitatio ipsa, vivido impetu et ardore animi peracta, infirmaverit oportet memoriam, multisque mutationibus causam dederit, &c.' In 264 and 265, this uncertainty becomes, in Wolf's mind, an irresoluble tension between the logic of his historical conclusions and his experience of the poems: 'Habemus nunc Homerum in manibus, non qui viguit in ore Graecorum suorum, sed inde a Solonis temporibus usque ad haec Alexandrina mutatum varie, interpolatum, castigatum et emendatum. Id e disiectis quibusdam indiciis iam dudum obscure colligebant homines docti et sollertes; nunc in unum coniunctae voces omnium temporum testantur, et loquitur historia.' Then, beginning a new chapter: 'At historiae quasi obloquitur ipse vates, et contra testatur sensus legentis. Neque vero ita deformata et difficta sunt Carmina, ut in rebus singulis priscae et suae formae nimis dissimilia esse videantur.'

[18] Parry comments in his lecture notes: 'Wolf was strong by his very vagueness. He made possible the large number of different theories concerning the composition, which appeared in the 19th century.'

[19] Parry comments in his lecture notes: '. . . the laws of mathematical probability should have prevented the first conception of Lachmann's theories.'

following the lead of G. Hermann (1772–1848),[20] played itself out in a series of hypotheses of an original nucleus by a single poet, which might as well have been a written text, and which underwent various expansions and transformations | at the hands of editors and 'reworkers'. First just the *Iliad*, then it and the *Odyssey* as well were treated in this way. For both poems the problem essentially reduced itself to the discovery of early and late *layers* of composition.

This was done by analysing the texts in order to establish discrepancies of plot, of historical and archaeological reference, of language, and of style. Thus G. Grote (1794–1871) in Chapter 21 of his *History of Greece* of 1846[21] argued that of the *Iliad*, 2–7, 9–10, and 23–24 were later additions to Homer's original *Achillêis*. (We note that from Wolf's point of view, the length of that *Achillêis* would have made it scarcely more conceivable as Homer's work than the whole *Iliad*; but the problem of illiteracy is of no real concern to Grote.) 2–7 must be intrusive because Achilles does not appear in them, and they are not connected with the story of his Wrath. 9 is inconsistent with the passages of 11 and 16 where Achilles appears not to know that he has already been offered compensation. 10 is alien in tone, and again unconnected with the Wrath, and likewise 23 and 24.[22]

One of the most influential of the discerping critics of Homer, and one of the last in their line[23] was Ulrich von Wilamowitz-Moellendorff (1848–1931), whose *Die Ilias und Homer* of 1916[24] is described by E. R. Dodds,[25] in what may be an excessively generous estimate, as 'one of the great books on Homer', although Dodds himself finds the pattern of Wila-

[20] *Opuscula* v (1832), 52 f.

[21] ii. 119–209 (1883 edn.).

[22] Despite this fierce analysis, Grote is so impressed with the coherence of the whole that he wonders (202) if the additions were not made by Homer himself. Grote expressed the prevailing opinion of his day in contending that the *Odyssey* is a unity. It was not until Kirchhoff's *Die homerische Odyssee und ihre Enstehung* of 1859 that the *Odyssey* became in its turn a victim of dissection.

[23] Not that the line has died out: see, e.g., Denys Page, *History and the Homeric Iliad* (Berkeley, 1959), app. on 'Multiple Authorship in the Iliad'.

[24] Wilamowitz dealt with the *Odyssey* in two books, *Homerische Untersuchungen* (1884) and *Die Heimkehr des Odysseus* (1927).

[25] In his valuable article 'Homer' in *Fifty Years of Classical Scholarship* (Oxford, 1954), 1–37, esp. 5.

mowitz's dissection unconvincing. Lesky's comment on this work could apply to most analytic Homeric scholarship since Wolf: 'No one who reads the concluding pages of Wilamowitz's *Die Ilias und Homer* (Berlin, 1916) with their summary of his intricate theory of the origins of the *Iliad* can conceive a process of such complexity without making the assumption of widespread literacy.'[26]

In his essay on Wolf of 1865,[27] Mark Pattison said, in pointing out the immense influence of the *Prolegomena*, that 'no scholar will again find | himself able to embrace the unitarian thesis'. Pattison also speaks (p. 381) of a 'crudity of conception' in Wolf's great work: 'The Homeric problem was too complicated to be capable of being thought out by the first mind which grappled with it. The question has been wrought out with much greater precision and fullness of detail since by Lachmann, Lehrs, Nitzsch, ... &c.' We can easily see one hundred years later that Pattison was looking into a very clouded crystal ball when he made the first of the comments. But the second may have been equally misconceived. Wolf's sense of the limitations of his own knowledge and his feeling for Homeric poetry, together with superior powers of logic, combined to keep him from making the errors of his successors. If he had little conception of what an oral tradition is, he did not at any rate put forth any theory of the poems which, by essentially assuming a literate tradition, would have undone the bases of his own theory. Nor would he ever have argued as Wilamowitz does, e.g. that the scene in Book 1 of the *Iliad* where the Greeks go to make amends to Chryseus must be a 'later addition', because the tone in that scene differs from the tone in the quarrel between Achilles and Agamemnon, as if the poet of the *Iliad* could command only a single tone or mood.

The assumption that in retrospect seems to have been common to all the analyst scholars, underlying all the erudition and ingenuity of their constructions, that Homeric poetry was essentially poetry like ours, only subject to peculiar distortion

---

[26] 'Wer etwa in dem Buche von Wilamowitz *Die Ilias und Homer* (Berlin, 1916) die letzten Seiten mit der Übersicht über die so komplizierte Entstehungstheorie der *Ilias* liest, kann sich so verwickelte Vorgänge nur unter der Voraussetzung reicher Schriftlichkeit vorstellen' (*Geschichte der griechischen Literatur*[2], 53–4).

[27] See above, n. 13.

and development in its transmission, was more harmful finally
to their work than the qualities for which they have been
frequently taken to task: their dogmatic representation of
guesswork, their revealing disagreements with each other. But
this assumption, implicit in their conjectures, became the
avowed principle of their opposition. For there was, by the
1920s, a substantial reaction to the criticism of the previous
century. E. R. Dodds says in his review of Homeric scholarship
of 1954:[28] 'It is now more than thirty years since the old logical
game of discovering inconsistencies in Homer was replaced in
public esteem by the new and equally enjoyable aesthetic game
of explaining them away.' The simple argument of the
Unitarians, which had not replaced the old business of layer-
hunting, but was much in the air when Parry was a student and
was forming his own explanations, was this: the Homeric
poems are works of art too great, their dramatic structure is
too perfect, their characterization too consistent, to have been
the more or less random conglomeration of a series of poets
and editors. Moreover, those who attempt to assign different
parts of them to different periods show their weakness by their
inability to agree. Therefore each of the two poems is the
unique and individual product of a great poetic mind. Such
was the feeling of men of letters from Goethe onwards, |
throughout the nineteenth century,[29] and such was the fundamental argument of those scholars who took up the unitarian
cause in the years before Parry wrote.

Their work, of which J. A. Scott's *The Unity of Homer*
(Berkeley, 1921) was the most eloquent, if not the soundest,
example, was more satisfying to our sense of poetry than any
of the edifices of the analysts. But its superficiality was most
apparent in the fact that it took largely the form of refuting
individual analyst arguments. It provided no concept of epic
poetry that could explain the difficulties which the analysts
exploited. As Parry says succinctly: 'Yet those who have thus
well refuted the theories which broke up the poems have
themselves given no very good explanation of how they were

---

[28] See above, n. 25.
[29] e.g. Andrew Lang, who argued the point in three books. See M. Nilsson, op. cit.
(above, n. 4), 22–3.

made.' 'What reasons', he adds, 'have they had for passing over
the fact pointed out by Wolf that a limited use of writing for
literary purposes, which is the most that one can suppose for
Homer's age, must have made for a poetry very unlike ours?'[30]
Ignoring the problem of literacy, omitting any close study of
Homer's language and diction, and unable to conceive clearly
of the formation of the poems they prized, they wished to
cancel the Homeric Question and return to the naïve view of
antiquity, that Homer was a poet like Aeschylus (or Virgil or
Dante), and that the *Iliad* and *Odyssey* were unique single
creations of an original poetic mind. The essential insight, that
Homer was a different kind of poet from the literary masters of
a later age, an insight which had been offered the world by
Robert Wood in 1767, was at least as absent from these men as
it was from the analysts whose want of poetic and literary
sensibility they justly deplored.[31]

But there was a third strain of Homeric criticism for Parry
to draw on, one more technical and less prominent than the
other two, but in the end perhaps more valuable. This strain
consisted of close study of the language of the Homeric poems,
and of its relation to verse-form. Those who began this study,
notably Ellendt and Düntzer,[32] were themselves analysts, at a
time when hardly any serious Homeric scholar was anything
else; but both men were able to forget their divisive study of
the text long enough to establish what became for Parry the
fundamental axiom of Homeric study: the dependence of the
choice of words and word-forms on the shape of the hexameter line. Düntzer in particular not only pointed out that, out
of a whole array of epithets, that particular | one would be
chosen which satisfied the metrical need of the moment,
regardless of its particular meaning, but also noted the more
striking phenomenon that, of a group of words and word-

---

[30] HS 75.
[31] A quite different, and not at all polemical, work of unitarian cast was C. M.
Bowra's *Tradition and Design in the Iliad*, which appeared in 1930 (Oxford), too late to
influence Parry's work, too early to have been influenced by it. Bowra more or less
assumes a single author of the *Iliad*, but stresses, as most unitarians did not, that
author's dependence on a long poetic tradition. Parry's work made it possible to give
precision to Bowra's conception.
[32] See the references in TE 5 nn. 1–6.

forms which in meaning could replace each other, there would exist only one for each metrical use. Thus of the epithets of wine, he remarks: 'All these forms are metrically distinct . . . and it is never the sense that determines the choice of one or another of them.'[33]

This observation of the economy of Homeric diction, elaborated and confirmed with a methodical rigour of which Düntzer never dreamed, was to become the core of Parry's explanation of Homeric poetry. But its relevance to the larger questions of Homeric criticism was missed by both Düntzer and his contemporaries. Nor did the slightly later scholars who examined the dialect-mixture of the poems perceive any such large relevance. A. Fick and his follower F. Bechtel,[34] observing correctly that the amalgam of early and late Aeolic and Ionic word-forms in the language of the Homeric poems precluded its ever having been spoken speech, tried to show that an original poem in Aeolic Greek had been translated into Ionic, only those Aeolic forms remaining in the final version which would have had to be replaced by forms metrically different. Metrical convenience thus made for conservatism, and conservatism made for the amalgam. But Fick and Bechtel, like Ellendt and Düntzer before them, were analysts, and they carried on their investigation in the service of analysis. They wanted to show that certain portions of the poems as we have them were composed at certain relative or absolute dates. Their work has been judged a failure, and Bechtel admitted this, because the dialect-mixture of Homeric poetry goes too deep: it is pervasive in the poems, and like Anaxagoras' elements, it seems to be found in the smallest units of them.[35] An attempt to find chronological layers in this way would lead to atomization. Yet these scholars were contributing to a body of knowledge about the language of Homer which would one day suggest a new insight. The failure of what they attempted

[33] *Homerische Abhandlungen* (Leipzig, 1872), 514: 'Alle diese Formen sind metrisch verschieden; dass bei der Wahl nie der Sinn den Ausschlag gab, lehrt genaue Betrachtung des betreffenden Gebrauches . . . etc.'

[34] See Nilsson, op. cit. (above, n. 4), 9 and n. 1; also HL 2–4.

[35] Cf. Nilsson 9; HL 40 ff., esp. 41, n. 1; now G. S. Kirk, *The Songs of Homer* (Cambridge, 1962), 192–210, although Kirk does, at the end of this chapter, try to reintroduce a kind of linguistic criterion for 'post-Homeric' passages.

showed the wrongness of assumptions they shared with other
scholars of their time. What they themselves did not do, but
helped others later to do, was to conceive of the kind of poetry
which would use such a language as they described.

Meanwhile the effect of linguistic examination was the
reverse of what had been intended by those who practised it:
instead of discernible layers | of language which would corres-
pond to fixed stages of composition, they succeeded in demonstrating the homogeneity of the dialect-mixture. Thus K.
Witte, who wrote the article on the language of Homer for the
Pauly–Wissowa *Real-Enzyklopädie* (1913), thought that
linguistic criteria would show 'early' and 'late' passages; but at
the same time drew the famous conclusion that 'the language
of the Homeric poems is a creation of epic verse'.[36] The two
notions are not compatible. For if the tradition created an
artificial language, that language, with its forms of diverse date
and diverse place, could have been used at one time by one
poet to create one work.[37]

In the early 1920s, when Parry wrote at the University of
California in Berkeley the Master of Arts dissertation which
contains in essence his new image of Homer, there was an
established, yet monotonous and infertile, school of analyst
critics; there was a growing unitarian reaction which shared
with analysis the assumption that the stages of composition,
whether one or many, represented the original wording of a
fixed text; and there was a body of linguistic examination
which had demonstrated the dependency of Homer's language

---

[36] Pauly–Wissowa viii. 2214.

[37] The foregoing remarks are not intended to suggest that the attempt to
investigate the formation of the language of Homer is a waste of time, or that
relatively early and late forms and constructions cannot usefully be discerned in that
language. Scholarship since Witte which has done exact and illuminating work of
this kind includes J. Wackernagel, *Sprachliche Untersuchungen zu Homer* (Göttingen,
1916); K. Meister, *Die homerische Kunstsprache* (Leipzig, 1921); Pierre Chantraine, *La
Formation des noms en grec ancien* (Paris, 1933); id., *Grammaire homérique* (2 vols., Paris,
1948 and 1953); Manu Leumann, *Homerische Wörter* (Basle, 1950); G. Shipp, *Studies in
the Language of Homer* (Cambridge, 1953); A. Hoekstra, *Homeric Modifications of
Formulaic Prototypes* (Amsterdam, 1965). But I believe that it is almost always the
easier and more reasonable hypothesis (it is explicitly that of Hoekstra) to regard the
inferred changes in Homeric language as having occurred before the composition of
the Homeric poems. In holding this view, I very much agree with G. S. Kirk's
comments on the studies of Shipp (*The Songs of Homer* [above, n. 35], 202–3), but not
with Kirk's own attempts (ibid. 204 f.) to discover 'post-Homeric phraseology'.

on the verse-form in which he composed. No scholar had succeeded in imagining any better than Robert Wood in 1767, or even so well, the kind of poet who would sing the kind of song we have in the *Iliad* and *Odyssey*.

This was Parry's great accomplishment. It explains and justifies his present influence among scholars of Homeric poetry and of all poetry in the improvising style. He ignored the barren controversy between analyst and unitarian, and concerned himself instead with the implications of the linguists' work. He saw that it presupposed a different kind of poetry from all that we are familiar with. This was for him no vague intuition. To a romantic feeling for another kind of world and art he joined a strong and sober historical sense, and with this a strict method of procedure. Hence he was able to conceive with some precision what kind of poetic tradition made a Homer possible, and to give his conception considerable dramatic force. |

It could fairly be said that each of the specific tenets which make up Parry's view of Homer had been held by some former scholar. Thus the dependence of the given word, especially of the ornamental adjective, on necessities of metre rather than considerations of meaning, had been observed by Heinrich Düntzer; Antoine Meillet had stated, though he had not set out to prove, that all Homeric poetry is made up of formulae; while the formulary structure of contemporary illiterate poetry had been stated by earlier researchers (e.g. A. van Gennep); so had the unfixed nature of illiterate poetry, its freedom from any true sense of verbatim repetition (M. Murko). Even the term 'oral' as applied to a kind of poetry, and a sharp differentiation of that kind of poetry from anything composed in writing, is to be found in Marcel Jousse.[38]

Parry's achievement was to see the connection between these disparate contentions and observations; to form from them a single consistent picture of what Homeric poetry was and of the conditions that allowed it to come into being; and to

---

[38] Düntzer, see above, pp. xix f.; Meillet, TE 8–9; van Gennep, Murko, and others, HL 6 f.; Jousse, see below, p. xxiii.

give substance to that act of imaginative understanding by demonstrating, with precision and the power of repeated proof, that it must be so. But this statement may give a misleading impression of the order of events in Parry's scholarly history. When we read his Master of Arts dissertation we discover that the initial impulse in his work was not the insights and suggestive theories of earlier scholars, but the text of Homer itself. There is no evidence that at the time he wrote that short thesis he had so much as heard of the scholars named above; yet it contains in essence his whole vision of Homeric poetry.

The historical positivism toward which Parry was himself inclined, and which would find the sources of intellectual creation in external environment, does little to explain the origin of the ideas in this remarkable essay. It sets forth with a clarity so quiet that apparently little notice of the work was taken at the time (and how can a Master of Arts thesis say anything important?) the view of Homer which, when developed in Parry's subsequently published works, was to render so much of earlier scholarship obsolete. The view itself was apparently arrived at by the reaction of an unusual mind to the text of Homer: nothing in Parry's background (middle-class, not particularly intellectual, Welsh Quaker origins), nor in the place where he was born and lived until he went to France in 1923 (Oakland and Berkeley, California, and the University of California in Berkeley) makes that reaction likely.[39] Parry's teachers in | Greek at the University of California included two of the finest Hellenists of their generation, George Calhoun (1886–1942) and Ivan Linforth (b. 1879). Both men knew Homer well and had a sensitive understanding of his poetry. But they were not the source of any of Parry's specific ideas. His work was as much a surprise to them as to

[39] Perhaps this is as good a place as any to correct a few errors in the perceptive and moving tribute to Parry, written shortly after his death, by his pupil Harry Levin, now Professor of Comparative Literature at Harvard University ('Portrait of a Homeric Scholar', *CJ* 32 (1936–7), 259–66). Parry did not 'cross the bay' to go to the University of California, since he was born and brought up in Oakland, which is contiguous to Berkeley. Nor did he come to Berkeley to study chemistry. His first science course, in his second year there, was Zoology. His adolescence was no more burdened, or 'overburdened', than is that of most of us. And while he was much impressed by Harvard, he did not 'recoil . . . from the tawdriness of California'.

the rest of the world. The mind that presented Homer to the world as the singer of traditional poetry was itself the product of no traditions.

The idea once conceived, Parry was quick to see how the work of earlier scholars related to it; and in this way he broadened and deepened his vision. His French *thèses* take full account of the work of strictly Homeric scholarship, and in particular, of the students of Homeric language whose work sometimes anticipated his own, sometimes provided an analogy to it: Düntzer, Ellendt, Fick, Witte, Meister. He had gone to Paris to study with Victor Bérard. Bérard's notions of Homer turned out to be far from Parry's own, and he did not wish to direct his work. The *thèses* were in fact written under the supervision of Aimé Puech, who was of great help to him in the composition of his work and who, after Parry's death, published a brief but affectionate testimonial to him.[40] He was supported and encouraged by M. Croiset (1846–1935), the author (with A. Croiset) of the famous *Histoire de la littérature grecque*. The professor at Paris whose ideas were most in harmony with Parry's own was Antoine Meillet (1866–1936), who was primarily a linguist, and as such more disposed to see the language of Homer as the product of a tradition than most straight Homerists. Meillet gave Parry confidence in following out his intuition that the structure of Homeric verse is altogether formulary; but he cannot be said to have vitally affected the direction of his thought. Nor did another scholar of note, who knew Parry in his Paris days and was one of the first to appreciate his work, Pierre Chantraine.[41]

Two other writers in French were of importance in Parry's thought at this time. The first was an anthropologist and student of psycholinguistics named Marcel Jousse, the influence of whose long essay 'Le Style oral rythmique et mnémotechnique chez les Verbo-moteurs',[42] marks the change

---

[40] *REG* 49 (1936), 87–8.

[41] Chantraine's review of TE and FM in *Revue de philologie* 3 (1929) contains an admirable summary of the arguments of these books, and shows him to have been the first scholar to acknowledge in print the value and importance of Parry's work. Parry's reference to the review in HS 74 makes it sound adversely critical of his own work; but that is far from the case.

[42] *Archives de philosophie* 2 (1924), cahier IV, 1–240.

of emphasis in Parry's thought from seeing Homer as a
traditional | poet to seeing him as above all an *oral* poet. A
second decisive intellectual encounter was at the end of his stay
in Paris when Meillet introduced him to Mathias Murko, a
collector and student of Yugoslav poetry. It may have been
Murko and his work that first suggested to Parry the possibility of finding in a living poetry an observable analogue to
the poetry of Homer.[43]

These influences helped to show Parry the implications of
his perception of the nature of Homeric verse, and may have
suggested to him directions of further study. But the perception itself seems simply to have been Parry's direct reaction to
the text of Homer, as it appears in his Master of Arts thesis. The
arguments of that document are too clear to need any summary; but a firm grasp of their central point is important, even
essential, to an understanding of the whole range of Parry's
work. It is from an aesthetic perception of the quality of
Homeric verse that the whole thesis develops. What Parry later
speaks of as the *historical method,* i.e. the attempt to explain the
specific product of an age by the unique conditions of life in
that age,[44] is necessary to the development. But the first thing is
the reader's experience of the style of the poem.

Parry first describes this as the 'traditional, almost formulaic,
quality of Homer' ('formulaic' here has not yet the technical
meaning which Parry was later to assign to it). He adds that
only investigation, i.e. statistical investigation, shows how
pervasive this quality is; and the M.A. thesis does provide a
little of the careful statistical study which is set forth with such
copious exactitude in TE. But the point is that the reader's
experience precedes the counting, just as it precedes the
historical explanation; and throughout Parry's work the appeal
to the experience of the reader is over and again the strongest
argument which he can adduce.[45]

In this, Parry's work differs from the most famous document of modern Homeric criticism, Wolf's *Prolegomena,* which
applies arguments of a historical kind to the Homeric poems.
Because such and such conditions, notably the absence of the

---

[43] CH i–ii.   [44] See TE 1, HM, and below, pp. xxx ff.
[45] e.g. TE 126 f.

art of writing, were true at the time of the composition of the Homeric poems, therefore the poems themselves must be of a certain nature. Parry's work moves from the tangible quality of the words of Homer to a whole vision of an art and even of a society.

The essentially aesthetic vision of Parry's work is more evident in the M.A. thesis than in most of his later pages. He here compares Homeric poetry to a kind of sculpture, and specifically to the Lemnian Athene of Phidias as he read of it in the appreciation of Furtwängler, where Furtwängler's own vision appears to derive from Winckelmann's *edle Einfalt | und stille Grösse*.[46] The beauty of the poetry, like that of the statue, is simple, clear, calm, and traditional. It is almost impersonal in its freedom from striving after originality or individual expression. It is designed to realize what appears, in these early pages, as the concept of a whole people elevated to an almost platonic Ideal. '[The repeated words and phrases] are like a rhythmic motif in the accompaniment of a musical composition, strong and lovely, regularly recurring, while the theme may change to a tone of passion or quiet, of discontent, of gladness or grandeur' (MA 427).

Parry nowhere else speaks in so extended and unguarded a fashion of the aesthetic basis for his judgement of Homer. But neither did he abandon the perception. In his latest printed work, the posthumously published article 'About Winged Words' (here WW), where he takes strong issue with his former teacher Calhoun on the meaning of the well-known recurrent line

$$\tau\grave{o}\nu \; (\tau\grave{\eta}\nu) \; \delta' \; \grave{a}\pi a\mu\epsilon\iota\beta\acute{o}\mu\epsilon\nu os \; (\text{-}\acute{\epsilon}\nu\eta) \; \check{\epsilon}\pi\epsilon a \; \pi\tau\epsilon\rho\acute{o}\epsilon\nu\tau a \; \pi\rho o\sigma\eta\acute{u}\delta a,$$

he says that the principal issue is how we read Homer and criticizes (417) Calhoun's concept as 'too little Phidian'.

Parallel to the Platonic notion of an art which can, e.g., describe Athene not as she is 'on [a] particular occasion, but as

---

[46] See now the most recent edition in English of Furtwängler's *Masterpieces of Greek Sculpture*, ed. A. N. Oikonomides (Chicago, 1964). Parry is presumably referring to chapter 1, on Phidias, in the original English edition of 1895. Here Furtwängler gives an account of his 'discovery' of the statue, and a careful description of it. He stresses both the traditional (13 f.) and the original (26) quality of the statue. It was his remarks on the former that caught Parry's attention.

she is immutably' (426), and to the romantic notion of an art
which was 'the perfection ... of the popular ideal' (425), is
Parry's sense of the directness and swiftness of Homer. He
refers to Matthew Arnold's judgement of Homer's 'rapidity of
movement' (428), an idea which in TE (126 f.) he develops into
a definition of the essential indifference of the audience to the
single word, and in HS (306) to a definition of the traditional
metaphor which 'found its place in the even level of this
perfect narrative style, where no phrase, by its wording, stands
out by itself to seize the attention of the hearers, and so stop the
rapid movement of the thought ...' In HG (241) he refers to
'the direct and substantial nature of Homeric thought'. The
experiential and aesthetic insight remains at the centre of
Parry's thought, although it is expressed more briefly and
guardedly in his later works. Corollary to that insight, at first
sight at odds with it, and appearing already in MA (427 f.) is
the practical and objective judgement of the utilitarian nature
of Homeric style. Words and even phrases, he showed us (for
to this aspect of his work he was able to give incontrovertible
demonstration), are chosen for their metrical convenience,
rather than for their appropriateness to the particular context
in which they appear. This | observation he could find in
Düntzer (above, p. xx), although he does not seem to have
derived it from him. In any case, nothing in Düntzer's rather
chaotic essays suggests the extent of what Parry calls the
'schematization' (HS 314) of Homeric style. When our examination of Homeric style reaches the level of the individual
word, Parry suggested (and demonstrated in the case of the
epithet), convenience is the operative determinant in choice.

   This, along with the emphasis on Homeric Poetry as *oral*, is
the best-known feature of Parry's work on Homer, and the
one that has aroused most disagreement, even antagonism, for
it has seemed to many to deny the poetry the possibility of
artistic expression. Such is the thrust of the oft-repeated remark
of Wade-Gery, that Parry, by removing the controlling hand
of the individual artificer, became the 'Darwin of Homeric
scholarship'.[47] To Parry himself, the opposition between art
and convenience was unreal. Both the rapidity and the rituality

---

[47] *The Poet of the Iliad* (Cambridge, 1952), 38.

of Homeric verse, the qualities he most loved, were, as he saw it, directly dependent on the utilitarian nature of the diction. From a negative point of view, it is only because the single word in Homer does not hold up the mind with an ingenious pregnancy of thought that 'the ... heroic language ... ever sweeps ahead with force and fineness ... [and] also with an obviousness which ... may deceive ... the best of critics' (WW 418). But there is a positive point of view also: the inappropriateness, the ritually repetitive quality, of the single word or phrase, because it is not chosen for its context, instead illuminates the whole heroic world. The fixed metaphor is a part of the completely utilitarian nature of Homeric diction because (TM 373) 'a phrase which is used because it is helpful is not being used because of its meaning'. But this traditional diction as a whole is 'the work of a way of life which we may call the heroic', and so the fixed metaphor is 'an incantation of the heroic' and 'every word of it is holy and sweet and wondrous' (TM 374).

We can see here that the historical scholar is the child of his own age. In a sense Parry is one of the lovers of the exotic of our century, and his admiration for a language formed by the clear exigencies of singing and directly expressive of heroic ideals reminds us of Hemingway finding courage and beauty in the vision of the Spanish bullfighter, or of T. E. Lawrence (one of Parry's favourite authors) finding a more satisfactory theatre of self-realization in the austere simplicities of Arab life. But from a purely aesthetic point of view, if we remember that the tens of this century, when Parry was growing up, and the twenties when he formed his ideas and began to write, were the years in which in the visual arts the concept of the 'functional' as a positive value became established, we can better appreciate his assessment of Homeric language. His historical sense | led him to distinguish sharply between Homeric poetic style and that of his own era; but what he found in Homer was not only the romantic possibility of a poetry expressive of a whole people, but also a quality of purposive directness which spoke strongly to the artistic sense of his own time.

Almost all of Parry's ideas on Homeric poetry can be found

in the M.A. thesis, but his emphasis there is mainly aesthetic. The emphasis of the doctoral thesis is on demonstration. The imaginative grasp of Homeric style here recedes, though it does not disappear, to make way for a stringently scientific and objective examination of the use of the ornamental epithet in Homer. Of the scholarly level of the argument, one can do no better than cite the estimate of Denys Page:

> It is not easy at first to grasp the full significance of Milman Parry's discovery that the language of the Homeric poems is of a type unique in Greek literature—that it is to a very great extent a language of traditional formulas, created in the course of a long period of time by poets who composed in the mind without the aid of writing . . . That the language of the Greek Epic is, in this sense, the creation of an oral poetry, is a fact capable of proof in detail; and the proofs offered by Milman Parry are of a quality not often to be found in literary studies.[48]

Not every point in the long *thèse* commands unquestioning assent, and some have argued that the total picture of Homeric poetry which it suggests is wrongly coloured; but the principal arguments themselves have never effectively been challenged. It is hard to imagine that they will ever be, since there are several of them, all crossing and reinforcing each other, each carefully worked out with accuracy and logic. The cumulative weight of all of them is overwhelming. They show that beyond a doubt the operative principle of Homeric style, at least in regard to the recurrent epithet, was a traditional pattern of metrical convenience rather than any sense of choosing the adjective appropriate to the immediate context. That this was so had been suggested earlier. Parry himself had argued it with some force in his M.A. dissertation. But the doctoral *thèse* demonstrated it beyond question and, what was more of a revelation, showed that there were whole systems of noun–epithet phrases fashioned with such *complexity* and with such *economy*[49] that it was all but inconceivable that the diction of the poems could be the creation of a single man, while the

---

[48] *History and the Homeric Iliad* (above, n. 23), 222–3, where Page gives a concise summary of the nature of Parry's proof.

[49] See TE 7, 16, etc. Parry varies his terms. In HS 276, he speaks of *length* and *thrift* in the same sense.

difference in this respect between Homeric style and that of literary epic, such as Apollonius and Virgil, was complete. The term 'traditional' had in the M.A. thesis represented an | intuitive and aesthetic perception. In the French *thèse*, it became an inescapable scientific inference. Only many singers, over at least several generations, could have produced the poetic language whose finely adjusted complexities these analyses revealed for the first time in an author whom men had known and read for two and a half millennia.

Those who discuss Parry's work, even those who have published comments on it, have rather rarely had a good knowledge of TE. This is partly because the work was written in French, while those who are naturally drawn to Parry's work are mostly in the English-speaking world; partly because it has long been out of print, and has been unavailable in many libraries; and partly because the thoroughness of the argumentation makes the work less attractive than many of the later articles,[50] and in particular the two long articles in *HSCPh* (HS and HL), which have been the source of Parry's thought for most scholars. Yet it remains Parry's basic work, and all the others are more or less specific applications of the conclusions which it works out.

A characteristic example of the subtlety of analysis which Parry in this work brought to bear on the problem of Homeric diction is the chapter on equivalent noun–epithet formulae. That in some cases there appear two or even three ways of expressing what Parry defines as an *essential idea* seems at first sight to show an incompleteness in the whole structure of formulary diction, and therefore might point to an area where individual style and particular choice operate. Parry shows that the contrary is true, since in most cases these apparent deviations from the economy of the system are themselves best explained by the sense of analogy which controls the system as a whole and indeed created it in the first place.[51]

The first work to apply the conclusions of TE was the supplementary French *thèse*, FM. Here, as in a number of later

---

[50] Chantraine says of TE in his review (above, n. 41): 'On serait tenté de reprocher à son livre sa sobriété, si cette sobriété n'en faisait aussi la force.'

[51] See TE 173 ff.

English articles, Parry took up an old Homeric problem and showed how the concept of the bard working entirely within a traditional poetic language set it in a new light, and for all practical purposes offered a solution. Departures from the standard metrical pattern of the hexameter in Homeric verse had previously been put down to the carelessness of early poetry or had been justified by the vague notion of poetic licence. Since such metrical flaws appear most often at certain distinct places in the line, descriptive 'rules' had been set up, and were represented as the 'causes' of the metrical deviations. Parry first applies his vigorous sense of language to the modern term itself, showing that these so-called rules were merely an incomplete set of observations, and could not meaningfully be spoken of as 'causes'. He then shows how most of the metrical flaws in Homer can be explained as | uncommon juxtapositions of traditional formulae, or analogous formations where a change of grammatical form introduced a variation in metre, and shows that the flaws could only have been avoided if the poet had been willing to abandon his traditional diction. What had seemed inexplicable aberrations of style were revealed as phenomena natural to the living operation of a complex, but not infinitely adjustable, system.

When, in the spring of 1928, having completed his work for the doctorate, Parry made ready to return to America with his wife and daughter and son, he had no position and no notion of where he was going to go, until at the last minute, by the offices of George Calhoun, an offer of a job arrived from Drake University in Des Moines, Iowa, which Parry instantly accepted. One of his colleagues there, 35 years later (and 28 years after Parry's death), remembered him as the man who had built the Classics library into something respectable. But Herbert Weir Smyth of Harvard University, as Sather Professor at the University of California, had taught Parry when he was an undergraduate. On learning that he was once more in America, Smyth suggested that Parry read a paper at the American Philological Association meeting in New York at the end of 1928. At that meeting Parry was offered and accepted a position at Harvard, on whose faculty he remained until his death in December 1935.

The paper which Parry read at that meeting, and which was printed in the *TAPhA* of that year, was HG. It was Parry's first published article in English, and it sets the pattern for a number of articles to follow: DE, TM (of which HM appears to be an early summary), and TD. In all these articles, as in FM, Parry took up an old Homeric problem and looked at it in the light of his demonstration of the traditional character of Homeric diction. The method followed in HG is typical. Parry begins by making between the terms *signification, meaning,* and *sense,* a distinction which is a good example of his lively, and twentieth-century, sense of what language is, and how the force of words is a function of their usage and context. He has an instinctive feeling for the operative definition. He then brings up the problem of the word of unknown definition in Homer, reduces it to its simplest terms, refers to the false philological method which had been used in an attempt to solve it, and then looks at the problem from the point of view of his own notion of two distinct kinds of poetry. The notion provides the solution to the problem: the gloss in Homer came into being because words were retained in formulae where the meaning of the entire formula was important for the narrative, but that of the single adjective in it was not. What the gloss possessed was not a relevant definition, but 'a special poetic quality': it added 'the quality of epic nobility' to the noun–epithet phrase. The conclusion of the article is characteristically aesthetic as well as scholarly: the decisive factor is | the way the auditor reacts to the word as he hears it; his thought passes 'rapidly over the ornamental glosses, feeling in them only an element which ennobles the heroic style'. To ask the old question of the signification of these words, it is implied, is to ask the poet and audience alike to 'perform an etymological exercise of the mind' which is alien to the essential style of the poetry.

In TM and TD similarly, an old question is answered from a new point of view, so that the old formulation of the question is shown to be irrelevant. TD is the most purely linguistic piece Parry wrote, but here too the insight at the centre of the argument is the way in which 'the traditional formulaic diction must have trained the ears of the singers and their hearers . . .'; that is to say, the historical discussion is marshalled round an aesthetic perception of style. TM, with its brilliant comparison of Homeric style to that of English Augustan verse, is perhaps the single most elegant statement Parry made of the way words are used in Homer and the way in which they should properly be understood by us. At the conclusion of this article, Parry appeals, as he so often does elsewhere, to the value of the historical method of criticism, which he feels can give us a true picture of the art of the past, free from the kind of misunderstanding of forms of art different from their own which men have made, especially in the case of Homer, from Aristotle onwards. Our greater understanding of Homer is due to the growth of this historical spirit. But the historical spirit itself, he adds somewhat surprisingly, has accomplished so much in our own day 'through a study of the oral poetries of peoples outside our own civilization'.

At the time he wrote these words, Parry's growing interest in oral poetry had caused him to modify his earlier concept of the historical method as a way of overcoming directly the barrier of time, as it is expressed in the quotation from Renan with which TE begins, and to conceive the possibility of returning to the world of Homer by studying at first hand the singing of living bards in another tradition. Of course the two ideas are not contradictory. Attainment of knowledge of the past by observation of living peoples who carry on a way of life which has disappeared elsewhere is a device exploited in one of the earliest applications of the historical method, the opening chapters of Thucydides' *History*.[52] For Parry it was a natural development, but a significant one. The emphasis on Homeric style as *traditional* shifts to the emphasis on Homer as an *oral* poet. The sense that by an imaginative perception of style and scholarly rigour of research one can free oneself from the presuppositions of the present and seize something of the different world of the past is modulated into the belief that there are two kinds of poetry, literary and | illiterate, corresponding to two kinds of civilization, and that one can still move from one of these to the other.

This change in emphasis is first clearly discernible in the two

[52] Thucydides I. 6. 2: σημεῖον δ' ἐστι ταῦτα τῆς Ἑλλάδος ἔτι οὕτω νεμόμενα τῶν ποτὲ καὶ ἐς πάντας ὁμοίων διαιτημάτων.

long articles which Parry wrote for *HSCPh*. They have been
the most widely read of Parry's writings, and they are in an
important way central in his work, because they look back to
the detailed study of the formula in TE and FM, and at the
same time point the way to the preoccupation with modern
improvising poetry which marked Parry's last years. They also
represent the most complete summary Parry gave of his work
on Homer. In the first, Parry answers some of the critics of
TE.[53] Their objections help him to define his own position.
They had argued that the fixity and traditional use of the
noun–epithet formula was an exceptional feature in Homeric
verse, so that the originality of Homer, lying in the remaining
elements of style, was equal to, and essentially of the same kind
with, that of later poets. Therefore Parry in his article widens
the scope of his discussion. In a comparative section he shows
that in later Greek poetry and in Elizabethan verse, there is
nothing like the recurrent and functional element of the
formula in Homeric verse. *Formula* here extends far beyond the
most obvious example of it in the noun–epithet expression. In
Homer, Parry demonstrates whole systems of, e.g., conjunction–verb phrases used over and again in the same way in the
same part of the line to express a like idea, systems which in
their length and in their thrift have no counterpart in any
literary verse.

As Parry saw as early as the M.A. dissertation (426), no other
system can reveal the traditionally elaborated pattern so well as
that of noun–epithet formulae. The reasons for this are fairly
obvious. No single element recurs in a heroic narrative with
the same frequency as the names of the principal persons and of
common objects. The proof of the traditional character of
Homeric style depends on the length and thrift of the systems
of formulae in it. But the length and thrift of a system depends
on the poet's need for it. The need for systems of formulae
capable of disposing frequently occurring proper names and
common nouns in the line will be greater than for any other
class of word, so that the systems involving these words will
have greater length and thrift than those involving other
words. At the same time, these names and nouns themselves
occur more often, and so the evidence attesting the systems in

[53] See HS 266–7.

which they occur is greater. Both factors made Parry's case more impregnable for the noun–epithet formula. Hence the arguments for the pervasiveness of the formula in general, which had been touched on in TE, but are given extended discussion only in HS and HL, are at once more telling in their implications and more open to criticism than the analyses of TE. Significantly, those who have tried to argue against the central thesis of | Parry's work (viz. that the style of Homer is so traditional throughout that originality of phrasing, as we understand the term, is a negligible factor (HS 137–8)) have been likely to concentrate their fire on HS.

The matter is clearly still open to debate, and a dogmatic pronouncement is futile here. It can be said, however, that at the date of writing of this introduction, the balance of scholarly and informed critical opinion finds Parry's central arguments convincing. This does not make them the last word on Homer: if we accept them, we have still agreed only that the poetry, in the narrowest sense of the way single words are put together within units of thought, is traditional and not the work of a single mind. Yet this is enough to determine our view of Homer in a radical way.

The wider scope with which Parry treats the formulary structure of Homeric verse in HS entails one other important factor: that of analogy.[54] This was a factor which had already been dealt with in TE, but again the discussion there is more confined to the noun–epithet formula than in HS. Analogy, the formation of new formulary expressions on the model of particular words and of the sound-pattern of old formulary expressions, is, Parry argued, the creative force in the formation of the epic style. Its importance in Parry's work has been much overlooked. Parry showed that the operation of analogy, while it exists in all poetries (HS 321), observably plays a vastly greater role in Homer. In later poetry the poet either consciously borrows a phrase (HS 290), or attempts to create a style peculiar to himself and so to avoid close modelling of phrase upon phrase which is the very life-force of the traditional language of the oral poet who, in order to attain fluency of improvisation, must yield himself entirely to this sort of play on word and sound. If we accept Parry's evidence on this

[54] See HS 323 and n. 1.

point, we not only have a further dramatic manifestation of the traditionality of Homeric style; we have also significantly extended the concept of what a *formula* is. Parry's original and tight definition (TE 13, repeated in HS 272) had been 'a group of words which is regularly employed under the same metrical conditions to express a given essential idea'. The remarks on analogical formation in HS suggest that patterns of grammar, word-length, and pure sound are themselves 'formulaic', so that two examples of a given 'formulaic element' may, in the extreme case, have no words in common at all. Such a view of what is 'formulaic' goes far beyond the discussion of analogy in TE. It suggests a greater flexibility for the epic language, but a flexibility always controlled by the tradition. At the same time, the mobility of such a concept of the formula has displeased some, who insist that only phrases consisting of the same words can meaningfully be said to repeat each other and so to attain formulary status. Debate on the | subject continues;[55] but possibly the most important aspect of the evidence of analogy in HS is not the minor problem which it raises of the definition

[55] The broad view of the formula is now best represented by J. A. Russo: see especially 'A Closer Look at Homeric Formulas', *TAPhA* 94 (1963), 235–47; also 'The Structural Formula in Homeric Verse', *YClS* 20 (1966), 219–40. A narrower and more sceptical view is that of J. B. Hainsworth, 'Structure and Content in Epic Formulae: The Question of the Unique Expression', *CQ* (N.S.) 58 (1964), 155–64; see also A. Hoekstra, *Homeric Modifications of Formulaic Prototypes* (Amsterdam, 1965), esp. chapter 1. Russo wants to regard structural patterns (or 'structural formulae'), such as τεῦχε κύνεσσιν and δῶκεν ἑταίρῳ (verb, – υ, followed by noun, υ – υ, at the end of the line), the similarity of which was noted as significant by Parry himself, as formulary in much the same sense as the noun + ornamental epithet phrases analysed in TE. Hainsworth would restrict the word 'formula' to groups of words frequently repeated in like conditions. W. W. Minton, 'The Fallacy of the Structural Formula', *TAPhA* 96 (1965), 241–53, has challenged Russo's analyses directly on the grounds that the patterns Russo finds in Homer also occur in writers of literary hexameter verse, such as Callimachus. G. S. Kirk, 'Formular Language and Oral Quality', *YClS* 20 (1966), 155–74, has challenged J. A. Notopoulos' extension of the term 'formulaic' to the Homeric Hymns and the fragments of the Cyclic epics (see below, p. xlvii) on much the same grounds. Russo, in 'Is Oral or Aural Composition the cause of Homer's formulaic style?', *Oral Literature and the Formula* (Michigan, 1976), 33–4, answers Minton's objection by pointing out that Homeric poetry set the pattern for later Greek poetry in this metre.

One question here is the definition of the word 'formula' in Greek hexameter poetry. It is a question of nomenclature, and therefore of limited interest. Parry in HS was careful to speak of 'formulaic element' rather than 'formula', where actual repetition does not occur, and some such distinction is undoubtedly useful. Another, and more important, question is how far Parry's 'formulaic elements' and Russo's 'structural formulas' are uniquely characteristic of Homer, and so, presumably, of early Greek oral poetry. A still more important question is whether examination,

of the formula, but the suggestion it makes of a poetry controlled by patterns of sound to a degree far beyond that with which we are familiar.[56]

The much more extended discussion of the effects of analogy in HS as compared with TE derives from Parry's increased awareness of the importance of sound in Homeric poetry, that the character of its style depended on its being a poetry of the spoken word. In HL he reviews the linguistic structure of Homer from this point of view. It was the linguists | who had first established that the language, i.e. the dialect-mixture and the morphology of Homer, was the 'creation of the hexameter', an artificial poetic language which was created over the course of generations for heroic song. In TE Parry saw his work on diction, i.e. the combinations of words, as paralleling this earlier work of the linguists. Now, in HL, he explains the formation of the artificial language as the product of an oral technique of poetry, in this way synthesizing his own and earlier work on Homeric language. At the same time, even more than in HS, there is a strong emphasis on living oral poetry as an observable manifestation of the processes by which Homeric language and diction came into being. Here for the first time he quotes Murko and Dozon on Yugoslav poetry, Radloff on Kara-Kirghiz, and others on Berber,

such as Russo's, of structural and grammatical patterns within the framework of the hexameter line will shed light on the composition of Homeric verse. In being sceptical about the pervasiveness and regularity of such patterns, Hainsworth and Hoekstra seem to want to reserve some originality or spontaneity of style for the epic poet, although they do not make it clear how such originality might express itself. By stressing them, Russo wants to do the same thing, but in a different, and possibly more fruitful, way. Thus Russo, in his *TAPhA* paper, shows that the first two words of the *Iliad*, μῆνιν ἄειδε (noun, – ∪, verb, ∪ – ∪) belong to a pattern common at the end of the line, rare elsewhere. The third word, θεά, is also unusual from this point of view. He concludes (242): 'What we have, then, in μῆνιν ἄειδε θεά is a rather unusual expression ... created ... specifically and almost self-consciously to open this carefully wrought prologue.' In the unpublished paper referred to above, Russo gives other examples of departure, for special effect, from the normal patterning of the line.

[56] A number of the peculiarities of Homeric *language* (as opposed to *diction*) become more explicable if we conceive of the poetry as something existing as *sound*, not as writing. So, e.g., the derivation of ἀταλός from ἀταλάφρων, pointed out by Manu Leumann (*Homerische Wörter* (Basle, 1959), 139 f.); or ὀκρυόεις from ἐπιδημίοο κρυόεντος (Boisacq, *Dictionnaire étymologique de la langue grecque*[4] (Heidelberg, 1950), s.v. ὀκρυόεις). Such processes of formation occur constantly in all languages, e.g. English *adder* from *a naddre*. Language in general even now consists fundamentally of *speech*. What is peculiar in the cases noted above and others like them is that the formation appears to have taken place within a body of poetry.

Finnish, Russian, and Afghan poetry (329 ff.). The process of improvisation itself begins to dominate his mind. All this makes HL a somewhat curious article, since it mixes a new concern with what we might call Comparative Epic Poetry and a largely traditional and philological approach to the problems of Greek dialects in Homer.

Plato says of music (*Republic* 401 D) that of all the arts it is 'the one which plunges furthest into the depths of the soul'. If we can extend the idea of music to include the art of the spoken (or sung) poetic word, we can say that a like conviction informs Parry's ideas on the character and value of oral poetry. Parry's work on the diction of Homer put the whole Homeric Problem in a new light and has significantly changed the way we read the lines of the *Iliad* and the *Odyssey*. But this is not the aspect of his work which has most caught the imagination. What has made him best known, and has most aroused interest in his writing, is his sense that all poetry is divided into two great and distinct realms, the literary and the oral, that each of these realms has its own laws of operation and its own values, so that each is almost a way of looking at the world; and finally that, of these two realms, the oral is in some way the more natural and the more satisfactory.

That the rhythms of oral poetry may be more natural in themselves because they correspond more closely to funda-mental physiological rhythms of the human body Parry was willing to accept from Jousse (see HS 270); and the idea clearly matched his own inclinations. But what most interested him in oral poetry, and what he was able so well to describe, one might even say to dramatize, in his later work, is the close connection between the specific form of oral poetry and the way of life that surrounds it and allows it to exist. In the preface to CH he speaks of the possibility a careful study of Yugoslav song offered him of seeing 'how a whole oral poetry lives and dies'. And then: 'Style, as I understand the word and use it, is the form of thought: and thought is shaped by the life of men.' The same is presumably true of the style, or styles, of written | poetry; but it is clear that Parry felt the relation between word and life to be more direct, and more observable, in the realm of oral song.

The best example of this belief is WF, which Parry wrote after his first summer in Yugoslavia. In this article he states simply and clearly that (377) 'the one part of literature is oral, the other written'. Homer belongs to the category of poetry which is, in a clearly defined way, *primitive*, *popular*, *natural*, and *heroic*. Parry defines these terms as they apply to oral poetry, and shows how they all, like the qualities of the *formulaic* and the *traditional*, depend on the oral nature of this kind of poetry. It follows that Homer could never be understood by those who looked at his poetry from a conventionally literary point of view. '[The] proper study [of the heroic element in early poetry] is ... anthropological and historical, and what Doughty tells us about cattle-lifting among the Bedouins is more enlightening, if we are reading Nestor's tale of a cattle raid into Elis, than is the mere knowledge that the theme occurs elsewhere in ancient poetry' (377).

Having then, so to speak, taken Homer out of the conventional context of 'Greek Literature', and placed him in company with singers of other lands who tell of the heroic way of life, Parry turns to his own special interest, Yugoslav poetry. 'When one hears the Southern Slavs sing their tales he has the overwhelming sense that, in some way, he is hearing Homer.' Most of the rest of the article sets out to give a specific illustration to those romantic words. Parry shows that large numbers of the most common whole-line formulae in Homer, those introducing speeches, marking the movements of the characters, or indicating the passage of time, have remarkably close parallels in Yugoslav verse. They are not only like in themselves: they also have a like function in the narrative. Then, at the end of the article, Parry tells from his travels in Yugoslavia an anecdote which is an epiphany of his own feeling for this kind of poetry. It is a simple story of an old Yugoslav singer who in telling the tale of his own life slips into the old formulae of his poetry (389–90). One senses in Parry's own careful words the excitement he felt at this example of a man who with a natural and unselfconscious pride saw his own life in terms of the traditional poetry which he sang. The slowness of change and the firm laws of the traditional formulary language, which could only exist in a culture of oral poetry, offered a closeness between life and art, and a satisfactoriness of self-expression, which struck Parry as a revelation.

Parry spent the summer of 1933 in Yugoslavia, and returned to that country in the early summer of 1934, to stay there until the end of the summer of the following year. During that time he travelled about the country, met singers and collected songs, some recorded on phonograph discs, some taken down by dictation. The article by his assistant, Albert Lord, who accompanied him on the second trip | (HPH) tells much of his purposes in going and of his field methods there. That article also includes the first few pages, all that Parry was able to write, of a book which was to be entitled *The Singer of Tales*, and was to report on his work in Yugoslavia and to apply its results to the study of oral poetry generally.

Dubrovnik, where Parry took a house and where his family stayed while he and his assistants travelled into the more remote lands where singing still flourished, was then, as it is again now, a popular seaside resort. But the country itself was wild in comparison with most of America and Europe. The language was difficult and little known. Costumes and manners were strange. Roads were poor. Milk had to be boiled to be safe for drinking, a source of distress to Parry's children (aged 6 and 10 in 1934). There were no rules laid down for Parry's investigation. He had to learn the language, which meant getting to know a good deal of dialect; to choose his assistants; and to evolve the best methods of approaching singers and prevailing on them to sing. The recording equipment, involving aluminium discs, he had built by a firm in Waterbury, Conn., and for power he depended on the battery of his Ford V-8 (1934), which he brought over to Yugoslavia with him. Banditry was not uncommon in the inland valleys, and an air of risk and adventure always accompanied Parry's several trips into the interior.

Often Parry brought his assistants to the house in Dubrovnik to work with him there. His principal Yugoslav assistant, Nikola Vujnović, was a dashing and intelligent (though occasionally irresponsible) man whose abilities as an interpreter and interviewer of singers (he could himself sing somewhat) proved invaluable to Parry and, after his death, to Albert Lord, who returned several times to the country to continue Parry's work. Nikola became a familiar figure in the household and a great favourite with Parry's children. Other helpers were more awesome to them. Once Parry announced that one of them, a Turk, would come for dinner that night. In response to his children's eager questions, Parry said that the Turk was 'a real hero', a man of immense strength and ferocity, whose hands were 'as large as dinner plates'. 'Did he ever kill a man?' Parry's daughter asked. Yes, many times he had killed men. The Turk was anticipated with fearful excitement, and actually turned out to be tall, but stoop-shouldered and exceedingly gentle, with a scraggly black moustache.

From his children's point of view, the sojourn in Yugoslavia (even if the milk did have to be boiled in a great blue pot, and thus rendered unpalatable, and ginger ale was hard to come by) was a great adventure. This picture was not wholly due to childish imagination. Parry himself loved to dramatize what he was doing. The photograph of him in native | dress costume (which he may have worn only on the occasion when the picture was taken) reveals a romantic and even histrionic side of himself which reminds one of T. E. Lawrence. Part of this was pure game; but part also derived from his convictions about poetry. Poetry, at least this kind of poetry, was valuable because it embodied life. To know it, to apply the true historical method in this modern but exotic setting, meant the ability to enter into the life of which the Yugoslav song was the expression. Parry was in a way romantic, but in another way, logical. If he had not been able to learn the language as well as he did, and to drink with the singers and their audiences in coffee-house and tavern, if he had not been able to take part in this society and win the respect of its members, he could not have carried on the work itself.[57]

[57] Harry Levin (see above, n. 39) says well: 'He loved to meet the contingencies of travel, to tinker with his recording machine, to visit the local pashas and exchange amenities, to ply his *gouslars* with wine and listen to their lies. He attained a native shrewdness in apportioning their pay to the jealous canons of village renown and in detecting stale or contaminated material when it was foisted on him. He not only spoke the language, he produced the appropriate gestures and inflections. He respected the hierarchical nicety with which his hosts handed out the different cuts of meat. Their outlook seemed invested with an order that he had not encountered among the schools and movements of the civilization that had formed his own.'

Parry used to improvise stories to his children, and did it rather well.[58] In CH (448), he uses his own experience as a story-teller in this way as an analogy to the use of recurrent themes by the narrative poet. Can one say he was mistaken in seeing this kind of parallel? He sought and attained, in his own life, something of the connection between art and living which made heroic song itself so valuable to him.

What his family were in no position to observe, and what is made so clear by the descriptions of his methods in HPH and CH, is the care and the scholarly control which Parry exerted over his interviews[59] and his field-work generally; and the discrimination, as we see it in CH, with which he drew from his knowledge of Yugoslav singers and singing conclusions applicable to Homer. The desire in some manner to relive the world of Homer did not detract from the sobriety of his scholarly judgement.

The concrete result of Parry's study in Yugoslavia was the collection now named after him, which has been of uncommon interest to students | of music, folklore,[60] and Comparative Literature as well as of Homer. It consisted of nearly 13,000 Serbocroatian texts, including those on more than 3,500 phonograph discs.[61] It has since been augmented by the assiduous work of A. B. Lord in the field. None of these texts were available to the public until 1953, when the songs from Novi Pazar were published by Lord.

What Parry himself regarded as the prize of the collection, the *Wedding Song of Smailagić Meho* by Avdo Međedović, has not yet been published. The importance of this song (taken

---

[58] I recall one episode in a favourite series in which the setting was Paris, especially the sewers of Paris, and Mickey Mouse was always the hero and Winnie the Pooh the villain. Two of Winnie the Pooh's henchmen had been captured by Mickey Mouse, who told them that he would count to ten, and then, if they had not revealed some vital bit of information, he would shoot them. 'One of Winnie's men smiled to the other. *They knew Mickey was only kidding.*' Mickey then counted to ten, and shot one of them through the heart. The other straightway 'talked', enabling Mickey once more to vanquish the Pooh.

[59] In *Serbocroatian Heroic Songs* (above, n. 3), A. B. Lord records from the Parry Collection a number of interviews between Nikola Vujnović and various singers. The kind of questions Parry instructed Nikola to ask reveals something of the skill of his field methods.

[60] See, e.g., *Serbocroatian Folk Songs* by Béla Bartók and A. B. Lord (New York, 1951).

[61] See *Serbocroatian Heroic Songs*, I, xiii.

down by dictation) lies partly in its quality, for Međedović was in many ways a superior singer, but mostly in its length: it is a single song telling a single story and has over 12,000 lines. Another song by Avdo, *Osmanbey Delibegović and Pavičević Luka*, was recorded on discs and comes to 13,331 lines.[62] The length of these songs (even if we make allowance for the shorter Serbocroatian line), sung in a series of creative performances in the traditional formulary style by an unlettered singer, seemed to Parry to offer the most striking proof he had yet found that the *Iliad* and *Odyssey*, poems of not much greater magnitude, could be the products of similar oral tradition in Greece in the eighth century BC. He had of course earlier argued from their style that they must be the products of such a tradition.

Judgement on the quality and coherence of these songs will have to await their publication. Such a judgement will be important, because it will affect the now debated question of the validity of the analogy between Yugoslav and Homeric poetry.[63] To Lord, possibly even more than to Parry, the analogy is clear and certain, although Lord admits freely the superiority of the Homeric poems.[64] To others, for example G. S. Kirk and A. Parry,[65] the analogy is far less sure. In the case of the long works of Avdo, Kirk points out that they were very much *tours de force*, being 'elicited by Parry's *specific and well-paid request* for the longest possible song' (Kirk's italics).[66] The real question may not be so much the occasion of the | songs (for who can know the occasion of the composition of the *Iliad*

---

[62] Lord, *Singer of Tales* (see below, p. xxxix), 288. In *TAPhA* 67 (1936) (see below, n. 67) Lord says (107): 'But one singer at Biyelo Polye in the Sanjak, Avdo Medjedovitch, though only a peasant farmer, is a veritable Homer; and he gave us songs of twelve and even fifteen or sixteen thousand lines.' Presumably Lord had not yet made a careful count.

[63] Cf. A. Lesky, *Geschichte der griechischen Literatur*² (above, n. 4), 34: '. . . wie stehen die homerischen Epen selbst zu dieser Welt von *oral composition*? Damit ist die homerische Frage unserer Zeit formuliert . . .'

[64] He believes the finer songs in the Parry Collection to be comparable in quality to the *Chanson de Roland*.

[65] G. S. Kirk, *The Songs of Homer* (Cambridge, 1962), esp. 83 f.; A. Parry, 'Have We Homer's *Iliad*?', *YClS* 20 (1966), 177–216, esp. 212 f. [ch. 10 in the present volume].

[66] Op. cit. 274. Cf. p. 329, 'One can see a limited degree of novelty even in the expansions of an Avdo Međedović, although the chief basis of these is the extreme and in my view often tiresome elaboration of detail.'

and *Odyssey*?) as their coherence and unity, which is certainly less than that of the Homeric poems, though how much less, and how significantly less, remains to be seen. Meanwhile, one can learn much of these and other songs and of the epic traditions of Yugoslavia in general from A. B. Lord's detailed and informative book *The Singer of Tales* (Cambridge, Mass., 1960). The title of that book is that of Parry's unfinished work.

The theoretic results of Parry's Yugoslav study exist only in the notes which he made on his researches in the winter of 1934 and 1935 and which he organized into a kind of unity and entitled *Ćor Huso* (here CH). Considerations of space have made it impossible to reprint these in their entirety, and so they are represented by extracts. Being notes only, they have not the refinement of thought we find everywhere in Parry's published work; and as a record of his investigations in the field, they certainly contain much that he would later have modified: the more interesting singers, for example, such as Avdo Međedović, Parry only came to know after the date of writing of the last of these notes.

In CH, Parry tells us much of how he came to know various singers, what they were like as persons, and how they sang. He discusses in detail different versions of the same song, both from different singers and from the same singer. Throughout this chronicle and this minute examination of texts we see always the generalizing power of his mind. He constantly searches in the Yugoslav data for material which will illuminate the nature of Homeric, and of all oral, poetry. He touches on many topics of the broadest interest within the study of this poetry: the relation between poetry and social conditions; the effect of the encroachment of literate civilization upon a society in which oral song has flourished; how songs, and themes within songs, change as one singer learns a song from another, or as one singer sings the same song on different occasions, and in different circumstances.

He sees some aspects of Yugoslav poetry as directly applicable to Homer. Thus from the point of view of what he observes in Yugoslavia, he argues forcefully against the notion that the 'books' into which the *Iliad* and *Odyssey* are now divided, or any other divisions which one might make, represent any intended divisions in the composition of the poems.[67] But again, his observations at many points make him aware of the distinctions to be made between the Homeric and the Yugoslav traditions: | he finds he must posit a far greater degree of professionalism in Greece of the Homeric era in order to account for the unity of style and for the transmission of the Homeric poems (especially 444 f.). He also remarks on the differences in the verse itself. The hexameter, he argues, was a far more rigorous prosodic form than the Serbocroatian decasyllable.[68] And the far greater use of enjambement in the former makes for a different kind of poetry, as Parry shows in a postscript on this subject.[69]

In reading the pages of CH one can share Parry's intellectual excitement as the idea becomes vivid to him that much of Homer which formerly could only be the subject of scholarly conjecture can now be understood by direct observation. What he actually says on many topics is often inconclusive. He clearly was waiting for more evidence and more time to work out its application to Homer. In what we have, we can appreciate the range of Parry's mind and the flexibility with which he regarded each question, a flexibility which contrasts somewhat with the almost rigid certainty of conviction of much of his published work. Any one of a dozen subjects adumbrated in the midst of his observations in these notes could have been the theme of an extended study which might have brought it to the level of cogent conclusion.

To speak of one such example, Parry deals in CH, as he had not since MA, with some of the aesthetic criteria of oral poetry. He talks (e.g. 453) of the 'fullness' of detail which is so

---

[67] CH 454 ff. These arguments were to be the basis of the article of which the title and the summary appear in *TAPhA* 66: HH. After Parry's death, Lord put forward some of these ideas in an article of the same title in *TAPhA* 67 (1936), 106–13. Some of the arguments are repeated, with illustrations from Cretan heroic poetry, in James A. Notopoulos, 'Continuity and Interconnexion in Homeric Oral Composition', *TAPhA* 82 (1951), 81–101.

[68] 445 f. On this point, Parry seems to agree with the impressions of Sir Maurice Bowra, 'The Comparative Study of Homer', *AJA* 54 (1950), 184–92, esp. 187, and G. S. Kirk, *The Songs of Homer*, 89 ff., as against those of A. B. Lord, review of Kirk, *Songs*, *AJPh* 85 (1964), 81–5, esp. 84. The trouble here is that those who have uttered opinions on the subject have so far failed to define what they mean by expressions such as 'the far greater rigour of the hexameter as a verse form' (Parry, CH 445), 'the formal looseness of the South Slavic decasyllable' (Kirk, *Songs*, 90).

[69] These observations too were published by Lord, in one of his better articles, 'Homer and Huso III: Enjambement in Greek and Southslavic Heroic Song', *TAPhA* 79 (1948), 113–24.

characteristic of Homer, but distinguishes (446) 'real fullness' from 'empty fullness', a distinction one wishes he had developed and illustrated. Detail, he argues elsewhere (454), is never included in oral poetry for its own interest. He speaks on the same page of the first four books of the *Odyssey* as an extended theme which has, however, no independence, but is entirely subordinate to the single plot of the poem. And he speaks generally (450) of *concision* and *diluteness* as aesthetic criteria, and wonders (461) about the social conditions which make for 'a more or less noble tradition'.

Many of his incidental remarks represent the distillation of his best thought; and some have a general critical authority: 'A popular poetry rises to greatness only in the measure that it shows a full understanding of the life which is portrayed or symbolized in its verses (and then, of course, only as that life itself is admirable), and it is the natural ability of oral poetry to show such an understanding that explains the high quality of| so much of it' (441). Or: 'the mind, since it cannot think in a vacuum, must necessarily carry over to its comprehension of the past the notions of the present, unless a man has actually been able to build up from the very details of the past a notion which must necessarily exclude the application of his habitual notions' (454. 5).

The very fact that one can disagree with many of his unsupported and unqualified statements in these notes, such as that 'no parts of the *Iliad* and *Odyssey* have any unity in themselves' (461), shows how fully Parry let his mind range in these notes. They give us a tantalizing sense of the value which full development of the ideas touched on here would have had, if he had lived to provide it.

It is impossible to know for sure exactly what direction Parry's work would have taken, if he had not been killed instantly, in Los Angeles, California, by an accidental gun-shot only a few months after his return from Yugoslavia in 1935. His close friend, Professor John H. Finley, Jr., of Harvard University, states in the introduction to Lord's *Serbocroatian Heroic Songs I* (above, n. 3) that Parry would never have done such detailed editing as *Serbocroatian Heroic Songs* represents, since 'he had said that he gathered the material "least of all for the material itself"'. Had he lived, Finley thinks, he would have gone on to 'the wider comparative studies that he planned'. On the whole, this seems right: the evidence is not only the generality of interest that we find in CH, but also the title and the opening pages of the unfinished *Singer of Tales*, quoted in full in HPH. On the other hand, his interest, despite his reported disclaimer, in the Yugoslav poetry itself is great, as again CH amply attests. We must remember that the finer singers, especially Međedović, and the longer songs were unknown to Parry when he wrote CH. His concern with Yugoslavia and its oral poetry would hardly have diminished.

The principal theoretic change in Parry's work, to judge from CH, in the last year of his life is the emphasis on the *theme* in oral poetry at the expense of the *formula*. The *theme* is a sort of basic unit of narration in an oral poem. It may be a unit of action: a single combat, the calling of an assembly, the arrival at a palace; or it may be a description, of arms, or a chariot, or a feast. It is clear that such themes recur often in the *Iliad* and *Odyssey*, indeed that the poems are to some extent made up of them. Parry, as we can see clearly in his review of Walter Arend (WA), and others after him, saw this as a distinguishing characteristic of oral poetry. Lord later took up the subject in an article, defining the theme as 'a recurrent element of narration or | description in traditional oral poetry',[70] and his *Singer of Tales* devotes a chapter to the subject.

[70] 'Composition by Theme in Homer and Southslavic Epos' in *TAPhA* 82 (1951), 71–80. It is uncertain how far the *theme*, as Parry and Lord use the term, can be said to be unique to oral poetry. It would not be difficult to illustrate 'composition by theme' in the 19th-century English novel; or still more in the modern detective story. Anyone who has read more than two or three of the works of Rex Stout or Ross Macdonald will recognize that these writers compose more completely in standard scene types, most of them fairly traditional at that, than either the *Iliad* or *Odyssey*. Of course this observation does not invalidate what Parry says in WA of the reasons for the dominance of typical themes in poetry from an improvising tradition.

In analysing thematic patterns, however, one must be careful not to overlook the individual qualities of the single scene. Lord, in the article cited, lists the appearance thrice of the theme of feasting in the *Odyssey*, implying that we have three appearances of the same thing, although the wording varies. Actually the three passages (1. 146–51, 3. 338–42, 21. 270–3) differ in content: only the first describes a feast, the other two describe drinking only; and the additional line which distinguishes the second from the third is an addition of content, occasioned by the unique situation of Book 3, where it occurs. The analyses in chapters 7 and 8 of *The Singer of Tales* are still more impaired by a tendency to blur differences. e.g. 195: 'In fact, Patroclus' mission to spy out the situation for Achilles is strangely like the mission of Diomedes and Odysseus in the Doloneia.'

Parry had apparently worked out a kind of morphology of themes, for in CH (see e.g. p. 446) he refers to *major* themes, *minor* themes, *simple* themes, *essential* themes, and *decorative* themes. These categories were clearly not mutually exclusive, and we can get some idea of their relation to each other: thus the *essential* and *decorative* themes were different kinds of *simple* theme. But unfortunately, Parry nowhere gives us a real definition of these terms.[71] None the less, concern with the idea of *themes* pervades CH. On p. 445, he speaks of the problem of the technique of the themes, of which much must be said later' as 'a . . . way of getting at the problem of the authorship of the Homeric poems through the Southslavic epos'. On p. 446, he suggests a kind of equivalence of *theme* and *formula*. There can be little doubt that, as Lord has indicated in conversation with me, this subject would have absorbed some of Parry's scholarly energies.

The reasons for his concern with *themes* and relative lack of concern with *formulae* in CH lie somewhat in the material itself The study of a living tradition of oral poetry offered virtually an infinite number of songs. Therefore the amount of repetition of theme that could be observed was vastly greater than what can be found in the limits of the *Iliad* and *Odyssey*. On the other hand, one of the striking facts which emerge from the study of Yugoslav poetry is the variation in phrasing of simple expressions. Even within the songs of one singer there does not appear to exist the same close economy of formula which Parry was able to demonstrate for Homer. And from singer to singer, and region to region, the | variation is far greater. It was partly his perception of this that led him (especially in the digression beginning p. 451) to pose somewhat different conditions for the composition and transmission of the *Iliad* and *Odyssey* from anything in Yugoslavia, and to reflect on the differences, in the two traditions, of metrical structure and poetic form.[72] Possibly the whole problem of the formula in a

[71] Lord refers to them in 'Homer and Huso II: Narrative Inconsistencies in Homer and Oral Poetry', *TAPhA* 69, 439–45, esp. 440, but postpones their definition. In *TAPhA* 82 (above, n. 70), he has reduced the system to *essential* and *ornamental* themes, which he tries to distinguish and illustrate.

[72] The degree of formulary thrift in Yugoslav poetry is itself a matter for further study. Lord in the third chapter of *Singer of Tales* argues that it is very great and that

tradition like the Yugoslav one is one with which Parry would have dealt separately at a later point.

Parry's reputation has risen steadily since his death. Even those who at the time of his death knew him best and admired his work most could hardly have augured the high repute in which he stands today. There has even been a temptation to regard him as a misunderstood genius, a prophet not without honour save in his own time. This is really not true. Parry's ideas and the force and clarity with which he set them forth won him considerable recognition in the 1930s, both before and after his death. The Second World War was a natural distraction from the problems of Homeric scholarship, but from the late 1940s onwards there is a continual increase of interest in Parry's published writings and their implications.[73]

But Parry's true reputation rests on his influence among scholars and readers of Homer, and of other heroic poetries. Much of the most valuable work on such poetry since Parry's death and even before has been influenced by his theories, its direction even determined by them. They appeared at a time when the old Homeric Question, deriving from the doctrine of Wolf, had worn itself out and become a repetitive and futile debate. Parry's work gave the whole study of Homer a new life. Its fertility was bound to become more and more evident as more and more of the dialogue concerning Homer was involved with his name and his published arguments. The position which his theories and the whole | problem of oral poetry occupy in the Homeric chapters of the latest edition of

apparent departures from it, in the case of a single singer, can be explained by considerations of rhythm and syntax. But even if one accepts the whole of Lord's explanations, we are still left with a freedom of word-order within the formulary expression which far exceeds the usage of Homer.

[73] Today, his name has almost won popularity, since Marshall McLuhan, on the first page of the Prologue to his *Gutenberg Galaxy* (Toronto, 1962), has hailed him as one of the first to explore the different states of the human mind entailed by the use of different media of communication. McLuhan seems to know Parry's work from the references in Lord's *Singer of Tales*, and there is no evidence that he has actually read Parry. Cf. also Walter J. Ong, S.J., 'Synchronic Present: The Academic Future of Modern Literature in America', *American Quarterly* 14 (1962), 239–59, esp. 247–8, who joins the names of Parry, Lord, and H. Levin (who wrote the preface to Lord's *Singer of Tales*) to that of McLuhan, and argues, not very convincingly, that 'Parry's special type of interest in Homer was made possible by the fact that he lived when the typographical era was breaking up'.

Lesky's *Geschichte der griechischen Literatur*, and in his new Pauly–Wissowa article,[74] is as good an index as any of his now established importance.

The influence of Parry's work has taken roughly five different lines of development. The first, in point of date, is the historical. Apart from the reviews of TE and FM by Shorey, Bassett, and Chantraine (HS 266–7), the first published notice of the significance of Parry's work by a scholar of international reputation was that by Martin Nilsson, whose astute judgement recognized its value in his *Homer and Mycenae* (London, 1933), when Parry was virtually unknown outside Harvard and the University of California. When he wrote his book, Nilsson had been able to read TE, FM, and HG. He speaks (179) of Parry's 'able and sagacious work' and finds in it 'the final refutation of the view that the poets composed their epics with the pen in the hand'. But Nilsson's special interest in Parry's arguments lies in the evidence they provided of the antiquity of the epic language. Parry had argued that Homer (or the Homeric poet) was entirely dependent on the tradition and that he added little or nothing of his own to the stock of epic formulae. Parry was not concerned with the dating of the tradition, but merely with the mode of its formation; but it followed from his arguments that that formation was exceedingly slow, so that much of the language itself of the poems must go back to an extremely early date. On the other hand, the preservation of ancient formulae immediately appeared as the best explanation of the bard's memory of artefacts, of political conditions, possibly even of religious and mythological beliefs, which had ceased to exist long before his own birth. The memory of these things was embedded in formulary expressions which the bards retained from generation to generation because such expressions possessed, as Parry explained it, both nobility and convenience in versification. There is a danger that this sort of argument may be unjustifiably generalized: the hypothesis of the antiquity of formulae offers an explanation of cases where Homer's memory of things before his own time is guaranteed by external evidence. Because this is possible, the presumed antiquity of the formulae

[74] See above, n. 4.

is itself used as evidence of the historicity of certain phenomena mentioned in Homer where no undisputed external evidence is available.

Nilsson cited Parry's work as one support among many for his theory of the possibility of extracting genuine knowledge of the Mycenaean world from the text of Homer. Denys Page, in his *History and the Homeric Iliad* (Berkeley, 1963), takes the argument for the historicity of the Homeric epics as far as it can reasonably go. Some of his conclusions concerning historical material preserved in ancient formulae derive from the precise | studies of Miss Dorothea Gray.[75] The case for the documentary value of Homeric expressions is perhaps strongest when it is applied to artefacts. When it is applied to the Catalogue of Ships, which Page wants to be a 'Mycenaean battle order', it is weaker. Parry's arguments lay the foundation for the historical case, since they stress the antiquity and stability of the formulary expression. But the conclusions of Page must ignore Parry's many arguments for the generality and interchangeability of the Homeric epithet. It is unlikely, that is to say, that Parry would have been sympathetic to the view that such adjectives in the Catalogue as 'steep' or 'stony' or 'many-doved' actually described any specific place at all.[76]

Other arguments of interest deriving from formulary analysis for the possibility of gleaning historical information from Homer's language can be found in T. B. L. Webster, *From Mycenae to Homer* (London, 1958), especially 183, 287. We must remember that Parry himself was quite unconcerned with the question of the historicity of information in the *Iliad* and *Odyssey*. Or rather, it was another kind of historicity which held his imagination: the way of life of the poet and his audience as it was reflected in the content and even more, in the style, of the poetry. What existed before Homer was of interest to him only in so far as it had become part of the living tradition which was a thing of Homer's own time. While there is no reason to think that Parry would have denied that fragments of information pertinent to the ages before Homer

---

[75] See D. Page, *History and the Homeric Iliad*, ch. 6 and notes.

[76] See especially TE 126 ff., 191 ff.; and cf. A. Parry and A. Samuel, review of D. Page, *History and the Homeric Iliad, CJ*, Dec. 1960, 84–8.

could be found in our texts, he clearly felt that the world depicted in these texts was, almost by the definition of poetry, that of Homer and his contemporaries. The implications of his view would run counter, for example, to the hypothesis of M. I. Finley, *The World of Odysseus* (London, 1954, revised 1965) (which discusses Parry's work in its second chapter), that the *Odyssey* depicts a society somewhere between the Mycenaean and Homer's own.[77]

These historical arguments have been, implicitly or directly, criticized from a number of points of view. C. M. Bowra, in 'The Comparative Study of Homer' (above, n. 68), and at greater length in *Heroic Poetry* (London 1952), especially chapter 14, exploits his impressive knowledge of other heroic and oral poetries to show just how small the degree of historical accuracy to be expected from such poetry is. (That these other poetries—the *Chanson de Roland*, the *Edda*, Achin poetry, etc.—can be seen in Bowra's book so clearly as belonging to a like genre with the *Iliad* and *Odyssey* is itself due in no small measure to Parry's work.) | The scholarly validity of historical argument from Homeric language was subjected to a characteristically strict and sober review by G. S. Kirk in 'Objective Dating Criteria in Homer'.[78] A more recent criticism which is especially germane to Parry's own arguments is A. Hoekstra's *Homeric Modifications of Formulaic Prototypes* (Amsterdam, 1965), especially the first chapter, in which it is forcefully argued that the linguistic structure of the poetic tradition changed more rapidly, and achieved the form in which we have it at a time much closer to Homer, than Parry wished to allow.

In the category of historical applications of Parry's work fall also the books of R. Carpenter and E. A. Havelock. Carpenter was one of the earliest scholars outside the Harvard circle to see the importance of what Parry had done, and he pays him a handsome tribute in the opening chapter of his *Folk Tale, Fiction and Saga in the Homeric Epics* (Berkeley, 1946). Carpenter applies the historical argument of the long 'memory' of

---

[77] Parry says in his lecture notes: 'It is not possible that the [poetic] tradition should have kept the details of the social existence of man at another epoch.'

[78] *Mus. Helv.* 17 (1960), 4, 189–205, see also his *Songs of Homer*, especially 179 f.

Homeric language to his attempt to find in Homer not only hitherto unseen archaeological information, but also patterns of European folklore underlying the *Odyssey*. If his book is, as E. R. Dodds in *Fifty Years of Classical Scholarship*[79] has said of it, 'a work which suffers from an excessive preoccupation with bears', it is also one of the most lively and entertaining books of our time on a Homeric subject.

E. A. Havelock's *Preface to Plato* (Cambridge, Mass. 1963) is the work of a philosopher as well as an historian, and belongs properly in the domain of intellectual history. Starting from Parry's concept of a specific way of life which corresponds to the peculiar form of oral poetry, Havelock develops with great insight and much illustration the idea of what he calls the 'Homeric state of mind'. He sets out to analyse the implications, psychological as well as social, of oral poetry as the central vehicle of communication in early Greek culture. His boldest stroke is then to go on to suggest that even after the waning of the epic tradition and the rise of specifically literary forms of poetry, this 'oral culture' substantially prevailed in Greece until the time of Plato, whose 'war against the poets' in the *Republic* and elsewhere is to be explained as an attack on the bases of this older civilization of the spoken word by the greatest representative of the new age of prose, science, abstract thought, and writing. Havelock's book has been much attacked, but many of the criticisms made of it seem trivial in comparison with the energy and scope of the work itself. It is a work which could hardly have existed without Parry,[80] although its conclusions certainly go beyond any held by Parry himself, who would hardly have admitted that an 'oral culture' could exist without the living tradition of oral poetry which determined its character. Havelock is aware of this difference and ably disputes this and other points in Parry's theory. |

Another line of development from Parry's work is the *comparative*. Of course the studies of Parry himself and later of Lord concerning Serbocroatian oral poetry and its relation to Homer are the origin of all such comparison. But it has been extended, in a less thorough and more hypothetical way, to include other bodies of poetry too numerous to mention.

[79] See above, n. 25.   [80] Op. cit., p. x.

Bowra's *Heroic Poetry* remains the best general study of this field. Lord has always been a student of Comparative Literature, and he includes useful comments on several kinds of medieval epic poetry from the viewpoint of formulary analysis in his *Singer of Tales*. J. A. Notopoulos has sought to do for modern Greek oral poetry, both by research in the field and by scholarly study, what Parry and Lord did for Serbocroatian.[81] He has at the same time tried to extend the purview of oral poetry within ancient Greece to include Hesiod, the *Homeric Hymns*, the Epic Cycle, and other poetry.[82]

The remaining lines of development from Parry's work comprise studies which deal more or less directly with the problems which concerned Parry himself, and in particular with the criticism of Homer. It can itself be divided into three parts. First there are Parry's disciples, those who have bent their scholarly efforts to defending, expanding, and publicizing his theories. They are A. B. Lord and J. A. Notopoulos. The mantle of Parry has especially fallen on Lord, who was his assistant in Yugoslavia, and who since his death has worked with the material collected by him, material which has been kept in the Harvard University Library and increased by Lord himself. In a series of articles which have been mentioned above, as in his book, *The Singer of Tales*, Lord adopted titles proposed by Parry himself and tried to follow out lines of investigation as he would have done. Lord is also engaged in editing a series of volumes of Serbocroatian texts from what is now called the Parry Collection.

[81] See especially 'Homer and Cretan Heroic Poetry', *AJPh* 73 (1952), 225–50.

[82] 'Homer, Hesiod and the Achaean Heritage of Oral Poetry', *Hesperia* 1960, 177–97; 'Studies in Early Greek Poetry', *HSCPh* 68 (1964), 1–77; 'The Homeric Hymns as Oral Poetry: a Study of the Post-Homeric Oral Tradition', *AJPh* 83 (1962), 337–68. Strong criticism of this procedure is to be found in G. S. Kirk, 'Formular Language and Oral Quality', *YClS* 20 (1966), 155–74.

Individual studies of other poetries as they are illuminated by Parry's concept of oral poetry include Francis P. Magoun, Jr., 'Bede's Story of Caedmon: the Case History of an Anglo-Saxon Oral Singer', *Speculum* 30 (1955), 49–63; and, with criticism of the analogy, W. Whallon, 'The Diction of Beowulf', *PMLA* 76 (1961), 309–19. How useful the analogy can be for literary judgement of epic poetry may be seen in Stephen G. Nichols, Jr., *Formulaic Diction and Thematic Composition in the Chanson de Roland, University of North Carolina Studies in the Romance Languages and Literatures* (Chapel Hill, 1961).

Like Notopoulos, Lord is an active and creative scholar in his own right. He has had more Yugoslav material to work with than did Parry, | and of course more time, so that, in his book especially, he has been able to take the detailed study of Serbocroatian texts in directions which Parry can only have had in uncertain contemplation at the time of his death. This is clearly true, for example, of a section like that on pp. 56 ff. of *The Singer of Tales*, where he attempts to find complex alliterative patterns in certain passages of Serbocroatian heroic verse. Lord's work on the function of the *theme*, adumbrated by Parry, is developed with many comparisons and examples, and applied to the *Iliad* and *Odyssey*. His concept of the theme there is in several places marked by anthropological speculation of a kind that never appears in Parry's work.[83]

Notopoulos in his studies of Cretan improvising poetry has significantly extended the base of the comparative study of orally composed verse. He has also[84] sought to revise the notion suggested by Parry's work that Homer is virtually our only source of ancient Greek poetry. His interests have moreover embraced certain philosophical and critical concepts related to the idea of oral poetry.[85] It remains true, however, that the primary action of both Lord's and Notopoulos's scholarship has been a reassertion of the fundamental theses of Parry's own work. They have insisted on the correctness and the revolutionary usefulness of Parry's views, have reiterated and publicized these views to student and scholarly audiences, and have zealously defended them against doubters and unbelievers. In the history of Homeric scholarship since 1935 they must appear largely as the Defenders of the Faith. A notable feature of Lord's book, *The Singer of Tales*, is its apparent assumption much of the time that the reader knows nothing of Parry's concept of oral poetry and consequently must be sedulously indoctrinated.

If Parry has thus found a succession of champions, he has also had his attackers. The earliest of these, apart from what

[83] e.g. 88.
[84] See above, n. 82.
[85] 'Mnemosyne in Oral Poetry', *TAPhA* 69 (1938), 465–93; 'Parataxis in Homer', *TAPhA* 80 (1949), 1–23.

criticism there was in the reviews of his *thèses*,[86] was Samuel Bassett's decorous but energetic criticism in his posthumously published *The Poetry of Homer* (Berkeley, 1938). Parry seems so much to have won the field, that Bassett's counter-arguments are little regarded today. It is true that the *originality* which Bassett wanted to save for Homer from Parry's doctrine appears in Bassett's own exposition to be of a conspicuously modern kind.[87] He appears to have | had little notion that the originality of a poet working within the tradition which Parry exposed would have to be something rather different. But it is also true that the principles which Parry formulated, taken at

[86] See p. xliv above.

[87] Discussing the influence of Parry's work in the introduction (xvii) to his valuable edition and commentary on the *Odyssey* (London, 1959), W. B. Stanford says: '. . . one last warning: the reader must lay aside all contemporary prejudices on the subject of "originality", that specious legacy from romanticism . . .; otherwise he may rashly conclude that Homer's rank as a great poet is being impugned when it is shown how much he owes to his predecessors.' This statement contains much truth. Only 'originality' is surely more than a 'specious legacy of romanticism'. The difficulty with many statements on this matter is that those who make them do not bother to clarify what they mean by 'originality'. When Macbeth says (II. ii),

> . . . No; this my hand will rather
> The multitudinous seas incarnadine,

part of the beauty and dramatic force of the line derives from the two protracted Latinate words, and the unexpectedness of their choice and collocation here. Antony says, of himself (*A & C*, IV. viii),

> he hath fought today
> As if a god in hate of mankind had
> Destroyed in such a shape.

Again the power of the words depends on our hearing them in this place for the first and last time and as a way of speech unique to the speaker. (We are meant to admire Antony's own imaginative use of words, and at the same time sense a hollowness in his hyperbole.) The effect of *Iliad* 22. 132 is quite different

Ὡς ὅρμαινε μένων, ὁ δέ οἱ σχεδὸν ἦλθεν Ἀχιλλεύς   131
ἶσος Ἐνναλίῳ, κορυθάικι πτολεμιστῇ   132

but not less powerful. Each word occurs in a traditional and expected position; cf., e.g., 1. 187, 18. 309, 22. 314, 5. 602, and, for the rhythm in the last part of the line (———πτολεμιστῇ), 6. 239; and the image in the line is one central to the experience of heroic poetry. The compound third word does not appear elsewhere in our texts, and may conceivably be rare; but it is clear that the force of the line does not depend on any such rarity, or rare use, of the single word; it is due rather to the slow and relentless concentration on the image of destruction at this turning-point in the action. To have composed (or simply *used*) such a line at just this point certainly shows *originality*, i.e. words used uniquely well in poetry, but originality of a very different kind from Shakespeare's.

their face value, seem to offer virtually no room for poetic originality of any kind or, for that matter, for any real development of the tradition itself; so that Bassett's objections ought perhaps not to have been set aside so lightly.

When Bassett wrote, Parry's ideas could still be considered radical. More recent critics have felt that they are challenging what has become an orthodoxy. A spirited statement of this point of view can be found in M. W. M. Pope, 'The Parry–Lord Theory of Homeric Composition',[88] an article which attacks Parry's central theses directly, and combines some valid points of argument with others more dubious. Pope's essay suffers besides from his restricted knowledge of Parry's own work, which appears to be limited to the two articles in *HSCPh*.[89] But in the present state of Homeric studies, it is hard not to feel some sympathy with such protests as that of Pope against the acceptance as established doctrine of all the conclusions of Parry and his successors.[90] |

The debate over the specific tenets of Parry's studies of Homeric style and their reassertion and extension by Lord, Notopoulos and others—that is, over such matters as the proper definition of the *formula*, the extent to which Homeric diction as a whole is *formulary*, to what degree formulary means *traditional*, and *traditional* in turn means *oral*, and if *oral*, how far this justifies close analogy with other oral improvising poetries—is now a lively one, and far from settled. The works cited of Lord and Notopoulos themselves, of Hainsworth, Hoekstra, Russo, Pope, and Kirk, are among the more valuable

---

[88] *Acta Classica* 6 (1963), 1–21.

[89] Thus the argument on pp. 12–13 concerning χρυσέη Ἀφροδίτη could certainly be answered by the chapter on *equivalent formulae* in TE.

[90] Other recent publications offering either challenge or qualifying criticism of Parry's concept of the formulary quality of Homeric poetry are: G. E. Dimock, Jr., 'From Homer to Novi Pazar and Back', *Arion*, Winter 1963; T. G. Rosenmeyer, 'The Formula in Early Greek Poetry', *Arion*, Summer 1965; two mathematically concentrated articles by J. B. Hainsworth, 'The Homeric Formula and the Problem of its Transmission', *BICS* 9 (1962), 57–68, and 'Structure and Content in Epic Formulae: The Question of the Unique Expression', *CQ* (N.S.) 14 (1964), 155–64, and now his book, *The Flexibility of the Homeric Formula* (Oxford, 1968), which shows by careful examination how much the formula could be modified, both in length and in its position in the hexameter line; and especially Hoekstra's *Homeric Modifications of Formulaic Prototypes* (above, n. 55). The first chapter of this last-named work contains what may well be the best criticism that has yet appeared of Parry's work on its own terms.

contributions to it. But to the scholar with literary interests, or to the student or lover of literature in general, the whole argument may well appear to be so narrowly technical as to miss somehow the fundamental issue, which is the poetry of Homer, and how Parry's work, and that of his successors, both its champions and its critics, will affect our reading of it. Criticism, in short, in the wider sense of the attempt to understand and evaluate works of art which for almost three millennia have aroused men's admiration and love, and have seemed to make the world more beautiful and more comprehensible, has throughout so much of the controversy taken second place; and the question which is of real interest to most intelligent and educated persons—what does all this tell us about the *Iliad* and *Odyssey*?—receives little answer, or none.

The work of criticism, in this humane sense, has not altogether been assisted by the efforts of philology. In the nineteenth and early twentieth centuries, in the age when the Analysts, those who looked for layers of composition, held sway, any attempt to find meaning in the relation to each other of different parts of the Homeric poems fell under the shadow of the possibility that these parts had found a place together in our texts by accident. For one need not be particularly an Aristotelian to realize that a work of art can present a clear vision of life only if one can assume it to be the product of deliberate human design. To believe this of the *Iliad* and *Odyssey* will not mean that one must believe every part of these poems to be equally with every other integral to the whole; any intelligent assessment of the *Iliad*, for example, will have to keep open the possibility that some parts of it, perhaps portions as substantial as the Tenth Book, have been added from an alien source. But if we are to conceive of the *Iliad* as a work of art, we shall have to be able to regard it as in the main the complex construction of the designing artistic mind. This was precisely what Analytic criticism, by | direct statement or by implication, denied; and as a result, humane criticism, rather than being helped by the work of the philologists, was so hindered by it that one could say that it existed, during this period of Homeric studies, in spite of them.

It was not the smallest accomplishment of Parry's Homeric

theory that it made the whole Unitarian–Analyst controversy, at least in its older and best-known form, obsolete.[91] The idea of distinct layers of composition, or of poetic 'versions' of diverse provenience imperfectly welded together, makes no sense in the fluid tradition of oral improvising poetry. 'Naïve' Unitarianism, as Dodds calls it, becomes equally untenable in the light of Parry's theory, since the poetic tradition, however much it changed, and however recently in relation to Homer's lifetime it took its final shape, was now shown to be the product of many men over many generations, and the dependence of the poems on the tradition was in turn shown to be so great as to rule out the kind of individual creation which some Unitarians, such as Scott,[92] were looking for. Parry himself was by instinct a Unitarian, and his observation supported his instinct, because he saw how the individual bard could, without anything like deliberate manipulation of pre-existent versions, be the repository of a whole tradition. Moreover, he was in Yugoslavia impressed by the unity of style in the *Iliad* and *Odyssey* as opposed to Serbocroatian poetry. This seemed to him strongly to suggest single composition of the Greek poems as we have them. Yet his Unitarianism, if it should be called that, was far indeed from the earlier concept of individual creation, for to him the tradition was more important than the poet who at any moment embodied it. That poet's virtue lay not in the ability to create, as the modern world conceives it, but in the ability to focus and transmit what is, in a sense more precise than Vico had imagined, the creation of a people.

Parry's Lecture Notes (see above, n. 2) show that he had not finally made up his mind on the matter of single composition of the poems. He presents in them a number of arguments in favour of unity, notably (1) the lack of repetition of character and incident in the *Iliad*: not only is each person given his proper place, actions also are not repeated: the duel between Ajax and Hector in 7 is quite different from that between Paris and Menelaus in 3; (2) the use of what he calls 'conscious

---

[91] Cf. Dodds in *Fifty Years of Classical Scholarship* (Oxford, 1954), 16–17; C. H. Whitman, *Homer and the Heroic Tradition* (Cambridge, Mass., 1958), 4 f.

[92] See above, p. xix.

literary devices', as an example of which he gives the return of Hector to Troy in 6, where the ostensible reason is Hector's desire to tell his mother to appease Athena, the real reason, the poet's desire to present a colloquy between Hector and Andromache; (3) the fact that, out of the vast body of oral poetry, only the *Iliad* and *Odyssey* survived, which shows these | poems to have been far superior to all others, and therefore probably the creation of one great poet (or conceivably of some closely organized guild). But then, at the end of the lecture, as if he were afraid that the conception of unity would detract from the importance he wished to assign to the Tradition, he says, 'I have spoken of a unity of conception of the story of the *Iliad*, but it would be wrong to suppose that this conception came to being in the mind of an individual poet. I do not think we shall ever know just how much of the *Iliad* was the work of Homer and just how much of his master and of the Singers who were his predecessors', and concludes by reverting to the traditional nature of the *style*: '. . . One who studies the traditional style . . . comes to see that it is a device for expressing ideas such as could never have been brought into being by a single poet. One's admiration of the poems increases as one realizes that we have here the best thought of many poets.'

The poet, then, is essentially subordinate to the tradition; and it never occurs to him to depart from it, or even to fashion it so as to produce any personal vision of the world (HS 323–4). This belief, which is at the very heart of Parry's thinking about Homer, led to a kind of paradox in the relation between his work and the critical understanding of Homeric poetry. He himself, unlike many, if not most, of his followers, possessed an acute sensitivity to the poetic values of Homer. His ideas on Homer, as we see from MA,[93] derive from an aesthetic insight; and his sense of the quality of the poetry, and of the relevance of that quality to our own lives, informs all his work, and is possibly the greatest factor in the indelible impression he has made on all his readers. But the strength of his feeling for the tradition as opposed to any single manifestation of it was such that we shall look in vain through all he wrote for any

[93] See pp. xxi ff.

comment on the *Iliad* and *Odyssey* as poems.[94] If they have a
unity, it is because the use of formulary diction in them is
consistent, not because they have a beginning, a middle, and an
end, or because they as dramatic narratives reveal any vision,
or embody any attitude, which we shall find of value roday.
Nor is there anywhere any suggestion of the criteria by which
we might distinguish a more effective portion of the Homeric
poems from one less effective, or in general distinguish good
epic poetry from indifferent or bad.

An example may serve to illustrate this judgement. At the
end of HS, his best-known article, after urging that 'the
question of originality in style means nothing to Homer', he
says: ' . . . in certain places in the poems we can see how very
effective phrases or verses were made. The wondrously forceful line:

κεῖτο μέγας μεγαλωστὶ λελασμένος ἱπποσυνάων
16. 776 = *Od.* 24. 40

is made up of verse-parts found in other parts of the poems:
κεῖτο μέγας (12. 381); μέγας μεγαλωστὶ (18. 26); λελασμένος
ὅσσ᾽ ἐπέπονθεν (*Od.* 13. 92); λελάσμεθα θούριδος ἀλκῆς (11.
313).' The quotation of what is indeed a 'wondrously forceful
line' makes for a moving conclusion to Parry's own essay; and
we may think we have here, if in a detailed perspective, some
hint of the artistic construction of the *Iliad* and *Odyssey*. But it
becomes clear on consideration that Parry is by no means
committing himself to the notion, may not even be suggesting,
that the combination of parts which made this line is Homer's
own work. It too, if we follow Parry's logic, is more likely to
be a product of the tradition. Moreover, we are given no
indication of what is in fact the case, that this line, like so many
others which are repeated in our texts, may be far more
'wondrously forceful' in one passage than in another, because
of the context in which it appears. The power of 16. 776
depends first on the structure of the small scene (765–76) which
it concludes. Here we find a simile and description in the first
ten lines which combine free energy with marked symmetry—

---

[94] If we except the few remarks in LS; see above, p. li.

there are five lines of simile and five matching lines of
description, each group beginning with ὡς—so as to catch the
precarious balance of violent forces in battle. This is enhanced
by the repetition of the reciprocal pronoun ἀλλήλων (765, 768,
770, followed by οὐδ᾽ ἕτεροι 771) and the anaphora πολλὰ δὲ
... πολλὰ δὲ ... in 772 and 774. The tremendous but frozen
turbulence continues till the middle of 775,

μαρναμένων ἀμφ᾽ αὐτόν· ὁ δ᾽ ἐν στροφάλιγγι κονίης,

where we abruptly leave the multitude of living men in
turmoil for the sudden still vision of the single man in the eye
of the storm who has left it all behind. It is because of its
position in this scene that the line acquires its condensed pathos,
and becomes a kind of symbolic representation of death in
battle. Nothing matches this power in the passage of Book 24
of the *Odyssey*, where the slight inappropriateness of the image
can hardly lead us to suspect interpolation or the like, but will
almost certainly prevent us from finding the line in question
memorable; especially since, unlike Cebriones, Achilles is not
chiefly noted for his horsemanship.[95]

What holds Parry's attention in all his writing is the
tradition, never the poems in themselves. His ideas concerning
the tradition were new and exciting, and he clearly felt that it
was his business in life to present these ideas to the world. So
that there is no cause to blame him for the limitations of his
view. Yet it is strange that those limitations have been so | little
remarked by his admirers and followers. They have often
cheerfully adopted his limitations along with his constructive
arguments, and the result has been a further inhibition of
intelligent criticism of the Homeric poems. Freed from the
shadow of Multiple Authorship, the critic now finds his way
darkened by the all-embracing Tradition and by the alleged

[95] The more determined traditionalist may conceivably want to argue that 16.
765–76 as a whole was given to the poet of the *Iliad* by the tradition. Proof of such an
argument lies afar. It would not in any case account for the obviously effective
position of the passage immediately before the lines that signal the shift in the battle
and Patroclus' own death. Such argumentation will lead to the absurd conclusion that
the whole of 16, and even of the *Iliad*, was ready to hand in the tradition.

rules of oral style. If he now tries to present an interpretation of the *Iliad* or the *Odyssey* involving the relation between one passage and another, he will have to fear the objection that the oral poet plans no such coherent structures, and that the occurrence of the passages in question is due to the fortuitous operation of the Tradition. It thus turns out that Parry's own sensitivity to the quality of Homeric poetry led ultimately to the erection of barriers to the understanding of the Homeric poems.

This was not so much the doing of Parry himself as of his followers. Parry avoided comment on the *Iliad* and the *Odyssey* themselves. He did speak more than once of the necessity of establishing an *aesthetics* of traditional or oral style.[96] But the readers of this volume will, I think, agree that his efforts in this direction were inconclusive. In fact the two places where he comes closest to confronting the problems of aesthetic criteria and literary judgement are the M.A. dissertation and some of the scattered remarks of CH, the two previously unpublished works. He himself took such pleasure in the idea of a noble diction created by a great popular tradition[97] that he never really troubled to define this nobility, much less to show how it operates in the monumental artistic structures of the *Iliad* and *Odyssey*.

He was, moreover, so concerned to urge his discovery of the functional role of noun–epithet combinations and of other formulary units, that he was inclined to see the meaning of such expressions as equivalent to the 'essential idea' to which he reduced them. Thus the essential idea of

$$\tilde{\eta}\mu o\varsigma \ \delta' \ \mathring{\eta}\rho\iota\gamma\acute{\epsilon}\nu\epsilon\iota\alpha \ \phi\acute{\alpha}\nu\eta \ \rho o\delta o\delta\acute{\alpha}\kappa\tau\upsilon\lambda o\varsigma \ \mathring{H}\acute{\omega}\varsigma,$$

which is the standard way of introducing a new day, is 'when

---

[96] e.g. TE 21 ff.

[97] A pleasure which, of all his essays, TM perhaps communicates most directly. Parry sent an offprint of this article to A. E. Housman, who replied in a letter dated 16 Feb. 1933: 'Dear Sir, I am much obliged by your kindness in sending me your paper on Metaphor. I agree with what you say about the diction of Homer and the 18th century, only I do not admire it so much as you do. Yours very truly, A. E. Housman.'

dawn came' (TE 13), and Parry implied that the Greek
expression had no imagistic power, no wealth of connotation,
to distinguish it from the English paraphrase other than the
undefined quality of 'nobility'. Therefore (it was implied)
there is no valid distinction to be made between this phrase and
a formula of like function from another tradition of heroic
song (WF 383):

Kad u jutru jutro osvanulo. |

(Parry's translation in WF, 'When on the morn the morning
dawned', is quite literal.) The richness of this and other Greek
expressions, rarely paralleled by anything in Serbocroatian
poetry, is thus removed from critical discussion.[98]

Again, to express the essential idea 'Hector answered',
Homer gives us

$$\tau \grave{o} \nu \; (\tau \grave{\eta} \nu) \; \delta' \; \mathring{\eta} \mu \epsilon \acute{\iota} \beta \epsilon \tau' \; \mathring{\epsilon} \pi \epsilon \iota \tau \alpha \; \mu \acute{\epsilon} \gamma \alpha_S \; \kappa o \rho \upsilon \theta \alpha \acute{\iota} o \lambda o_S \; \mathring{E} \kappa \tau \omega \rho.$$

Having once uttered the standard expression for 'answered' in
the first half of the line, Parry showed, the poet had virtually
no choice in naming the subject in the last half. Hector had to
be $\mu \acute{\epsilon} \gamma \alpha_S \; \kappa o \rho \upsilon \theta \alpha \acute{\iota} o \lambda o_S$. No one can deny this; but Parry's
implied conclusion, that $\mu \acute{\epsilon} \gamma \alpha_S$ and $\kappa o \rho \upsilon \theta \alpha \acute{\iota} o \lambda o_S$ mean nothing
in themselves, does not follow of necessity; and is certainly
restrictive to the critic who tries to explain the effect of
Homeric verse.[99]

---

[98] Whitman, op. cit. (n. 91), 7, rightly objects to this procedure in the case of the
phrase 'winged words' (see WW).

[99] There are really two distinct questions here. (1) Parry seemed to believe that the
ornamental epithet had virtually no meaning at all; it was a sort of noble or heroic
padding in the noun–epithet formula (see especially TE 145 f.). Accordingly, he
never concerned himself with the problem of what individual epithets mean, and was
content, for example, to accept the time-honoured, but essentially indefensible,
translation of $\mathring{\alpha} \mu \acute{\upsilon} \mu \omega \nu$ as 'blameless' (TE 122). See the forthcoming monograph of A.
Amory, *Blameless Aegisthus* [Leiden, 1973], which, accepting the formulary nature of
such epithets, argues cogently that they, like other words in Homer, had none the less
precise and ascertainable meanings. (2) $\kappa o \rho \upsilon \theta \alpha \acute{\iota} o \lambda o_S$ is what Parry called an *épithète
spéciale*, or *distinctive epithet*: the poet awards it only to Hector. Parry recognized a
class of such epithets (TE 152 f.), but attributed little importance to them. It took
independence of judgement on the part of W. Whallon ('The Homeric Epithets',
YClS 17 (1961), 97–142) to argue that there is a significant connection between this
epithet of Hector and the scene at the Scaean Gate in 6, where Hector's helmet plays

The negative case for any criticism of Homer dealing with
the single word has been well put in the thoughtful article of F.
M. Combellack, 'Milman Parry and Homeric Artistry'.[100]
Combellack concludes: 'For all that any critic of Homer can
now show, the occasional highly appropriate word may, like
the occasional highly inappropriate one, be purely coinciden-
tal—part of the law of averages, if you like, in the use of the
formulary style.' Whether this statement is open to challenge
or not (the 'highly inappropriate' words in Homer seem
considerably more 'occasional' than the 'highly appropriate'),
it is much in the spirit of Parry's own argument, as we can see
in WW. Parry's followers, especially Lord and Notopoulos,
however, went considerably beyond this restriction on the
criticism of the single word. Thus Notopoulos in 'Parataxis in
Homer: | A New Approach to Homeric Literary Criticism'[101]
argues that Parry's placing Homer in the category of oral
rather than of literary poetry makes it possible and even
necessary to understand Homer in a new way. The old
standards of poetic art, deriving from Aristotle and his philosophic predecessors, must be replaced by new aesthetic standards
appropriate to this kind of poetry. For Homeric poetry is
'inorganic' and 'paratactic', as is oral verse in general, and must
be judged as such.

The difficulty is that these new standards of art appear to be
mainly negative: one must not look for any real coherence in
the Homeric poems, because they are by nature episodic; nor,
by the same token, for any relevance of part to whole, or of
part to larger part, either in the case of single words, or in that
of entire scenes. At the end of Notopoulos's article we may

---

so dramatic a role. Parry was willing to see κορυθαίολος as *distinctive*, because it is
never, in our texts, used of any hero other than Hector. He was unwilling to put, e.g.,
ἄναξ ἀνδρῶν into this category (TE 149) or to see in it any meaning besides 'hero', or
any dramatic function at all, because, although principally used of Agamemnon, it
occasionally qualifies the names of other heroes, including some of little note. Here
again, the critic will have to assert himself to insist on what the good reader
recognizes without meditation, that ἄναξ ἀνδρῶν is particularly appropriate to
Agamemnon, and underlines the public role he plays in the *Iliad*. (See Whallon, op.
cit. 102–6.)

[100] *Comparative Literature*, 1959, 193–208.
[101] See above, n. 85.

wonder what we are left with that could enable us to make an intelligent criticism of the *Iliad* or *Odyssey*.[102]

The effect of Lord's 'Homer and Huso II: Narrative Inconsistencies in Homer and Oral Poetry'[103] is much the same. Lord (like Notopoulos) argues against those who would dissect the *Iliad* and *Odyssey* into shorter poems on the basis of inconsistencies of narrative. For the poet composes in *themes* as well as *formulas*, and both of these are fixed. When the poet has embarked on a *theme*, he must go through it in the traditional manner, whether or not its content makes good sense in relation to other themes in the poem. The poet does not care whether or not he makes sense in this way, because his attention is wholly taken up with the theme he is composing at the moment. Lord closes his argument with the example of the alleged inconcinnity in the depiction of Diomedes in Books 5 and 6 of the *Iliad*. In the former, mist is removed from his eyes so that he can perceive the gods, and at Athene's instigation he does battle with two of them. In the latter, he meets Glaucus, son of Hippolochus, and tells him that he will not fight with him if he is a god, 'since I would not battle with the gods of heaven'. The critic, or indeed the simple reader, unless warned away from it, might be inclined to understand this change by the course of the narrative in Books Five and Six, and the difference in tone between them. Diomedes rises ro momentary greatness only with the direct help of Athene; without her, he is a fairly ordinary hero.[104] Book Five is one of the most martial and clangorous in the *Iliad*; Six, with the | conversation between Diomedes and Glaucus, and the then between Hector and Andromache, one of the most explicitly peaceful. The contrast is characteristic of Homer, and to have made it

---

[102] Cf. the thoughtful remarks of Norman Austin in 'The Function of Digressions in the *Iliad*', *GRBS* 7 (1966), 295–312, esp. 295: 'An important result of the studies of Milman Parry on the nature of oral composition is that scholars are more cautious about imposing their own aesthetic bias on Homer and making anachronistic demands of him ... A danger of this new receptive attitude, however, is that while Homer may be vindicated as a historical personage, as an artist he may be merely excused ... The suggestion implicit in the oral approach is that we must recognize that there is after all no artistic unity in Homer, just as many Analysts claimed; moreover, we must learn not to look for any.'

[103] See above, n. 69.

[104] Cf. Whitman (above, n. 91), 167–8.

through the medium of a Diomedes placed in two different
situations, and behaving, as is his wont, with perfect correctness in each, might seem a result of the poet's deliberate art—if
the extension of the Parry theory to criticism did not tell us
that this is wrong, and that we should instead recognize that
the poet is dealing successively with two themes, of which 'the
first is used, completed, and forgotten, and then the second
comes in. This is just the sort of thing we found in [Serbocroa-
tian] oral poetry.'[105]

It is quite impossible to know how Parry would have
reacted to this extension of his theories to the criticism of the
Iliad and *Odyssey* themselves by his successors. We can only say
that his own work provided the impetus for theirs and that,
although one may miss in their publications the intense feeling
for the *style* of Homer which pervades his work, no explicit
statement of his runs counter to their contentions. We must
recognize, at all events, that the effect of these contentions has
been to form a barrier to the sophisticated attempt to explain

[105] Lord, op. cit. 444. Parry himself could be averse to seeing obvious relationships
in the poems. He says in his lecture notes '. . . because of the circumstances of oral
recitation [the] story must be told in episode[s] . . . When the poet within a certain
episode makes some reference to another part of the legend, it is not one of the
previous lays of the Singer which occurs to the mind of the public, but rather the
simple legend. For instance, when Pandarus is slain by Diomedes in the fifth book of
the *Iliad*, Homer makes no mention of the fact that this was the man who but recently
was responsible for the breaking of the truce; the evidence of course has been taken as
showing that the poet of the fifth book was not that of the fourth book, but Homer,
when he told of the slaying of Pandarus, though he only a few hundred lines before
had been dealing with him at length, treats him simply as a well-known character of
the legend, not as one of the personages who had a place of especial sort in his own
particular poem.' Parry's observation that Homer is likely not to make cross-
references, and that this reticence is characteristic of his style, is correct. But he
appears to assume further that the proximity of Pandarus' death in 5 to his actions in 4
is a coincidence, unless he wanted to suggest that the connection was in the mind of
the poet, but would not exist for the audience. In either case, the assumption is based
on an undemonstrable theory of the circumstances of composition of our *Iliad*, and,
like theories of Lachmann (see above, n. 19), should have been 'prevented [by] the
laws of mathematical probability'. The death of Pandarus in 5 is no more the effect of
chance than is the death of Euphorbus in 17, where the connection with that hero's
role in 16 is equally implicit.

Cross-references are by no means entirely absent from the Homeric poems. Those
in the *Iliad* are well examined by W. Schadewaldt, *Iliasstudien, Abhandlungen
der sächsischen Akademie, Philologisch-historische Klasse* 43. 6 (Leipzig, 1938). For the
Odyssey, see e.g. K. Reinhardt, 'Homer und die Telemache' and 'Die Abenteuer der
Odyssee' in *Tradition und Geist* (Göttingen, 1960).

the greatness of the Homeric poems. Hence while formulary
analysis and the concepts of traditional diction and oral
improvisation have of recent years seemed to offer exciting
new approaches to poetry other than Homeric, application of
these methods and concepts to Homer himself had become
increasingly technical, and what poetic criticism we have had
of Homer has often ignored, for practical purposes, the
implications of Parry's work. |

   This is certainly true of non-academic criticism, such as that
of Simone Weil, in her justly celebrated essay 'L'*Iliade* ou le
Poème de la Force'.[106] But it is largely true as well of such a
work as C. Whitman's *Homer and the Heroic Tradition*,[107] the
most ambitious work on Homer in recent years, and one of the
few to treat Homer, with seriousness and imagination, as a
poet. In a somewhat abstract chapter, 'Image, Symbol and
Formula', Whitman discusses the artistic function of the
formula, and tries to work out an aesthetics of Homeric style.
But much of the critical examination, some of it most perceptive, of the *Iliad* which follows this chapter has little to do with
formulary analysis, or with any concern for Homer as a
composer of oral verse. On the other hand, like all work in
German until recently, Karl Reinhardt's subtle and illuminating *Die Ilias und ihr Dichter*[108] knows nothing of Parry or his
work. The most balanced and scholarly recent treatment of
Homeric poetry, G. S. Kirk's *The Songs of Homer*,[109] is more
concerned with historical than poetic matters. Where it does
concern itself with poetic criticism, it seems to accept the
prohibition imposed by Parry's successors. Thus Kirk states
(337) that the *Iliad* would have greater dramatic impact if it
were considerably shorter, but adds: 'Yet one cannot say that
such a contraction would seem desirable by the *completely
different canons of oral poetry* . . .' (italics mine). The finest critical
passages of Kirk's own book (e.g. his eighth chapter, 'Subjects
and Styles') owe their merit partly to his willingness to

---

[106] *Cahiers du Sud*, 1940–1, and again 1947; reprinted in *La Source grecque* (Paris, 1953).
[107] See above, n. 91.
[108] Edited by Uvo Hölscher after Reinhardt's early death (Göttingen, 1961).
[109] See above, n. 35.

examine the poetry without regard to this principle, which would seem to render any literary discussion impossible.

There will always be criticism of the *Iliad* and *Odyssey* that treats these texts as if they were contemporary poems, and some of this, like Simone Weil's essay, may be of the very best. But for the more scholarly- and historically-minded critic, the revelations of Parry, and the attempt by some of his successors to derive from them principles of criticism, will pose a problem as well as offer an insight. The problem is really one manifestation of the fundamental problem of all historical method. If the historical method requires that we try to abandon our own natural judgement in order to grasp the conceptions, and adopt the standards, of a culture essentially different from ours, then the question of the value, or the relevance to us, of that culture, or of any product of it, arises. And together with the question of the value of the undertaking will be that of its feasibility: if the other culture is so thoroughly *other* as the historical approach appears to insist, will it ever, in any meaningful way, become comprehensible to us? The almost uniformly negative character of the | artistic principles enunciated by Parry's successors appears to make an affirmative answer to this question doubtful in the case of Homer.

That Parry himself was aware of these problems, which his successors have mostly ignored, we know from his subtle if inconclusive article, originally a speech to the Overseers of Harvard College, in the Harvard Alumni Bulletin of 1936, 'The Historical Method in Literary Criticism', here HC. Parry there states:

I can ... see nowhere in the critical study of literature anything to check his ever accelerating concern with the past as the past. But when one trained in this method, ... while still staying in the past, turns his eyes back to his own time, he cannot prevent a certain feeling of fear—not for the fact that he has become a ghost in the past, but because of what he sees in the person of his living self. For in the past, where his ghostly self is, he finds that men do the opposite of what he has been doing: they by their literature turn the past into the present, making it the mirror for themselves, and as a result the past as it is expressed in their literature has a hold upon them which shows up the flimsiness of the hold which our past literature has upon ourselves.

Parry's discussion, which follows this statement, of Robert Wood's anecdote about the 1st Earl Granville illustrates the dilemma, and perhaps reveals an excessive strictness in Parry's own conception of the historical method, since one may wonder if Lord Granville's situation in eighteenth-century England was after all utterly different from that of Sarpedon in the world of the *Iliad*. In this very strictness, we can again see, together with the relentless logic which was characteristic of him, Parry's sensitivity to the spiritual directions of his own time: for it is observable in HC that he stresses the historical approach partly as a defence against the propagandistic treatment of past literature as it was being practised by the political extremists of the 1930s. Nor does he quite succeed in resolving the dilemma which he formulates so accurately. He says in his conclusion:

In the field with which I have been particularly concerned here, that of the literatures of the past, unless we can show not only a few students, but all those people whose action will determine the course of a whole nation, that, by identifying one's self with the past, with the men, or with a man of another time, one gains an understanding of men and of life and a power for effective and noble action for human welfare, we must see literary study and its method destroy itself.

Possibly there was a quality in Parry's own life and in his use of words which goes some way toward realizing this requirement. But no explicit statement of his shows how it can be done, nor did he ever fulfil his stated wish to articulate a distinct aesthetics of traditional improvising poetry.

A distinguished scholar of Medieval History[110] once said that the historian must ideally possess the Then and the Now, and must at the same | time sit at the right hand of God. To interpret the past to the present, that is, he must understand what is unique, and uniquely valuable, in the past; he must know and be able to respond to his own time, both so that he can prevent the concerns of the present from distorting the

[110] The late Ernst Kantorowicz, in conversation with the writer.

image of the past, and so that he can know what the present needs and can use from the past. But with all this, he must be able to conceive an Olympian perspective which embraces them both; or in less theological terms, he must maintain a sense of what is in some measure universally human.

Such a thought can be applied with particular relevance to the current state of Homeric studies. It is because we now, as others have done for so many centuries in the past, respond with such directness, such instinctive immediacy of understanding, to the greatness of the Homeric epics, that all this work of archaeology and scholarship continues to take place. It would be perverse if the effect of our scholarship were to deny the validity of the spontaneous judgement which provided the impetus for that scholarship in the first place. One of Parry's strongest arguments for the central point of his Homeric theory—the ornamental nature of the epithet—was, we remember, the reader's own experience of the way words are used in the poems. In developing this kind of argument, Parry was making criticism fulfil its truest purpose: that it should tell us what we already know, only we did not know that we knew it; or less paradoxically put, that it should make clear and articulate what we had apprehended dimly and intermittently. The historical weight of his Homeric studies, their emphasis on differences that lie between the poetry of the *Iliad* and *Odyssey* and poetry deriving from later and literate traditions, has changed our picture of Homer and increased our understanding of his verse: we shall never read it in quite the same way. But the historical argument can only illuminate our understanding if it derives from, and eventually adds to, a conception of Homer itself not based on a purely historical perspective, but on a recognition, in the *Iliad* and the *Odyssey*, of an artistic order and a human significance not limited to any time or any place.

It has been stated by more than one scholar that Milman Parry's work began a new era in Homeric studies. That there is a large measure of truth in this judgement few would now deny; and I have tried here to show why and how this is so. It is equally true that some of the limitations of his work still need to be clearly recognized, while some of its positive value has yet to be perceived. It is unlikely, although some of his followers appear to make this claim, that his approach alone will ever provide the basis of a full criticism of Homer's art. He himself, it must be repeated, almost never discussed Homer, that is, the author or authors of the *Iliad* and the *Odyssey*, as opposed to the tradition in which Homer worked; nor did he ever demonstrate, although at times he seems to | assume it, that Homer was himself an oral poet.[111] His discoveries about the style which Homer employed will perhaps be best exploited when we learn how to combine them with ways of criticism which we know already. His work has suffered from the attempt to make it an exclusive key to the understanding of the epic.

One thing that is surely needed is criticism which can use

[111] Parry manages generally to avoid stating this assumption; but in his lecture notes he says: '. . . it can be shown from the style of Homer himself that the poet composed orally.' The text on this page of the lecture notes, however, is lacunose, and it is not absolutely certain what is meant by 'the poet'.

Dorothea Gray puts the matter with characteristic neatness and accuracy in her edition of J. L. Myres, *Homer and his Critics* (see above, n. 4), 241, where she speaks of 'Milman Parry's proof that Homer's style is typical of oral poetry'. This is in fact what Milman Parry proved. That Homer himself, i.e. the composer or composers of the *Iliad* and *Odyssey*, or of either of these poems, or of any substantial connected part of either of them, was an oral poet, there exists no proof whatever. Otherwise put, not the slightest proof has yet appeared that the texts of the *Iliad* and *Odyssey* as we have them, or any substantial connected portion of these texts, were composed by oral improvisation of the kind observed and described by Parry and Lord and others in Yugoslavia and elsewhere. Hence the statement of Nilsson quoted above (p. xliv) is, strictly speaking, false; so is that of Dodds (*Fifty Years* [above, n. 25], 13): 'the decisive proof that the [Homeric] poems are oral compositions'; and of Lord (*Singer* [see above, p. xxxix], 141): 'There is now no doubt that the composer of the Homeric poems was an oral poet.'

What has been proved is that the style of these poems is 'typical of oral poetry', and it is a reasonable presumption that this style was the creation of an actual oral tradition. But it is still quite conceivable, for example, that Homer made use of writing to compose a poem in a style which had been developed by an oral tradition. This notion, first argued at length by H. T. Wade-Gery (*The Poet of the Iliad* [Cambridge, 1952], 39 f.), was challenged by Dodds (*Fifty Years*, 14), on insufficient grounds, in my opinion, and has more recently been defended as a possibility by Lesky (*Geschichte*[2] [above, n. 4], 56–7), and A. Parry ('Have We Homer's *Iliad*? [above, ch. 10], esp. 210 f.). Cf. Whitman, op. cit. (above, n. 91), 79–80. Such an argument would hold that the composer, or composers, of the *Iliad* and *Odyssey* made use of writing (either directly or through a scribe), but did so at a time when no literary tradition had been able to develop; that the products of this composition are dependent on the oral tradition not only for their diction, but for many other distinctive features as well, such as the reticence in cross-reference noted above (n. 105); but that they owe to their use of writing both their large-scale coherence and their subtlety, qualities in which no known oral poem has begun to equal them.

Parry's insights into Homeric style to understand more of how the Homeric poems are put together. It may be sobering to our belief in scholarly progress to note that one of the few essays which answer this need is the short monograph published in 1933 and entitled 'Homeric Repetitions' by Parry's teacher at the University of California, G. M. Calhoun.[112] Here Calhoun makes use both of Parry's three earliest publications (TE, FM, HG) and of the much earlier work of C. Rothe.[113] By way of explaining how Homer is able to vary some of the traditional elements at his disposal to produce the proper emotional effect of a given scene, Calhoun notes (6 f., especially 9) that the epic poet composes in formulae and whole lines as '. . . freely [and] readily as does the modern poet in words'. | Calhoun's essay thus began a kind of criticism which could show in a precise way how the poet was free to manipulate the materials given him by the tradition in which he so closely worked.

But criticism such as this, which tried to grasp both the existence and the poetic effect of the formula, and then to show how it becomes part of an artistic construct, has been rare since Calhoun's essay.[114] Among the few examples of it are J. Armstrong, 'The Arming Motif in the *Iliad*',[115] N. Austin, 'The Function of Digressions in the *Iliad*' (above, n. 102); some of the remarks of C. Whitman in chapter 10 of his book on Homer;[116] A. Parry, 'The Language of Achilles',[117] and 'Have We Homer's *Iliad*?'[118] In this last article, an attempt is made to show how the poet (of the *Iliad*) can choose the disposition of traditional formulae, how the quality of a scene can depend on this disposition, and how such organization in detail is related to the larger economy of the poem.

The rarity of this kind of criticism, now more than thirty

---

[112] *University of California Publications in Classical Philology*, vol. 12.

[113] *Die Bedeutung der Wiederholungen für die homerische Frage* (Berlin, 1890).

[114] So A. Amory, in the introduction to her monograph on Homeric epithets (see above, n. 99): 'Parry had not the time and most of his successors have lacked the inclination to take up the task of analyzing the interplay between formula as a device for oral composition and formula as a vehicle of meaning.'

[115] *AJPh* 79 (1958), 337–54.

[116] See above, n. 91.

[117] *TAPhA* 87 (1956), 1–7; reprinted in G. Kirk, *The Language and Background of Homer* (Cambridge, 1964) [and in the present volume, ch. 1].

[118] [Ch. 10 of the present volume.]

years after Parry's death, should remind us not only that our knowledge of Homer is far from complete, but also that the implications of Parry's own work are in many ways yet to be realized. We want to see ever more clearly Homer in his own time, to grasp more fully the sense of language, the rhythms of thought, in which he composed. We want also to understand better what we have always known, that the poet (or poets) of the *Iliad* and *Odyssey* was not the representative of his tradition merely, but its master. By continuing the work of Parry, and of other scholars and critics who have extended his discoveries, with some such goal as this, we shall be taking heed of Parry's own warning (HC 413): 'I have seen myself, only too often and too clearly, how, because those who teach and study Greek and Latin literature have lost the sense of its importance for humanity, the study of those literatures has declined, and will decline until they quit their philological isolation and again join in the movement of current human thought.'

16

The Idea of Art in Virgil's *Georgics*

EDMUND WILSON says, a little clumsily but suggestively, in *Axel's Castle*: 'How can the *Georgics*, the *Ars Poetica*, and Manilius be dealt with from the point of view of the capacity of their material for being expanded into "pure vision"?'[1]

Wilson singled out T. S. Eliot's phrase 'expansion into pure vision' for attack. Eliot meant that poetry should present the sort of unencumbered aesthetic experience that we find in music. If a poem seemed to urge a particular set of values, or way of life, or philosophy, it was merely presenting that prosaic material as an 'object of contemplation'. But the philosophy or doctrine itself must be 'capable of expansion into pure vision'. Lucretius falls down, Eliot suggested, because his Epicureanism was too poor in feeling ever to become 'pure vision'.

To this Wilson made the common reader's objection that great poetry often does in fact attempt to persuade us of something; we cannot really separate the philosophy from the poetry; or assess the degree to which a given philosophy can be 'expanded into pure vision'. The three Latin works he cited seemed to him to clinch his case. His question was rhetorical, and the implied answer was No Way.

Wilson therefore applied to the *Georgics* a simple historical relativism. The modern reader finds the agriculture puzzling and tiresome; the Roman reader must have liked it. But if we keep the notion of 'vision' while enlarging it to mean more than a purely aesthetic experience, we may find that Eliot was after all closer to the truth. By this I mean some picture of life, some dramatized attitude, that may work through such

[1] New York 1943, p. 120.

Arethusa, 5. 1 (1972), 35–52. Reprinted by permission of the State University of New York at Buffalo.

vehicles as agriculture, prescriptions for good writing, or astronomy, but finally rises above the details to be valid for all activities and for men of all periods of history. Such a vision would be a product of all the disparate parts of complex literary works; it would at the same time shed light on and explain the literary work as a whole. It has seemed to me that the Virgilian vision in the *Georgics* was finally one of the function and value of human art. And this paper is an attempt to define the idea of art in the *Georgics*.

Of the three works acknowledged as authentic of the greatest of the Roman poets, the *Georgics* may well contain the finest expression of Virgil's poetic art. At the same time, it is, and is likely to remain, the least popular and the least accessible. The pastoral poems, or the *Eclogues*, have the advantages of simplicity and brevity. The | *Aeneid* too, the most sweeping and powerful of Virgil's poems, is simpler in expression, possibly even simpler in thought, than the *Georgics*. That the *Aeneid* is heroic narrative, telling a continuous story, in a manner deriving directly from the most perspicuous and accessible of all poets of the Western tradition, Homer, and a narrative moreover centrally concerned with the Roman historical experience, ensures it a popularity which the more difficult and didactic poetry of the *Georgics* will certainly never attain. The *Aeneid* is perhaps not a less subtle and complex poem than the *Georgics*. Its strange amalgam of triumph and sadness, of confidence and nostalgia, of the martial tones of Roman and Augustan achievement, and of the poignant notes of personal loss and renunciation, has hardly been fathomed by ancient or modern criticism.[2] But the *Aeneid*, like the *Eclogues* in a very different way, can be read and enjoyed in a simple fashion. It tells a story, a compelling story, a story of travel and experience, of sacrifice and victory. The poem can be read for that story, and what most of us would regard as its deepest meanings, its characteristic mode of taking back what it gives, of mixing personal regret with Roman hope, can, without a positive distortion of perspective, be left to be apprehended by

---

[2] See e.g. A. Parry, 'The Two Voices of Virgil's *Aeneid*', *Arion* 2 (1963), reprinted in *Virgil, a Collection of Critical Essays*, ed. Steele Commager (Englewood Cliffs, New Jersey, 1966) [and as ch. 8 in the present volume].

the reader, as, shall we say, overtones and undertones, further vistas to be felt almost subliminally, not requiring, in order for us to reach any adequate understanding of the poem, to be fully analysed and raised to the level of conscious and explicit critical exposition.

Not so the *Georgics*. Portions of that work, many of the individual descriptions of natural phenomena, or the constantly recurring digressions, mythological and otherwise, such as the one on the Roman Civil Wars at the end of Book 1, or the praise of the farmer's life at the end of Book 2, even the baroque invocations to the gods, to Augustus, and to Maecenas at the beginning of Books 1 and 3, can be appreciated in a simple way for and of themselves, particularly since in these and other passages the sheer art of words reaches a height and a degree of finish greater than anything else in Virgil. As a collection of pieces worthy of inclusion in anthologies, the *Georgics* offers no difficulty. It is the total thrust and meaning of the work, the way in which all the many descriptions and short narratives, natural, mythological, historical and philosophical, the expressions of different attitudes, sometimes in apparent contradiction to each other, are orchestrated into a complex but unified vision of the world, which is hard for us to grasp. The *Georgics* is ultimately about the life of man in this world, about a kind of art in living which can confront the absurdity and cruelty of both nature and civilization, and yet render our existence satisfactory and beautiful. All its rich and diverse in|gredients are expressed and arranged to conduce to this vision, a vision less cosmic and explicitly philosophical, but finally as comprehensive and purposeful as that of the poem which may, more than any other, have been Virgil's model, the *De Rerum Natura* of Lucretius.[3]

Virgil's three works are each ostensibly modelled on one of the classical (to Virgil and his audience) works of Greek poetry, the *Eclogues* on the bucolic poems of Theocritus, the *Georgics* on the *Works and Days* of Hesiod, and the *Aeneid* on the Homeric epics. As the *Georgics* is in itself the least accessible of Virgil's works, so the *Works and Days* is the least accessible of

[3] See most recently L. P. Wilkinson, *The Georgics of Virgil* (Cambridge, 1969), 163–5.

his three models—or at least if Theocritus seems harder than Hesiod, the greater difficulty is mostly due to the former poet's use of dialect. It is not merely that the Hesiodic poem and its Virgilian descendant belong to a genre of literature which seems strange to the modern world: the didactic poem. It is also the didactic content itself. The pastoral vision of nature has still its living manifestations: the willed simplicity and the waving fields of grain in *Bonnie and Clyde* may have more of a connection with Theocritean and Virgilian pastoral vision than at first sight occurs to us. But a long poem—four books of between 500 and 600 lines each—in an intricate language dealing with the details of agricultural practice in 35 BC: that is something far enough from us to give us pause, and to incline us, following the lead of so many classicists, to talk about the consummate art of Virgil's *Georgics* when the occasion arises, but to confine our reading and effort to understand to a few splendid anthology plums. The agricultural mode does not seem to us a natural expression of poetry. To show that the idea is not wholly alien, I suppose one could cite something like Knut Hamsun's *Growth of the Soil*, a work which does not spare the laborious details of the farmer's life, and yet—like Virgil's—contains a vision which goes beyond the mere assertion of the superiority of the life attached to the living earth and far from the corruptions of metropolitan, or even small-town, civilization.

I cite the difficulty and the remoteness of subject of the *Georgics* primarily to explain why one of the finest poems of all antiquity is one of the least known; and further to indicate how much worth study it is: the deepest beauties of the poem are a little like the value of the natural existence itself, as Virgil defines it: man's livelihood is hidden in the earth by the harsh but wise counsel of Jupiter—*curis acuens mortalia corda* (1. 123)—to be gained therefrom by labour and assiduous art.

One feature of both the excellence and the complexity of the *Georgics* it shares with the other great didactic Latin poem, that of Lucretius. What we may, for lack of a better term, call the *philosophi|cal* aspect of the poem, the network of ideas and attitudes which makes up its essential vision, is not separable from the detailed description of natural phenomena or from the specific instructions offered to the hypothetical tiller of the soil to whom the poem is addressed. The characteristic Virgilian attitudes toward labour and joy, toward all the conditions of existence, in fact, pervade the poetry; and many of the most beautiful passages are also the most didactic. Book 1. 160–8 may illustrate this double aspect of the poem:

> Dicendum et quae sint duris agrestibus arma,
> quis sine nec potuere seri nec surgere messes:
> vomis et inflexi primum grave robur aratri,
> tardaque Eleusinae matris volventia plaustra,
> tribulaque traheaeque et iniquo pondere rastri;
> virgea praeterea Celei vilisque supellex,
> arbuteae crates et mystica vannus Iacchi.
> omnia quae multo ante memor provisa repones,
> si te digna manet divini gloria ruris.

This passage comes shortly after an important section, to which I shall refer later, in which the poet urges the necessity of constant, intelligent labour in extracting life from the soil. The next step, and here Virgil follows Hesiod fairly closely, is to say something of the tools which make this labour possible. The corresponding passage in the Greek poet is, at least in part, simply practical: Hesiod there tells us what tools we need and actually how to make them. But Virgil talks of tools to illustrate his point: they represent the element of *art*, an element largely absent from the Hesiodic original.

Now I must tell you what arms are possessed by the tough farmer, without which the crops cannot be sown, and cannot grow: the ploughshare first, and the heavy tree of unbending plough; the slow rolling waggon of the Eleusinian mother; sledges, harrows, and the unbalanced weight of the hoe; the humble wattle-work equipment of Celeus as well; wooden hurdles, and the mystic winnowing-fan of Iacchus. All these things you will take thought to lay up beforehand, if the godlike glory of the fields is to be yours, and you worthy of it.

The tone of this short paragraph is firmly didactic: the gerundive *dicendum* and the second person future *repones* for the imperative at the end of 167 catch the business-like and peremptory note of the true agricultural treatise. Note also the expressiveness of 164: the alliteration and the double *-que* construction give the sense of a Greek | catalogue, and of the specialization of the farmer's tools: the heavy last part of that line is more than a conventional or ornamental circumlocution; the hoe or mattock is unusually and harshly heavy; it is also unbalanced, with the weight in the head.

The farmer's equipment thus set out in detail is modest: the *vilis supellex* carries an implicit contrast to the expensive furniture of the town-dweller. At the same time these things are *arms*: the farmer has the manly virtues of the soldier, and like his plough, he must be hard: *duris agrestibus arma*. But these arms, the emblems of his austere and honest life, all have divine connections: Demeter had the first waggon; and such a waggon is used in her rites at the Eleusinian mysteries; Celeus, father of Triptolemus, was taught by her, and taught the rest of mankind how to till the fields; the winnowing-fan is simple and practical, but it is also prominent in the mysteries of Bacchus: the phrase anticipates the celebration of the magic of the vine in Book 2. The penultimate line, though maintaining the stern instructive tone, has the alliteration and assonance which accounts for so much of the amazing music of Virgilian poetry. Then the last line gives, so to speak, the reward of all the labour implied by the preceding. It reveals the vista of godlike splendour and satisfaction that makes the drudgery worth it. The release and assertion of this line surprises, but is made to follow naturally from the enumeration of the farmer's means, and that is one of the principal points of the poem.

Another passage: this time from Book 3, the book devoted to livestock (49–71):

> Seu quis Olympiacae miratus praemia palmae
> pascit equos, seu quis fortis ad aratra iuvencos,
> corpora praecipue matrum legat, optima torvae
> forma bovis cui turpe caput, cui plurima cervix,
> et crurum tenus a mento palearia pendent;
> tum longo nullus lateri modus: omnia magna,
> pes etiam; et camuris hirtae sub cornibus aures.
> nec mihi displiceat maculis insignis et albo,
> aut iuga detrectans interdumque aspera cornu
> et faciem tauro propior, quaeque ardua tota
> et gradiens ima verrit vestigia cauda.

aetas Lucinam iustosque pati hymenaeos
desinit ante decem, post quattuor incipit annos;
cetera nec feturae habilis nec fortis aratris.
interea, superat gregibus dum laeta iuventas,
solve mares; mitte in Venerem pecuaria primus,
atque aliam ex alia generando suffice prolem. |
optima quaeque dies miseris mortalibus aevi
prima fugit: subeunt morbi tristisque senectus
et labor, et durae rapit inclementia mortis.
semper erunt quarum mutari corpora malis:
semper enim refice ac, ne post amissa requiras,
ante veni et subolem armento sortire quotannis.

Whether a man raises horses out of admiration for the Olympian victor's palm, or whether he raises stout bullocks for the plough—let him select above all the bodies of the mothers. The best shape of a cow is fierce, an ugly head, an oversize neck; and all the way down from her chin to her lower legs hang her dewlaps; long flanks, the longer the better; everything big, even the hoof. And beneath her twisted horns, shaggy ears. No objection to one with white spots, or one that resists the yoke, or now and then threatens with her horns, and looks more like a bull than a cow, and standing high all along her length, sweeps as she walks her hoofprints with the end of her tail.

The age of the goddess of childbirth, and of lawful union, is over at ten years, and begins after four. Outside these limits, she is neither capable of reproducing, nor strong enough for the plough. Meanwhile, then, while their rich prime of life still lasts, let loose the males; be the first to give over your cattle to Venus, and by generation replenish the stock of the young. The best days for unhappy mortal beings are the first to flee away; then come sickness, and unhappy age, labour and trouble; and the unpitying hardness of death takes all away. There will always be some whose shape you want to change; keep replacing them; don't wait till the stock is gone and you miss them; be beforehand, and choose each year the young to breed for the continuation of the herd.

The picture of the ideal cow here is particularly fine. The sharp edge of Virgil's observation expresses a connoisseur's professional delight in the best product of its kind. The massive ugliness of the beast—*turpe caput ... plurima cervix ... nullus lateri modus*—defies the simple-minded aesthetics of the inexperienced and suggests a new and more functional feature. The single points are made with almost staccato urgency, but with no less expressiveness for that: note how after the first three brief features, the fourth, describing the extravagant length of the animal's dewlaps neatly fills a whole line (53), and how another whole line, of more formal and heroic structure, appears | at the end of the description (59), as if, like a judge, Virgil has looked over the animal and now lets it grandly walk off.

No Latin poetry is more inventively and variously expressive of external phenomena than the *Georgics*; but what I want to bring out in this didactic passage is the *suite* of thought: 'For both horses and cows, breeding is important. Breeding depends especially on the physical characteristics of the mother; e.g., here is the best kind of cow. There is also a right age for breeding; it is soon over. So take advantage of that short period, and keep up the numbers and the standard of the breed.' That is the bare argument of the passage as a whole, but Virgil's feeling for animal life, and its relation to human life, leads him to introduce an entirely different, and tragic, note. The phrase *laeta iuventas* in 63 appears conventional, a formulary phrase, when we first come to it, but it and the mention of Venus, the divine source of love and generation, again a convention that becomes more than that, lead to lines of sudden poignancy, 66–8, lines which, as readers of Boswell remember, Dr Johnson said were the saddest of all ancient poetry. The limits of the proper age for breeding, a practical matter of husbandry, is transformed by Virgil's quick thought into a melancholy reflection on the transience of happiness and life itself. But this again is no mere ornament or indulgence; the clear thought that underlies the transformation is the characteristic contrast between the evanescence of the individual and the continuation of the race. Though contrasted, there is an indissoluble connection: if the race, the process of life, is to continue, the individual must be selected, and the brief moment allotted must be seized and exploited; otherwise the race itself will decline or disappear. But even if it does not disappear, the irreducible value of the individual life is lost, and that loss is inconsolable. Hence the famous lines 66–8 are not really comprehensible outside their context, where they are at once a warning and a lament.

The *Georgics* begins with a series of indirect questions (1. 1–5):

> Quid faciat laetas segetes, quo sidere terram
> vertere, Maecenas, ulmisque adiungere vitis
> conveniat, quae cura boum, qui cultus habendo
> sit pecori, apibus quanta experientia parcis,
> hinc canere incipiam.

What makes the crops grow rich, under what constellations to turn the earth, Maecenas, and to graft the vine to the elm, what care we should give to oxen, what is the management of cattle, and what sort of experience is needed for thrifty bees: these are the subjects of my song. |

These words in fact outline the subjects of the four books as we have them: Book 1 concerns the seasons of the year and how they are marked by the heavenly bodies: it is the book closest in subject to Hesiod; Book 2 deals with care of trees and the grapevine; Book 3 with livestock; and Book 4 with bees. These subjects, however, offer only a small idea of the true contents of the books. Thus Book 1 includes the fundamental passage concerning the fall of man from the Saturnian Golden Age, and it moves at the end from the signs given by the stars to the farmer to the apocalyptic warnings in the heavenly bodies of the Roman Civil Wars. The Civil Wars are over, but wars elsewhere continue: *saevit toto Mars impius orbe* (1. 511).

It is a dark and pessimistic book, ending with a Lucretian sense of inevitable movement into disorder and destruction. Book 2 is the happiest book and ends with the famous praise of the Italian countryside. Book 3, dealing with animals, and hence closer to the human condition, is yet darker than 1. Unlike trees, animals are subject to passion, and passion leads to tragedy. At the beginning of a passage describing the destructive and irresistible power of *amor*, Virgil specifically includes mankind (3. 242–4):

> omne adeo genus in terris hominum ferarumque
> et genus aequoreum, pecudes pictaeque volucres,
> in furias ignemque ruunt: amor omnibus idem.

The whole race of men and of beasts, and the race of fish and cattle and varicoloured birds—all rush into frenzy, into the fire of passion—*amor* is the same for them all.

Specific directions towards the end of the book concerning the treatment of disease in livestock bring Virgil to speak of a great plague of cattle which destroyed vast herds. The model here is the despairing end of Lucretius' poem, which is in turn based on Thucydides' description of the plague at Athens. The plague is inevitable, Virgil says, and its effects final. All the animals died, and the bodies were contaminated, so that those who wore clothes made from their skins contracted the sacred disease, and were themselves devoured by it. The transition at the end from animals to men makes explicit the symbolic purpose of the whole book. All existence, including human, is doomed, despite even our best labours, to annihilation. And in the face of the nightmare destructions of the race, the whole subject of the poem, the value of care and ingenuity and work, is called into question (3. 525–6):

> quid labor aut benefacta iuvant? quid vomere terras
> invertisse gravis? |

What does this labour avail them now? What their services? What reward for having turned the heavy earth with the plough?

It is left to Book 4 to find a resolution of the tension of the first three between the joy and beauty of nature and its ultimately destructive power, a tension which we saw in the passage concerning the breeding of cows, in the brief turn from directions for the maintenance of the herd to the irreparable loss of individual happiness and life. In that passage, collective survival seemed at least a counterbalance to individual loss. The relation of such an equation to the feelings of a Roman in the closing years of the Civil Wars need not be elaborated. But that melancholy assessment, as I said, is replaced by the vision at the end of the book, of the total destruction of the herd, of life itself.

The solution in Book 4 appears at first to be the unique character of the bee. The description of the life of bees filling the first half of the book is done with—on the whole— accurate observation and felicitous humour. The bees, unlike plants, have intelligence and motion. They have, more than any other animal, a complex social structure, a structure which Virgil's description makes sound analogous to that of men. But

within the analogy, Virgil shows his awareness of the differences. On the one hand, bees are small and in a way
insignificant: 'I hold up for your admiration, Maecenas, the
spectacle of the small beings', *admiranda tibi levium spectacula
rerum* (4. 3). And as they lack passion, so they lack power: lines
67 and following describe with a delightful mockery of human
effort, a battle of bees. They circle around the *praetoria* (75) of
their leaders, and there is martial clangour imitating that of the
war trumpet. Yet if you toss into their midst (86–7) a small
handful of dust, the battle instantly ceases, and these great
combats are reduced to quiet. On the other hand, bees are
superior to all other animals, including men, because of their
collective instinct, their completely social orientation. Each
does his job without protest or reluctance. Their obedience is
perfect. They are a faultless social organization, offering in this
sense a corrective model to the chaos of human society. This
perfect subordination of the individual to the group and
specifically to the *rex*, the apian analogy to the *princeps*, seems
to bespeak a divine influence.[4] In one passage Virgil speaks, in a
manner anticipating the Pythagorean statements of *Aeneid* 6, of
how the movements and life of bees show us the pervasive
presence of god in nature. This presence is not confined to bees:
divinity moves through all nature, and from it 'cattle great and
small, men, all wild beasts—each one as he is born, acquires the
subtle elements of life' (4. 223–5):

> hinc pecudes, armenta, viros, genus omne ferarum, |
> quemque sibi tenuis nascentem arcessere vitas.

The passage is especially noteworthy because in Book 1,
observing the remarkable sense of weather to be observed in
some birds, Virgil had denied that this was due to any
particular infusion of divine intelligence. Here we find that all
living things derive their being from a divine source; but bees
are especially symbolic of this, because of their devotion to the
collective good. Since they and other living things derive from
a living source and return to it after individual death, the sum
of life appears constant: 'Nor is there any place for death; ever

---

[4] See Brooks Otis, *Virgil, A Study in Civilized Poetry* (Oxford, 1963), 184–5.

alive they fly off to take their place among the throng of heaven' (4. 226–7):

> nec morti esse locum, sed viva volare
> sideris in numerum atque alto succedere caelo.

Using Lucretian language, *viva volare*, Virgil has adopted a Platonic idea—see the first argument of the *Phaedo*—to assert a collective immortality. But though this idea is here phrased in general terms, it is still centred on the image of bees. And does even their ideal social instinct ensure them continuing life? The Platonic-Pythagorean generality does not really solve the essential problem posed in the *Georgics* as a whole: partly because death is an observed fact, partly because bees are ultimately not an adequate image for humanity.

Virgil continues the argument by reverting to the didactic: *labour* is necessary here too, and we learn how to care for the hive and keep it clean and how to remove the honey, the honey which Virgil refers to in the first line of Book 4 as 'the heavenly gift of honey', *mellis caelestia dona*. But bees too are subject to sickness. The analogy here‘is first to the diseases of livestock so vividly described in Book 3. The poet, however, passes over this logical step, and equates the sickness of bees directly to that of men: 'But if (for life brings to bees our sufferings too) their bodies languish with dread sickness . . . ' (4. 251–2):

> Si vero (quoniam casus apibus quoque nostros
> vita tulit) tristi languebunt corpora morbo . . .

Remedies for various maladies follow. The course of thought retraces that of Book 3, ending with the possibility of the incurable sickness that destroys the entire hive: 'But if all at once the entire race is extinguished, if there is nothing left from which to create a new generation' (4. 281–2):

> Sed si quem proles subito defecerit omnis
> nec genus unde novae stirpis revocetur habebit . . . |

Here the poem takes a final and unexpected turn. An ancient tradition recorded by Servius tells us that Virgil's poem originally ended with the praise of his friend Cornelius Gallus, the one who is the subject of the tenth *Eclogue*. Gallus was appointed prefect of Egypt at about the time when Virgil is supposed to have finished the *Georgics*—30 BC. Afterwards, he fell into disfavour with Augustus, and in 26 BC committed suicide. After his disgrace, Servius tells us, Virgil replaced the original ending of the work by the Aristaeus story which now ends it—presumably in a second edition if the tradition of first publication is correct. The question is important aesthetically as well as for literary history, because if the story is true, then the judgement of those who have considered the Aristaeus episode an inorganic addition to the body of the *Georgics* appears to receive some support.

The story, at least taken at its face value, seems, almost certainly, to be untrue. Several arguments beside the problem of dating can be brought against it. One is simply that if it was true, and if the original Book 4 was approximately as long as the preceding three books, the praise of Gallus must have occupied more space than all the passages praising Augustus and Maecenas put together. This would have been most tactless, and Virgil as well as Gallus might have had to commit suicide. Another reason is that the Aristaeus episode, or something like it, is necessary to the sense of the work. What may have happened, and what could account for the story, is that originally Virgil included, at the beginning of the episode, a few lines in praise of Gallus which were afterwards excised.[5] That would help to explain the emphasis on Egypt with which the episode begins. Having posed the question of total extinction of the race, the poet says that 'it is time now to set forth the discoveries of the Arcadian shepherd'—i.e. Aristaeus. Then he says that this *art* is practised in Egypt. The country of Egypt is richly described in a climactic sentence of 8 lines, ending with the words, 'this whole land finds sure salvation in the *art* of which I am about to tell you' (4. 294):

> omnis in hac certam regio iacit arte salutem.

The description of Egypt and its exotic neighbours (287 f.) is functional, as Klingner saw,[6] in that it helps to remove the

---

[5] That is essentially the opinion of Wilkinson, op. cit. 108 f. and app. IV. The Servius story is accepted as it stands by e.g. E. de Saint-Denis, *Virgile, Géorgiques* (Budé edition, Paris, 1960), pp. xxxvi f.

[6] F. Klingner, *Virgil* (Zürich and Stuttgart, 1967), 327.

focus of attention from the real world of Italy to the mythical and magical realm where life can be created from death. But the choice of Egypt in the first place may well be due to Gallus.

The Aristaeus episode is an ἐπύλλιον, and while it plays a vital part in the economy of the whole work, it is obviously written in a different key from the rest of the poem, and has a considerable degree of autonomy. Lines 281–314, which purport to describe the actual process |of βουγονία, the generation of bees from the rotting carcass of a young bull, form a transition from the didactic poem to the rich involution of the myth of the ἐπύλλιον proper. It is doubtful that Virgil actually believed in this remarkable process; but it would make little difference if he did.

The βουγονία must take place in spring, and as such it is symbolic of the regeneration of life in springtime. Virgil stresses this in three unusually beautiful lines (4. 305–7):

> hoc geritur Zephyris primum impellentibus undas,
> ante novis rubeant quam prata coloribus, ante
> garrula quam tignis nidum suspendat hirundo.

This is to be done when the West wind first begins to move the waves, before the fields come out in new colours, before the talkative swallow hangs her nest on the roof-beams.

The way in which the two conjunctions *ante . . . ante* are placed at the beginning and end of 306, leaving 307 clear for the swallow to hang her nest, with the adjective that gives life to the scene, *garrula*, in the emphatic position at the beginning of the next line, which ends with the sonorous name of the bird, is true Virgilian art.

The ἐπύλλιον proper begins with line 315. The perfection of its form and its truth to type reminds us of how much Virgil was, after all, a neoteric poet. The shepherd Aristaeus has lost all his bees by sickness. He implores his mother, the sea goddess Cyrene's help. She tells him to go capture the elusive deity Proteus. When Proteus is finally pinned down, he tells Aristaeus that he must expiate his fault in causing the death of Eurydice. Proteus then tells the whole elegiac tale of Orpheus' descent into the underworld to regain Eurydice, his losing her a second time, his return to earth to mourn her continuously until his own death at the hands of jealous Thracian women. Proteus then vanishes, Cyrene takes over, giving her son specific and ritual directions how to create a new hive of bees from the rotting carcass of a cow, after having made expiating sacrifice. Aristaeus does as he is told, a new hive appears, and the poem ends on the note of renaissance and pastoral paradise regained.

The first lines contain the epic invocation of the Muses and the inquiry after the divine causation. This introductory passage embodies, in transformation suitable to the tone of the work, scenes from both *Iliad* and *Odyssey*: the lamenting Aristaeus, complaining that he is denied even that little honour due him, is Achilles of *Iliad* 1, while the catalogue of nymphs matches those who rise from the sea to mourn Patroclus in *Iliad* 18. The capture of Proteus as the story continues is | of course taken from Book 4 of the *Odyssey*. But the allusions are not merely Homeric. Aristaeus visiting his mother in the world under the sea is from Bacchylides' dithyramb on Theseus, where that hero is welcomed into the depths by Amphitrite. And Eurydice, in the story within a story, as she says farewell to her imprudent husband and is drawn back to death, is the dying Alcestis of Euripides' play.

The language is epic. *At*, the Latin equivalent of ἀτάρ or αὐτὰρ ἔπειτα, marks the steps of the narrative again and again, cf. lines 333–4:

> At mater sonitum thalamo sub fluminis alti sensit
> Then did his mother in her chamber under sea hear his cries

or 360–1:

> at illum
> curvata in montis faciem circumstetit unda
> then did the waters rise up like a mountain about him

and in a line like 320:

> multa querens atque hac affatus voce parentem
> lamenting much, and with these words addressing his mother

with its archaic and religious verb *affari*, is pure heroic diction.

The Hesiodic and Alexandrian quality of the ἐπύλλιον appears particularly in the catalogues which fill the poem. Besides the muster (336–44), we have the sources of the great rivers in the underworld (367–73), a theme which Eric Havelock[7] has ingeniously shown to derive from early Greek geographical speculation, and to have echoes as late as Coleridge:

> In Xanadu did Kubla Khan
> a stately pleasure dome decree,
> where Alph the sacred river ran
> through caverns measureless to man,
>   down to a sunless sea.

And the Orpheus-Eurydice story itself contains two more geographical catalogues: 460–3, at the beginning the Thracian lands that weep for Eurydice, and matching it in a corresponding position at the end, 517–19, the lands traversed by Orpheus, lamenting her now twice lost. |

> solus Hyperboreas glacies Tanaimque nivalem
> arvaque Riphaeis numquam viduata pruinis
> lustrabat . . .

These, and the many other geographical references in the episode, are designed to express both the universality and the exotic splendour of the story. The sort of rich and rare and highly wrought language of these passages in particular shows Virgil in his most lyrical, and his most Alexandrian, mood. The theme of the Orpheus story—more than in any other ancient version of the myth—is lamentation and death. The *imperium* of art here is entirely in the threnodic mode. But this dense expression of grief, these unremitting plangent notes, are elevated into what is almost a symbol of music and art by Virgil's employment in them of every possible lyrical device: the magical and exotic notes of Greek names and Greek metres, of anaphora, alliteration, and assonance. Consider 460–6:

---

[7] E. A. Havelock, 'Virgil's Road to Xanadu', *Phoenix* 1/1 (1946), 3–9; 1/2 (1946), 2–7; suppl. to vol. 1 (1947), 9–18.

At chorus aequalis Dryadum clamore supremos
implevit montis; flerunt Rhodopeiae arces
altaque Pangaea et Rhesi Mavortia tellus
atque Getae atque Hebrus et Actias Orithyia.
ipse cava solans aegrum testudine amoreum,
te, dulcis coniunx, te solo in litore secum,
te veniente die, te decedente canebat.

The chorus of Dryads, of her age, filled with their cries the mountain tops; the summits of Rhodope wept, and lofty Pangaea, and Rhesus' land of Mars, and the Getae, and the Hebrus, and from the Attic cliffs, Orithyia; while he went solacing his bitter love with the hollow lyre, singing you, sweet wife, alone on the shore by himself, singing you, as the day arose, you, as the day declined.

Line 463 contains at least three elements alien to Latin metrics: the hiatus of the first syllable of the second dactyl; the long syllable in Hebrus before -br-; the unlatin short second *a* in Actias; and the drawn-out spondaic ending in the single final word, Orithyia, with its unlatin vowels. Note also the *tour de force* of assonance in 465–6: the alternating *o*'s and *u*'s of 465 succeeded by the repeated *e*'s of 466. The language, that is to say, in fairly direct fashion, presents Orpheus as the essence of poetry as well as of grief.

The construction of the ἐπύλλιον too shows a conscious symmetry to a degree beyond anything else in the *Georgics* or the rest of Virgil's poetry. The form of the ἐπύλλιον often depends on an αἴτιον: that is, the poem purports to be an explanation of the origins of a custom, | usually a religious rite still continued in the poet's day. This is expressed in the first line, 315: 'What god, Muse, fashioned this art for us?':

Quis deus hanc, Musae, quis nobis extudit artem?

The whole passage ends with the first successful practice of the art of βουγονία. The concluding lines, 556–8, by mentioning weather, trees, and grapes, as well as the bees born from an animal, also sum up the subjects of the whole poem: 'The bees buzz in the animal's belly, but burst forth from its broken ribs, are borne aloft in a huge cloud, and settle in supple branches in a grape-like cluster':

liquefacta boum per viscera toto
stridere apes utero et ruptis effervere costis
immensasque trahi nubes iamque arbore summa
confluere et lentis uvam demittere ramis.

The matching of beginning and end is the outer layer of an
intricate nested structure: 317–32, the pastoral god Aristaeus
complaining to his mother of the loss of his bees, corresponds
to the scene at the end, Aristaeus successfully carrying out the
βουγονία.

The passage from lines 333 to 424 is dominated by Cyrene,
receiving her son into her watery kingdom and telling him
how to discover the cause of his affliction from Proteus. It
corresponds to 530–47, where Cyrene tells Aristaeus after the
speech of Proteus how he can expiate his fault. Lines 425–52
describe at length the appearance of Proteus surrounded by his
faithful seals, and Aristaeus' capture of him. That corresponds
to two lines, 528–9, where Proteus goes back under the
foaming sea. Proteus' oracle occupies the centre, 453–527, and
this oracle is itself chiastically disposed: beginning with the
death of Eurydice, ending with the death of Orpheus, those
sections, 453–63 and 516–27, are followed and preceded respectively by the first and second scenes of Orpheus' lamentation (465–6 and 507–15); and these in turn by Orpheus
regaining Eurydice (467–84), and losing her again (485–506).
The turning point, where Orpheus looks back a moment too
soon to see if Eurydice is still there, and so loses her forever, is
in almost the exact centre of the Orpheus-story, marked by
two aorist verbs in 490–1—*restitit* and *respexit*. 'He stopped,
and right there, at the edge of the world of light, oh god,
forgetting himself, yielding to desire, he turned back to look.'[8]

What is the point of this art of such extraordinary richness
and intricacy? Klingner properly compares Catullus 64,[9] the
long poem on the marriage of Peleus and Thetis. Here too we
find the Alexandrian device of a story within a story, two
myths joined by an ingenious and | unexpected, if somewhat
arbitrary, transition. Moreover, the two myths in that poem,
like those of our text here, are joined closely in form, but

[8] St.-Denis, op. cit. p. xxxix, has a useful chart of the structure of the episode.
[9] Op cit. 353 f., esp. 359.

embody contrasting moods. The marriage of Peleus and Thetis is all joy and success: human life virtually raised to the level of the divine. The Ariadne story embroidered on the wedding present is tragic. Two aspects of existence are symbolized by the two intertwined myths of the poem. But, Klingner points out, the poem also contains a suggestion that the outcome of the two myths is the same, and not in contrast.[10] The marriage of a mortal, Peleus, to a goddess, Thetis, is paralleled by the suggestion at the end of the poem of the marriage of a mortal woman, Ariadne, to a god, Bacchus.

Klingner wants to find a similar higher resolution of the two contrasting stories of our poem, the regeneration of life for Aristaeus, and the inevitability of death for Orpheus. He tries to find this resolution in the Pythagorean sense of the continuation of all life on a super-terrestrial plane as a compensation for the tragic loss of individual life, an idea which, as we saw in the passage on the breeding of cows (39–40 above), occurs elsewhere in the poem. Klingner misses this passage, but compares *Aeneid* 6, where the individual deaths of Icarus and Marcellus at the beginning and end of the book are balanced by the sense of unending Roman history merged with religious philosophy in Anchises' speech.[11] The resolution here in the *Georgics* and in the *Aeneid* is less clear than that in Catullus' poem, and perhaps this refusal to work out a comforting equation is itself characteristic of Virgil.

But I should like to suggest another sort of resolution. In Book I of the *Georgics*, Virgil takes over from Hesiod the idea of the fall of man from the Golden Age as Jupiter's punishment of man's intellectual challenge to the gods (I. 121–3): 'The father himself wanted for men no easy way of agriculture, and he was the first to see that art must move the fields. . .'

> pater ipse colendi
> haud facilem esse viam voluit, primusque per artem
> movit agros. . .

But Virgil varies the Hesiodic theme. In Hesiod, Zeus' punishment is ultimately a good thing, because it necessitates *work*, and work is man's way of achieving justice, or δίκη. For

---

[10] Ibid. 355.   [11] Ibid. 361.

Virgil, work, *labor*, is important, but not so much because it embodies justice. Rather, because it in turn necessitates *art*. Virgil's whole poem, in contrast to that of Hesiod, stresses the beauty and variety of human experience raised to the level of art. Art, *ars*, is at once an intellectual and an aesthetic achievement. To continue the passage in Book 1. 125 ff.: |

> ante Iovem nulli subigebant arva coloni;
> ne signare quidem aut partiri limite campum
> fas erat: in medium quaerebant, ipsaque tellus
> omnia liberius nullo poscente ferebat.
> ille malum virus serpentibus addidit atris,
> praedarique lupos iussit pontumque moveri,
> mellaque decussit foliis ignemque removit,
> et passim rivis currentia vina repressit,
> ut varias usus meditando extunderet artis
> paulatim. . .

Before Jupiter was, no farmers tilled the soil. Nor was it right to mark the fields or part them with boundaries; all sought life in common, and earth herself bore freely all, so that no one needed to ask. Jupiter added poison to serpents, ordered wolves to prey on other animals, made the sea a place of storms; he struck the honey-dew from the leaves, took away fire and stopped the flow of wine in rivers, *so that* need and experience would *fashion* the several arts, little by little. . .

The whole process of agriculture in which Virgil takes such delight is one of the *variae artes* which men have fashioned by experience and design. Note the key word *extunderet* in 133 — 'hammer out' — a word which properly suggests the sculptor's creation of a statue. The sculptor's, or metalworker's craft, I suggest, is a bridge between the farmer's and the poet's art.[12] Nature, for which Virgil feels so vivid a passion, is to be understood and appreciated through art, and the assimilation of the farmer's lore and the poet's song is furthered by their common subject, the natural world.

At the beginning of the ἐπύλλιον, line 315, these key words

---

[12] In a curious footnote (op. cit. 327) Klingner, to argue against Norden's idea that Aristaeus somehow represents Zeus, denies any connection between the occurrences of *extundere artem* in Book 1 and in Book 4. Norden's point is not to be taken seriously, but Klingner's argument here appears both literal-minded and perverse.

reappear. And the verb is repeated in Aristaeus' lament (4. 326–8):

> en etiam hunc ipsum vitae mortalis honorem,
> quem mihi vix frugum et pecorum custodia sollers
> omnia temptanti extuderat, te matre relinquo.

See how this one honour of my mortal life, which with such pains my care of crop and herd, after long experience, had *fashioned*, is taken from me, though I am your son!

He is angry because his art, which he has fashioned with such effort, appears to be of no avail.

The lesson which Aristaeus must learn to make his art viable, to attain by it a kind of immortality, is a lesson of poetry. Proteus in fact does not tell him how to regain his bees: that practical matter, the | religious rite of expiation, is left to Cyrene. Proteus instead sings a song, and the song is about the singer *par excellence*, Orpheus. The grief of Orpheus is irredeemable, despite the regeneration of bees that follows it, except that it becomes the subject of song; and the song in turn becomes the condition for the recreation of life. The grief is elevated to the highest art, and in that art, the epitome of all human art and craft, lies the true immortality of the poem, the resolution of man's confrontation with the absolute of death.

17

Thucydides' Historical Perspective

THUCYDIDES' *History of the Peloponnesian War* is an intensely
personal and a tragic work. A careful reader feels this from the
very first sentence: '. . . I began writing the History from the
moment the War broke out; I expected it to be a great war and
more worth a λόγος than any war that had preceded it.' This
tone is maintained throughout the work. Even if we leave aside
the dozens of personal judgements in the *History*,[1] its intensity
of feeling everywhere reminds us of Thucydides' personal
involvement.

We learn from ancient criticism that Thucydides was
admired as the historian of πάθος, as opposed to Herodotus, the
historian of ἦθος. The sense of the tragic, which exists as a fine
suffusion in parts of Herodotus' work, dominate the
whole *History* of Thucydides. This sense of the tragic is

[1] A number of these are well-known to everyone at all acquainted with
Thucydides; e.g. the comment on war as βίαιος διδάσκαλος in the account of the
Corcyrean revolution, 3. 82. 2. More personal, and less well-known, is the comment
he makes later in the same passage on the moral degeneration which revolutionary
activity brought upon all of Greece, 3. 83. 1: οὕτω πᾶσα ἰδέα κατέστη κακοτροπίας
διὰ τὰς στάσεις τῷ 'Ελληνικῷ, καὶ τὸ εὔηθες, οὗ τὸ γενναῖον πλεῖστον μετέχει,
καταγελασθὲν ἠφανίσθη. No less personal, in a different way, is the unexampled
rhetorical question in 7. 44. 1, where the historian, with all the impatience of an
intensely orderly mind, comments on the essentially confused nature of all military
actions, even those that take place in broad daylight, where even the individual
participant, like Tolstoi's Captain Tushin, hardly (μόλις) knows what is going on
around himself, and then asks how anyone can have any clear knowledge (πῶς ἄν τις
σαφῶς τι ᾔδει;) of a battle by night, such as Demosthenes' attempt to storm Epipolae.
A valuable close study could be made of the revelations of personal attitude in the
History. Such a study would comment on the author's tendency to use distancing,
'scientific' language at points where emotion is strongest, as in οὕτω πᾶσα ἰδέα etc.,
above; and on the reasons for the many evident exaggerations in the text; e.g. 5. 26. 3:
the prophecy that the war would last thrice seven years was the only one in the whole
course of it that came true; or the estimate (3. 98. 4) of the 120 hoplites who fell in the
battle against the Aetolians in 426.

Yale Classical Studies, 22 (1972), 47–61. Reprinted by permission of Cambridge
University Press.

something quite different from the clinical objectivity which
has been so often, and often so thought|lessly, ascribed to him.[2]
His very reluctance to speak of himself, his way of stating all as
an ultimate truth, is, if we must use the word, one of his most
subjective aspects. When you can say, 'so-and-so gave me this
account of what happened, and it seems a likely version', you
are objective about your relation to history. But when,
without discussing sources, you present everything as αὐτὰ τὰ
ἔργα (1. 21. 2), the way it really happened, you are forcing the
reader to look through your eyes, imposing your own assumptions and interpretations of events. To say all this is of course
not to cast doubts on Thucydides' veracity or on the validity of
his methods of inquiry, little as we know of them.

The reasons for Thucydides' personal involvement are
evident enough. He was a passionate admirer of Periclean
Athens. 'The devout disciple', Wade-Gery calls him.[3] When he
has Pericles say (2. 43. 1), 'You must each day actually
contemplate the power of the city (τὴν τῆς πόλεως δύναμιν
καθ' ἡμέραν ἔργῳ θεωμένους) and fall in love with her (καὶ
ἐραστὰς γιγνομένους αὐτῆς), and when you grasp the vision of
her greatness (καὶ ὅταν ὑμῖν μεγάλη δόξῃ εἶναι)', and so on;
when he has Pericles speak in this vein, can we doubt that he
was, and as he writes is in retrospect, one of those who heard
Pericles' words with willing ears? Now compare this passage
with a famous one from the 'Archaeology'. This is 1. 10. 2,

---

[2] The notion of Thucydides as the passionless scientific gatherer of facts goes back
to the positivistic interpreters of the nineteenth century (e.g. Gomperz, *Griechische
Denker*[2], 1, 401 f.), and despite protests like those of F. M. Cornford (Preface to
Thucydides Mythistoricus, p. vii: 'Xenophon, I suppose, is honest; but his honesty
makes it none the easier to read him'), has become the standard handbook view,
much enforced by the double and doubly dubious equation Thucydides = ancient
medical writers minus modern medical research methods. See A. Parry, 'The
Language of Thucydides' Description of the Plague', *University of London Institute of
Classical Studies Bulletin* 16 (1969), 106–17 [ch. 13 of the present volume]. Wade-Gery
in the *Oxford Classical Dictionary* shows an ambiguous attitude characteristic of much
modern judgement. On p. 904, he says finely: '[Thucydides] uses a language largely
moulded by poets: its precision is a poet's precision, a union of passion and candour';
but shortly thereafter falls into a slightly sentimental version of the old cliché:
'Thucydides would no doubt prefer to substitute, for those great names [W.-G. has
compared him to Shakespeare and Marlowe], the practice of any honest doctor.' The
conception of the self-effacing doctor is as alien to Thucydides' aims as to his practice.
One need only consider the sense of rivalry which informs his attitude to Homer as
well as Herodotus.

[3] In *PCPhS* (1953).

where Thucydides interrupts his account to speculate on the |
possibility of utter destruction of Athens and Sparta. Later
generations would not guess, from her meagre foundations,
how powerful Sparta had once been; and if the same thing
were to happen to Athens— Ἀθηναίων δὲ τὸ αὐτὸ τοῦτο
παθόντων—her power would be judged to be twice what it is,
from the evident appearance of the city. Once again we have a
vision of Athens—*circumspice!*—but in how very different a
perspective. If Thucydides, as I believe, wrote 1. 10. 2 along
with the rest of the 'Archaeology', after 404, or if he wrote it
earlier but let it stand in his final version,[4] he is not only
making a good logical point: he is also indicating the perspective from which he is writing the *History*. Although not in the
literal sense envisaged in 1. 10. 2, Athens has been destroyed,
her greatness has vanished. The transition from the first
passage, Pericles' words in 2. 43. 1, to the vision of destruction
in 1. 10. 2 marks Thucydides' experience of the Peloponnesian
War. For him it was the end of the world, after the world had
reached its high point. This experience must be seen as the basis
both of his dramatic presentation and of his theory of history as
he had worked these out in the text we have.

Thucydides' final theory of history is one which he can only
have evolved after the defeat of Athens. This is evident enough
if we read 1. 23. 1–3, where he sums up the conclusions of the
'Archaeology'.

Of former actions, the greatest was the Persian (τῶν δὲ προτέρων
ἔργων μέγιστον ἐπράχθη τὸ Μηδικόν) and yet this in two battles by
land and two by sea had a swift conclusion. But of this war the
duration was great, and disasters to Greece took place in it (παθήματά
τε ξυνηνέχθη γενέσθαι ἐν αὐτῷ τῇ Ἑλλάδι) such as no others in an
equal space of time. Never were so many cities captured and made
empty of their inhabitants, some by the barbarians, some by the
Greeks themselves as they fought against each other; and there were
those that changed their populations on being captured. Never were
there so many exiles and so much slaughter, slaughter in battle,
slaughter in civil war. Things which formerly had been known by
story only, but had been | rarely attested in fact, now ceased to be

[4] The two alternatives come, practically speaking, to the same thing. The
'analytic' view becomes of interest only if it can be argued that the text was put
together without the author's intent.

incredible, earthquakes, which were at once the most extensive and the most violent of all history, and eclipses of the sun, which came with a frequency beyond any recorded in earlier times, and in some places droughts, and from them famine, and that not least worker of harm and in part utter destroyer, the death-dealing Plague. And all these things were the accompaniments of this War.

Thucydides' vision of history is of greatness measured by war, and greatness of war measured by destruction, or πάθος.[5] This vision is a product of Thucydides' own experience. Unlike all the other great historians of the ancient world, he writes of the events of his own lifetime. He is so strongly concerned with this experience, that he has by modern scholars been accused of having no understanding of the past, or of regarding it with contempt.[6] To some degree, that is so; but we can conjecture from the *History* itself that he did not begin with this perspective.

It is likely, on the contrary, that Thucydides, as a young man, began with a genuine interest in the past, and did researches in the Herodotean manner. Witness the vestiges of these researches in the Cylon, Themistocles, and Pausanias episodes and in the excursus on Harmodius and Aristogeiton.[7] When the War breaks out, he decides to devote the time he can spare from the affairs of the City either largely or wholly to recording its progress. He does so because he sees that it will be a great war or the greatest of wars, | but he does not yet see it in terms purely of destruction. On the contrary, he must have

---

[5] The end of Book 7 illustrates this principle again. 'The Sicilian ἔργον [87. 5] was the greatest of this war, and of Greek history, most glorious to the victors, most unfortunate to the defeated.' The two poles of glory and suffering seem balanced, for a moment, and we might have a Herodotean view. But Syracusan triumph has throughout Book 7 been pale next to Athenian grief; and here in the splendid final sentence that follows the one quoted, he talks only of the extent of the disaster for the vanquished.

[6] Cf. Collingwood, *The Idea of History* (Oxford, 1946), e.g. p. 30.

[7] A. Momigliano (*Memoria d. R. Accad. d. Scienze di Torino* 68 (1930)) seems to have been the first to make the suggestion that there are early essays incorporated into the text. In the work we have, all four 'digressions' play an important part in the structure of the whole. See recently, on Harmodius and Aristogeiton in Book 6, Stahl, *Thukydides* (Munich, 1966), ch. 1. Stahl, however, sees in this episode only a general point. The obvious relevance of the story in the *History*, *pace* Dover, *Historical Commentary on Thucydides*, ii (Oxford, 1970), 329, is to enforce the historian's point that the Athenian δῆμος was characteristically fatuous and self-destructive in rejecting Alcibiades.

been hopeful of its outcome—or at least that is the mood of 431 as he dramatizes it in the *History*. We may even go so far as to suppose that he includes himself in the wry comment in 2. 8. 1 about the many young men both in the Peloponnese and in Athens who because they had no experience of it were not reluctant to make themselves part of the War. He was young, he believed Pericles, and he was a keen professional soldier. The long course of the War changes his mind. By its end he has become convinced that this war is so final a version of the historical process as to supersede all preceding events, and that the greatness of historical events is measured by their power to destroy. He might have said of it, as he does of the Plague, εἰ δέ τις καὶ προύκαμνέ τι, ἐς τοῦτο πάντα ἀπεκρίθη—'all earlier disasters ended in, were subsumed by, this one'. He therefore can see its structure only when it is past, when all there is to lose is already lost. Then he can write of the loss and of what was lost for the benefit not of his contemporaries, but of men of some later civilization who will thereby be better enabled to understand the destruction of their own (1. 22. 4). So he does write about the past, but his own, the experienced past; and he has no great interest in earlier events, because they after all only led up to this one.

The purpose of the 'Archaeology', that is, of chs. 1–22 and their summary in ch. 23 of Book 1,[8] is to state and develop his theory of history and thereby to justify his exclusive concern with the Peloponnesian War. For all his famous obscurity, Thucydides' style is such as to make the patterns of his thought very evident. Taking over the devices of the Sophists and turning them to an individual use, he writes an exposition in which ideas and events are strongly marked by key terms. These key terms are semi-abstract nouns and verbs designed to distil the elements of experience into an articulate pattern. He establishes the relation between judgement and fact in the first sentence, a first example of that pervasive contrast of λόγος and ἔργον which dominates | his work and is what we might call its

---

[8] The 'Archaeology' is of course not Thucydides' term. As modern critics have used it, it refers strictly to chs. 2–19 of Book 1. But 23 in fact refers back to 1. 3, elaborates 18. 2 in the Persian Wars, and makes in final form the point of the whole beginning of the work, so that 1–25 is an obvious unit.

central metaphor.⁹ The great fact is the War, τὸν πόλεμον. It is the supreme ἔργον; as in 1. 23. 1, Thucydides often uses the word ἔργον as a synonym of war or battle.¹⁰ The other side is the judging intellect, the intellect that can give a conceptual shape to events, and that is expressed in the word ἀξιόλογον: I expected this war to be ἀξιολογώτατον τῶν προγεγενημένων. This contrast is maintained throughout the 'Archaeology'. ἐκ δὲ τεκμηρίων ὧν ἐπὶ μακρότατον σκοποῦντί μοι πιστεῦσαι ξυμβαίνει (1. 1. 3). The facts are now the evidence, and the intellectual judgement is expressed by σκοποῦντι, σκοπεῖν being one of Thucydides' favourite terms. Analogous terms of intellectual discernment occur 23 times up to the end of ch. 21. The historical facts which make up the object of intellection appear primarily as words meaning *power*. History in fact is movements of power. Thus forms of the word δύναμις occur 10 times to the end of ch. 21 and if we include synonymous expressions, e.g. δυνατός, δύνασθαι, ἰσχύς, βιαζόμενοι, ἐκράτησαν, κρεισσόνων, etc., we have a count of 35. Two other terms which by repetition assume a special function are, first, compounds of ἵστημι, usually in the middle voice, which are regularly used to mark significant qualitative changes in the historical situation, and second, the adjective μέγας, which, in contrast to Herodotus' glorifying use of it—ἔργα μεγάλα τε καὶ θωμαστά—means size of power as measured by size of war.

He argues that the earlier Greeks—meaning by this, I believe, all earlier generations down to the Peloponnesian War—did not have comparable greatness, 'either in wars or in other matters' (1. 3). Note that in this sentence greatness in every other sphere of life is made subordinate to greatness in war. There was no greatness because there was no power, and it soon becomes apparent that power means order, because he at once begins a description of earliest times (2. 2), where with disconcertingly inconsistent syntax he describes the disorganization of that period. He dramatizes this state by his style, moving rapidly from genitive absolute to a series of nominative plural participles, to a nominative absolute to another genitive absolute, with subordinate clauses of | varying kinds in

⁹ Cf. my Harvard doctoral dissertation (1957), 'Λόγος and Ἔργον in Thucydides'.
¹⁰ e.g. 1. 23. 1: τῶν ... προτέρων ἔργων ... τούτου δὲ τοῦ πολέμου; 1. 49. 7.

between, before he finally comes to the main verb ἀπανίσ-ταντο, the imperfect of ἴστασθαι to mark the frequency of change of historical situation which prevented any order from being established. The sentence contains several key terms. There was no *communication*, no *capital*, 'so men *shifted* their dwellings easily, and so had *strength* neither in the size of their cities nor in other *material means*'.[11]

This is the beginning of history; one might almost say, man in a state of nature. Thucydides describes it in negative fashion, listing those appurtenances of civilization which were lacking, things deriving from, and adding up to, power. The effect of power and resource (παρασκευή) is first to create order. Minos was first to *get a navy* (4. 1) (ναυτικὸν ἐκτήσατο), and thereby he *got power* in the Aegean Sea (τῆς νῦν Ἑλληνικῆς θαλάσσης ἐκράτησε); he *established* his sons (ἐγκαταστήσας) as rulers in the islands, and set about clearing piracy off the seas (τό τε λῃστικὸν ... καθῄρει) so that he could get *revenue* (τοῦ τὰς προσόδους μᾶλλον ἰέναι αὐτῷ). This establishment of order by removing piracy made possible the accumulation of *capital*: 'The establishment of Minos' navy made it possible for those living close to the sea to *amass money* and achieve security, and some began to surround themselves by *walls*.'

The missing elements of civilization begin to be filled in. And so on through the 'Archaeology' a series of civilizations, as power, are established, each with its elements of ships, capital, walled cities—the features, obviously, of Athens in 431—and, the final transformation, war. Thus Agamemnon (9. 1) is superior in power (δυνάμει προύχων), and this, not the legendary oath to Tyndareus, enabled him to gather the expedition against Troy.[12] The way in which the Mycenaean rule came to be inherited by Agamemnon is then described in 9. 2. The foundations of that rule were laid by Pelops, who by having a *supply of money* (πλήθει χρημάτων) *built up power for*

---

[11] Παρασκευή is often translated by specific words like 'equipment'; in Thucydides' system it assumes a much larger meaning.

[12] This famous interpretation of Homer is often cited to illustrate the historian's critical powers. It shows the boldness of his thought, but also how limited a reader of Homer he was. Homer does not mention the oath to Tyndareus; and the *Iliad* makes it clear that the Greek warrior-princes fought primarily for booty and to maintain their position in society.

himself (δύναμιν περιποιησάμενον). In 9. 3 we read that |
Agamemnon took over this rule (ἃ ... Ἀγαμέμνων παρα-
λαβών), acquired more *naval power* than anyone else (ναυτικῷ
... ἐπὶ πλέον τῶν ἄλλων ἰσχύσας), and so was able to make the
attack on Troy.

And the Tyrants (13. 1): 'As Greece became more powerful
(δυνατωτέρας) and more and more engaged in the acquisition
of money (τῶν χρημάτων τὴν κτῆσιν), tyrannies began to be
established (καθίσταντο) in the cities; revenues (τῶν προσόδων)
increased; and Greece began to provide herself with navies
(ναυτικὰ ... ἐξηρτύετο).'

And finally, Sparta and Athens (18. 1–2): 'The Spartans,
having power (δυνάμενοι, used absolutely), established (καθίσ-
τασαν) governments in other cities; while the Athenians
equipped themselves (ἀνασκευασάμενοι; cf. παρασκευή), took
to their ships and became a sea power (ἐς τὰς ναῦς ἐσβάντες[13]
ναυτικοὶ ἐγένοντο).' Then after the Persian Wars (18. 2) all of
Greece ranges itself on one side and the other (διεκρίθησαν).
'For these were greatest in power; the might of the one was on
land, the other in ships.'

Thucydides sees these establishments of order and power as
admirable, and his style communicates this admiration to his
readers. The severe impetus of that style, where words for
power and *force* continually spring up to dominate the order of
the sentence, enforces the sense that the creation of this sort of
dynamic sovereignty is the most serious pursuit of man. I say
creation, because each of these civilizations, these complexes of
power, is seen as an order imposed by human intelligence. The
notion that civilization is a product of the human mind, rather
than of institutions and laws vouchsafed to man by the gods, is
a characteristic Sophistic concept, and no one expresses it more
clearly than Thucydides. Sea-power in particular, as we shall
see, is an aspect of the intelligence, and the growth of this,
culminating in the Athenians' becoming entirely nautical
(ναυτικοί), parallels the development of civilization in general.

---

[13] The phrase ἐς τὰς ναῦς ἐσβάντες looks like a standing slogan in Athenian
imperial apology, the kind of thing Pericles says he will not indulge in in the Funeral
Speech, 2. 36. 4. Cf. the Athenians at Sparta, 1. 74. 2. From such pat Athenian self-
advertisement Thucydides constructed his historical system.

But there is so far one essential point missing in Thucydides' account of history. That is, destruction, πάθος. The reason lies in the importance he attributes to Athens and to the Peloponnesian | War. He presents us, in the 'Archaeology' as a whole— that is, including ch. 23—with two historical curves, two lines of historical development. One is the rise and fall of a series of civilizations. The Empire of Minos had to dissolve before the Empire of Agamemnon could be established, and Agamemnon's had ceased to exist by the time of the Tyrannies, while these in turn were variously undone to make way for fifth-century Athens and Sparta. In terms of this historical curve, which could be represented as a periodic curve on the graph of history, the Empire of Athens (let us for the moment forget, as Thucydides often does, about Sparta) is but one term in an endless series, perhaps the largest term so far, but still not a unique point, not the convergence of all history. The second historical curve is a line of continuous development, ignoring minor ups and downs, from earliest times (1. 2. 2) to Athens in 431 BC, when, with what has been blamed as exaggeration,[14] Thucydides says that her individual power exceeded that of Athens and Sparta together when their alliance at the time of the Persian Wars was at its height. It is this second curve which makes Athens and the fall of Athens into what I have called the final version of the historical process. The rise and fall of earlier empires must accordingly be seen as steps upward, and so he stresses their rise only, casually alluding to such matters as the confusion and faction attendant on the return of the Greeks from Troy, which could have been presented as the calamitous dissolution of Agamemnon's realm. Rather than this he stresses the creativeness of the early empires, presenting all history as a single trajectory, reaching a height in Periclean Athens, and coming to an end with the close of the 27-year war. The ruin of all empires is subsumed under that of Athens.

1. 23. 1–3 is the inevitable and fitting summary of the scheme of history which Thucydides has developed through-out the 'Archaeology'. Civilization is the creation of power and is splendid and admirable, but it inevitably ends in its own destruction, so much so that this destruction is virtually the

[14] Cf. Gomme, *Historical Commentary on Thucydides*, i (Oxford, 1945), 134.

measure of its greatness. And all this is a pattern which Thucydides finally worked out *after* the defeat of Athens in 404, and it expresses his personal experience.

But some questions remain. Is the process absolutely inevitable? And if so, how are we to regard the historian's presentation of | Pericles and Pericles' policy? And how does Sparta fit into the scheme?

The answer to the first question seems to be Yes. For one thing, Thucydides has Pericles himself say so in a beautiful passage from his last speech (2. 64. 3):

Know that Athens has the greatest name among all men because she does not yield to disasters, because she has expended most labour and lost most lives in war, because she has acquired the greatest power in all history; and the memory of that power will be left eternally to succeeding generations, even if we should now sometime give way; it is the nature of all things to decline. They will remember that as Greeks we ruled over most Greeks, that we fought against others singly and all together in the greatest wars, and that we had the city richest in all things, and the greatest.

$\Delta \acute{\upsilon} \nu \alpha \mu \iota \varsigma - \mu \epsilon \gamma \acute{\iota} \sigma \tau \eta - \pi \acute{o} \lambda \epsilon \mu o \varsigma - \mu \acute{\epsilon} \gamma \iota \sigma \tau o \varsigma - \dot{\epsilon} \lambda \alpha \sigma \sigma o \hat{\upsilon} \sigma \theta \alpha \iota -$ $\mu \nu \acute{\eta} \mu \eta$ (a memory only)$-\mu \epsilon \gamma \acute{\iota} \sigma \tau \eta$: here we have all the essential elements of the Thucydidean scheme, and expressed by the statesman who, Thucydides tells us in the next chapter, had such justified confidence in victory.

Other considerations too enforce this sense of inevitability. First the comparison with the Plague. Strong verbal echoes confirm our sense that the Plague is presented as a kind of concentrated image of the War.

The word $\dot{\epsilon} \pi \iota \pi \epsilon \sigma \epsilon \hat{\iota} \nu$, 'to fall upon violently', which is used of the Plague (2. 48. 3), is used again of the inevitable effects of war in the description of the revolution in Corcyra: 'things many and terrible befell the cities of Greece in the course of revolutions, things which happen and always will happen as long as the nature of man remains what it is' (3. 82. 2). And he uses in his description of the Plague the same word $\sigma \kappa o \pi \epsilon \hat{\iota} \nu$, 'to discern', that he had used in 1. 22. 4 of his description of the whole War, where he expresses the hope that his work will be judged useful 'by those who shall want *to see clearly* what happened in the past and will by human necessity (κατὰ τὸ ἀνθρώπινον) happen in the same or in similar fashion in the future'. Of the Plague he says, in 2. 84. 3, 'I shall confine myself to describing what it was like [instead of offering either explanation or cure], and to putting down such | things as a man may use, if it should strike again, to *see it clearly*, and to recognize it for what it is'.

Finally, the connotations of Thucydides' basic terms imply the inevitability of the process his *History* describes. I have spoken already of his use of λόγος and ἔργον and of equivalent terms as a kind of fundamental metaphor in his historical presentation. I use the word *metaphor* advisedly, if we recall Aristotle's statement that the use of metaphor involves the perception of similarity in apparently dissimilar things. Following out his notion that the course of civilization and thereby of all history is man's imposing an intellectual order on the world outside him, Thucydides continually makes a division in his presentation of history between words meaning judgement or speech or intention, etc., and others meaning fact, thing, resource, power, etc.

On one side of the constantly repeated and endlessly varied opposition, we find e.g. λόγος, γνώμη, διάνοια, ἀκοή, ὄνομα, ἐλπίς, διδαχή, μέλλοντα; on the other, ἔργον, παρασκευή, μελέτη, δύναμις, βία, φύσις, ὄντα, παρόντα. The point of this whole terminological system is to present history as man's constant attempt to order the world about him by his intelligence. Each actor within the historical drama attempts to formulate, present, and enforce his own interpretation of external events and situations. This is done in speeches, and accordingly Thucydides, in 1. 22. 1–2, divides his whole work into 'what [the participants in the War] said *in speech*' (ὅσα λόγῳ εἶπον: λόγῳ dative singular of category) and 'the reality of what was done' (τὰ ἔργα τῶν πραχθέντων). But both of these as seen from Thucydides' own point of view are past actions and hence both fall under the heading of αὐτὰ τὰ ἔργα of the War at the end of the preceding chapter. And the λόγος that matches αὐτὰ τὰ ἔργα, in its widest sense of all actions and speeches in the War, is, of course, Thucydides' own *History*.

Of all the words on the fact–external reality side of the

opposition, ἔργον is by far the most common and has the widest range of meaning. It is a fundamental Greek word and means anything *wrought* or *done*, *work* being its obvious cognate. Thus it can mean *deeds of war* (the Homeric πολεμήια ἔργα), or, very commonly in Thucydides, a *battle*, or the whole business of a war, as in 1. 23. 1, where he begins τῶν δὲ προτέρων ἔργων, and then continues | τούτου δὲ τοῦ πολέμου; or 1. 80. 1, where Archidamus, urging the Spartans to caution, says that he and his contemporaries have too much experience of wars for anyone to long to be in one: ὥστε ἐπιθυμῆσαί τινα τοῦ ἔργου.

But then there is a slightly different direction in the meaning of ἔργον, whereby it stands for *fact, reality*, the thing that was *actually done*. This is the nuance of meaning that makes it appropriate for the common fifth-century idiom wherein λόγος and ἔργον are distinguished: 'He *says* such and such, but *actually* . . .' Some of this stretch of meaning is in the English word *deed*. We can speak of 'deeds of war' and at the same time have an adverb 'indeed'; and the sinister nursery rhyme 'A Man of Words and not of Deeds' joins the two meanings. The man does not perform deeds and in some way he lacks reality.

So Thucydides, by using the word ἔργον in a great variety of contexts, and in associating with it, by a series of antitheses, other words such as δύναμις and πόλεμος, is indicating, building the notion into the structure of his language, that power and war are simply aspects of reality. War is the final reality. There can be no civilization, no complex of power without war, because the one word implies the other.

If it be objected that I am playing a word-game here, the answer is that it is Thucydides' own word-game, and that he uses it to express an interpretation of history that he makes explicit in other ways as well.

The other unanswered questions are the role of Pericles and that of Sparta. They can be answered together. Throughout the *History*, Thucydides presents the Athenian character as dominated by λόγος and the Spartan character as dominated by ἔργον. In two Spartan speeches, those of Archidamus and Sthenelaidas in Book 1 (80–5 and 86) and in one Corinthian speech, that at the Second Congress in Lacedaemon, also in

Book 1 (120–4), the Spartans and their allies are characterized as distrusting the intellect and putting their faith in fact. 'We are trained to believe', Archidamus says in 84. 3, 'that the chances of war are not accessible to, cannot be predicted by, human reason' (οὐ λόγῳ διαιρετάς: the last word has interesting philosophic associations).[15] 'And so', he goes on, 'we put our faith in strict | discipline' (μελέτας 85. 1). Sthenelaidas is more brutal: τοὺς μὲν λόγους τοὺς πολλοὺς τῶν Ἀθηναίων οὐ γιγνώσκω, he begins in 86. 1. He means (1) 'I choose to ignore the protracted speech of the Athenians in defence of their Empire', and (2) 'As a Spartan I reject the use of speech and reason and urge immediate recourse to fact; that is, to war'.

By contrast the Athenians and Pericles in particular urge that reason is the indispensable preliminary to action. 'We differ from other men', Pericles says in 2. 40. 2, in the Funeral Speech, 'by not believing words harmful to action, but rather that harm lies in not working out beforehand in words what must actually be carried out.' As everywhere in his great speech, Pericles worries the distinction for all it is worth. The Athenians' τέχνη and ἐπιστήμη, intellectual words, are several times contrasted with the Spartans' μελέτη and ἀλκή, words of institution or instinct. The contrasted speeches of the Peloponnesian generals and of Phormio in 2. 87 ff. are another good example of such contrast.

The implications of this much-elaborated opposition between the two national characters are something like this. Inasmuch as civilization is the successful imposition of intelligence on the brute matter of the outside world, Athens, not Sparta, represents civilization. The Athenians in fact are the moving force throughout the *History*. They, from the moment they followed Themistocles at the time of the Persian Wars and took to their ships—this itself an act of the creative intelligence: 1. 18. 2 διανοηθέντες ἐκλιπεῖν τὴν πόλιν καὶ ... ἐς τὰς ναῦς ἐσβάντες—from that moment on it is they who have created an Empire far greater than those of Minos and Agamemnon. It is they who have changed the map and the character of Greece. They, in the formation of history, are a sort of second cause alongside of intelligence, as Plato, in the

[15] Cf. Diels, *Fragmente der Vorsokratiker* III, s.v.

Timaeus (47 E 2 f.), added Necessity or the Wandering Cause to the Demiurgic Mind. They are that incommensurate, irrational factor in reality which makes it sure that ultimately you can never win; what corresponds in the large scheme of history to what Thucydides calls παράλογος, that which the keenest intelligence cannot foresee.

In Thucydides' scheme therefore, there is only one civilization in 431 BC, that of Athens. Its power, created by intelligence, inevitably becomes war: 'As the Athenians became great, the Spartans were compelled to war' (1. 23. 6), and eventually this | war destroys the civilization that brought it about. The Spartans are merely the external agents of this destruction. Pericles is the essence of Athenian intelligence. The word constantly attached to him is γνώμη. He is that aspect of intelligence which will not yield to the pressures of external reality. The reason the Athenians could have won the war if they had followed his judgement throughout is that this judgement is presented as transcending the vicissitudes of actual events. 'I continue to hold to the same conception (γνώμη),[16] citizens of Athens, not to yield to the Peloponnesians . . .' So begins his first speech, the first words he speaks in the *History*.

The same assertion of unwavering judgement in the face of the παράλογοι of reality dominates Pericles' last speech after the Plague. In general, men's conceptions, and the words they use to express them, vary with events, and alter with every alteration they find. Pericles alone is above this, and hence Thucydides attributes to him an almost superhuman judgement, which he asserts could have carried the Empire through to victory if the Athenians, who to Pericles are part of the recalcitrant matter of history, had been able to follow it.

It may fairly be objected that Thucydides has no right to insist on the inevitability of the fall of empire on the one hand and on the invincibility of Periclean policy on the other. It is the great paradox of his work and a point where his system seems to break down. Two considerations should modify this criticism. One is the peculiar nature of the Athenian Empire as

[16] e.g. 1. 140. 1; 2. 12. 2; 34. 6; 43. 3; 59. 2 (comments on Periclean γνώμη by implication); 62. 5 (possibly the most revealing instance); 65. 8.

Thucydides has Pericles conceive it. The Athenians do not
merely use sea-power to build their realm: they become almost
entirely identified with sea-power: they 'became nautical'.
Inasmuch as sea-power is especially the creation of the intelligence, we have here a vision of the Athenian Empire as pure
product of the mind, and consequently inaccessible to those
elements in the world which the mind cannot control. 'Consider', Pericles tells his countrymen in his First Speech (1. 143.
5), 'if we were islanders, who could be more safe from the
enemy? ... You must *approach this conception* as close as
possible, and let your homes and lands go.'[17] This vision of
Athens | as a power so completely created by the intellect as to
be proof against the waywardness of reality is almost fantastic;
and yet perhaps true to the historical Pericles' own imagination.

The second consideration is deeper. It is the foreknowl-
edge of Athens' defeat which Thucydides attributes to Pericles
in his Last Speech. There the historian suggests that there is a
valid sense in which it does not matter whether Athens falls or
not, because the quality of her memory will remain; and that
Pericles was clearly aware of this sense.[18] As conception in the
present becomes fact in the past, so fact in the past, in this case
the uniqueness of Athenian power at one moment of history,
stays alive in the present as concept. In this way, Periclean
Athens does escape the grim system which Thucydides develops as the intellectual foundation of his narrative. Because
Athens under Pericles remains an ineffaceable image in the
mind, the city is truly invincible, and to fix this image is
precisely the purpose of Thucydides' account.

---

[17] ὅτι ἐγγύτατα τούτου διανοηθέντας. Cf. 1. 18. 2 οἱ Ἀθηναῖοι διανοηθέντες ἐκλιπεῖν
τὴν πόλιν καὶ ... ἐς τὰς ναῦς ἐσβάντες, and cf. n. 13 above.

[18] Esp. 2. 64. 3, the great expression of imperial heroism. The same note is struck in
2. 43. 3, where some of the language, especially ἀνδρῶν γὰρ ἐπιφανῶν πᾶσα γῆ τάφος
is rhythmical, metaphorical, and lapidary to a degree that, historically or not,
distinguishes Pericles from other speakers in the *History*, and sounds almost like a
quotation. The prophetic element in these and other passages of Pericles' last two
speeches has been judged obviously anachronistic by scholars like Kakridis (*Der
thukydideische Epitaphios* (Munich, 1961), 5 f., etc.). We shall never know what
Pericles really said on such occasions. But it does not seem to me impossible that he
said something like what Thucydides has attributed to him. Shall we confidently
deny Pericles both foresight and a tragic sense of life? If Kakridis and others are right,
they must not only reject the historicity of these statements in the text as a document;
they must also find no verisimilitude in the text as a dramatic work.

18

Language and Characterization in Homer*

Meaning in the Fixed Epithet

THE so-called oral theory of Homeric poetry has raised serious problems for anyone concerned with critical analysis of the *Iliad* and *Odyssey*. Of the rigorous investigations of Milman Parry, the author of the modern form of this theory, I will not speak here, and I will assume that his examination of noun-epithet formulae and his proof of the remarkable economy of these formulae are known to you. His conclusion, that only a language designed for composition in the process of recitation would show such an economy, has at the least an extremely high degree of probability; and his own and subsequent studies of living oral poetries have on the whole corroborated this conclusion. But it is worth while to state again what Milman Parry's investigations have proved: viz., that the *style* of the *Iliad* and *Odyssey*, as the poems have come down to us, shows many features of a style originally created for oral composition. That the poems themselves, as we have them, were in fact composed in the process of improvising recitation has not been proved, and probably cannot now be proved. Such a notion must remain, for the many scholars who now hold it, a plausible speculation only.

But regardless of the question of the specific conditions under which the *Iliad* and *Odyssey* were composed, the detailed

* This article is printed from the typescript of a lecture which the late Adam Parry delivered in London in 1969. He did not live to remove a few phrases indicative of its original form or equip the article with footnotes incorporating references and elaborating suggestions in the text, but in submitting it for publication in *HSCPh* he had expressed his intention of doing so. *Editor's Note*

Harvard Studies in Classical Philology, 76 (1972), 1–22. Reprinted by permission of Harvard University Press.

exposition of the way in which their style depends on set
phrases which tend conservatively to be used whenever they
can, raise serious critical problems. These problems centre
round the elusive and yet vital question of the *consciousness* of
Homer's audience. That is, we must be prepared somehow to
answer the query, of how much significance should the ideal |
member of Homer's audience be aware? Homer's audience, for
the purposes of this question, will include us as well as the
poet's contemporaries. Do the set phrases in which the poetry
so largely consists have a meaning dependent on the individual
words which are their ingredients? Or does the formulary style
preclude such meaning, so that these phrases are in operation
equivalent to single words? Such questions are hard to answer,
but they are vital, because the answers to them will determine
the entire way in which the poetry will be read; they will
determine the meaning of the poetry itself.

Milman Parry's demonstration of what we might call for
short the oral affinities of Homeric style does not in itself
answer these questions. To them, however, Milman Parry
himself and many of his successors have offered an answer: that
the words within those expressions which can, by the repetition of their use, be established as set phrases, or formulae, do
not bear an individual meaning. They are inseparable parts of
the whole phrases, and the consciousness cannot focus on them.
The question then arises, how does the phrase composed of
these words differ from a phrase composed of other words, or
how does it differ from a single word which could be used to
paraphrase it?

In the seventh line of the *Iliad*, the poet names two men in
the nominative case to be the subject of the dual verb
διαστήτην in the preceding line. Line 7 reads:

Ἀτρεΐδης τε ἄναξ ἀνδρῶν καὶ δῖος Ἀχιλλεύς.

In what sense do the words of this line differ from the words
Ἀτρεΐδης τε καὶ Ἀχιλλεύς? The obvious answer, that the actual
words of line 7 form a hexameter, is trivial, because in that case
a meaningless sound could have replaced the epithetic expressions. Milman Parry's answer was that they differ only by
invoking the world of heroic poetry. To call Agamemnon

ἄναξ ἀνδρῶν and Achilles δῖος tells us nothing about these men; it tells us only that this is epic song, that this is an imaginative world where all men are entitled to epithets of this kind. Milman Parry supports this view by pointing to the large number of heroes, some major, many minor, who are qualified by δῖος and to the smaller number, but again including minor characters, who are called ἄναξ ἀνδρῶν. Since these epithetic expressions are not used of these men alone, they cannot serve to tell us anything unique about them. If they tell us nothing unique about them, they tell us nothing about them at all, but only that we are listening to Homer, and are hence in a heroic and kingly world. |

Milman Parry's argument was an appeal to experience. δῖος occurs ninety-one times in the *Iliad* and qualifies ten different heroes, although in the great majority of cases it is used of Achilles and Odysseus. Long before we have heard or read this word ninety-one times, he argued, we have ceased to inquire into, or even to be aware of, its meaning. Similarly we learn to assign no particular meaning to ἄναξ ἀνδρῶν long before we have encountered its fifty appearances, of which forty-five are for Agamemnon, the other five, one each, to five other heroes.

There may be a danger, however, in using experience to contradict experience. It is true that a reasonable man who has read all of the *Iliad* is not going to base a critical interpretation of the poem solely or chiefly on any occurrence of a word so commonly and freely used as δῖος, or of an expression like ἄναξ ἀνδρῶν. He learns not to stop at these expressions and not to pay them undue attention. But he will, I think, no matter how many times he has read Homer, have some feeling on beginning the poem and coming to line 7, which grandly ends the first sentence and the invocation, that Agamemnon and Achilles are being rather well described. He will, despite the strictures of this school of modern scholarship, think a bit about ἄναξ ἀνδρῶν and δῖος. He will here already to some degree be aware of the contrast between the two on which so much of the quarrel depends: Agamemnon, the man of political power and the public figure, complete with much of the insecurity which an aspirant to the highest public position can possess; and Achilles, the man whose values are more

purely heroic and individual, and become increasingly personal and private as the story progresses.

Of course the reader will not spell this out as he reads the
beginning of the poem, as I have done. But he will be aware
that the adjectives are right for each man. For Achilles, the
superb individual hero, the simplest heroic adjective, δῖος,
whether it means 'godlike' or 'bright'; for Agamemnon, the
jealous king, the most political of heroic epithets, ἄναξ ἀνδρῶν.
This impression will not be much affected by his occasionally
finding ἄναξ ἀνδρῶν used of other men, of Aeneas and
Anchises, and even Augeias, Euphetes, and Eumelus. This kind
of experience does not require chemical purity: the impression
will only be confirmed by more reading, in the course of
which he will find the poet describing Agamemnon as ἄναξ
ἀνδρῶν thirty-seven times, and other characters calling him this
eight times. Agamemnon is in fact the only character to be so
addressed in the poem.

There is another argument, besides that from experience of
repetition, to dissuade us from the awareness of the meaning of
these words. | δῖος cannot be used easily of Agamemnon, nor
ἄναξ ἀνδρῶν of Achilles, for metrical reasons. δῖος is used most
often of names metrically equivalent to Ἀχιλλεύς: Ὀδυσσεύς,
Ἀγήνωρ, Ἐπειός, etc., and ἄναξ ἀνδρῶν is always used of names
metrically equivalent to Ἀγαμέμνων at the end of the line: ἄναξ
ἀνδρῶν Ἀγχίσης or Εὔμηλος, or the like. Although both
Achilles and Agamemnon of course elsewhere receive different
epithets, neither can ever receive the epithet used for the other
man here, and given the position of the names in this line, no
other epithet will do for either man. The epithets are, then,
chosen for metrical convenience. If they are so chosen, the
argument goes, they cannot be chosen for their meaning.
Hence they cannot have meaning.

But it does not follow. The epithets do indeed possess
metrical convenience, and the demonstration of this does add
to our knowledge of Homer. But it simply does not follow
that they lack meaning. The argument is based on faulty
inference. The demonstration of the metrical convenience of
the epithets, together with our natural experience that they do
have meaning, may lead us to admire the convergence of these

phenomena: but not to deny the one in favour of the other.

It is more reasonable to ask how it comes about that the one person in the story who can, shall we say, most profitably be qualified by ἄναξ ἀνδρῶν is also one of the few persons whose names are of a metrical form which makes combination with this epithet possible. Is that merely the poet's good luck?

Two kinds of explanation seem to me possible, and both must remain speculative. If we assume that the character Agamemnon, two shorts and two longs, existed in the tradition as the powerful king and public figure before the phrase ἄναξ ἀνδρῶν was invented (because it was invented, at *some* point), then we can say that, out of a number of possibilities of epithets expressive of kingship which we naturally do not know, one was chosen which could go with Ἀγαμέμνων. Once it was chosen, no other was necessary. At that point a great poet could develop the character of the self-conscious ruler with the depth which we find in the *Iliad*, and when he did so the epithet ἄναξ ἀνδρῶν, enabling him to make the second half of the line out of ἄναξ ἀνδρῶν Ἀγαμέμνων, could stand him in good stead; although, not being a purist, he would not hesitate to use the phrase casually of a minor figure, too minor to be characterized by it, when this was convenient for him.

If on the other hand we do not assume that Agamemnon, as an epic character, something like what we see in the *Iliad*, preceded his most common epithet, then we may be led along a slightly different path of speculation. We are perhaps misled by the more romantic of archaeolo|gists, those who show us Nestor's throne and Clytaemnestra's and Agamemnon's bath-tub, into thinking that the characters of the *Iliad* and *Odyssey* are much older than the language used to depict them in the poems can possibly be. The Linear B tablets, to be sure, offer us an A-KI-RE-U, but one far removed from the passionate leader of the Myrmidons; and no Agamemnon at all. Could he have been developed to his present importance in the poem because his name fits so well with ἄναξ ἀνδρῶν? The possibility cannot be ruled out.

The important thing is that we see that the epithets like those in line 7 of Book 1, however convenient metrically and however often repeated, can have the kind of meaning we naturally find in them, can help to define the characters and to tell the story that depends on those characters. But before we leave the line and the metrical utility of the phrases in it, something else should be pointed out. The usefulness of ἄναξ ἀνδρῶν lies in its being able to precede a name beginning with a vowel of the shape of Ἀγαμέμνων or Αἰνείας and so to form an expression completing the line after the trochaic caesura in the many cases where the first half of the line ends with a vowel— often the third singular past tense of a verb. But in 1. 7, the phrase is Ἀτρείδης τε ἄναξ ἀνδρῶν and it fills the first half of the line. This is the only occasion in either poem on which ἄναξ ἀνδρῶν stands in the first half of the line, although ἄναξ alone frequently occurs in the same position in the line as here.

The poet has constructed an unusual line here, and if we stop to think we can see that the conditions are unusual. For how often would one have a line consisting of only the name of Agamemnon plus epithet and the name of another, plus epithet? But if we go a step further and ask why these conditions themselves exist, we enter new realms of the mystery of Homeric composition.

J. Russo has pointed out that the opening phrase of the *Iliad* is metrically unusual. μῆνιν is natural at the beginning of the line, occurring six times in the *Iliad*. But the combination of trochaic noun and amphibrachic verb belongs at the line end, as in ἄλγε' ἔθηκε at the end of the second line. Here the poet has made—or, if you want, inherited, in which case some other poet made—a special phrase to mark the theme of his story, a phrase remarkable in metre as well as significance, for μῆνις, the noun, is used of the gods and only for Achilles among men. The poet of course does not seek the unusual for its own sake; but he is sufficiently concerned with what he is saying to depart from the patterns which he usually follows to say it with the greatest effect.

The unusual structure of lines 6 and 7 shows us a similar departure from pattern. The first half of line 6, ἐξ οὗ δὴ τὰ πρῶτα, is made up of | two set phrases both occurring in their most common position; the verb διαστήτην and the participle ἐρίσαντε with which the line ends appear to be in a normal position, though parallels are neither close nor common enough to justify our speaking of formulae. The unusual thing is line 7, where each phrase can be paralleled, but the position of ἄναξ ἀνδρῶν is special.

The poet clearly wanted an effective close to the first sentence of the poem in a line entirely devoted to the two contestants in the quarrel. He also wanted what becomes a thematic epithet for Agamemnon. He therefore prepares line 6 so as to contain the conjunction and the predicate and to lead to line 7, where to fit in Agamemnon and his epithet he makes a slight displacement of a familiar formula, and the grand resounding line leaves us looking at the two enemies, properly qualified.

Describe the process leading to line 7 in this way, and you make it sound a little like someone fiddling with written lines on a piece of paper. But this would be quite wrong. The perfection of the structure of lines 1–7, like that of so many parts of the *Iliad*, comes from Homer's having sung this part of this story over and over again. In doing so, he held to familiar patterns where he could, and varied the patterns where, as he sang, something occurred to him that would make it more effective. Each change might of course necessitate other changes. If we make the reasonable assumption that those elements which we find effective correspond to the poet's own purposes, then the epithets of line 7 were part of what he wanted in order to make the opening of the poem as strong as possible. Conceivably he—or others—had sung versions of this quarrel in which the names of the two men were mentioned differently and not with these epithets. Eventually our version evolved, a version in which the epithets do not depend merely on the blank spaces in the line in which they occur, but on the structure of the whole passage, which could have been formed as much for them as for anything else.

Among scholars who do not reject the oral theory of Homer out of hand, there have recently been a few who have questioned the rigid exclusion of meaning from the fixed epithet as it is urged in the original statement of the theory in *L'Épithète traditionnelle dans Homère*. W. Whallon in *Yale*

Classical Studies 1961 argued for a greater general relevance to character in the distinctive epithets, those reserved largely or entirely for one man, such as πολύμητις. Hoekstra in *Homeric-Modifications of Formulaic Prototypes* (Amsterdam, 1965) cites as an example of a formula which bears an irresistible meaning for its immediate context *Odyssey* 3. 352–3:|

οὔ θην δὴ τοῦδ' ἀνδρὸς 'Οδυσσῆος φίλος υἱὸς
νηὸς ἐπ' ἰκριόφιν καταλέξεται,

where the common formulary periphrasis for Telemachus, 'Οδυσσῆος φίλος υἱός, is joined to the specific words τοῦδ' ἀνδρὸς in the first half of the line in such a way that one must think of the name 'Οδυσσῆος in the formula.

An example at least as striking is the use of πεπνυμένος of Antilochus in Book 23 of the *Iliad*. The adjective is used, falling between the masculine caesura and the bucolic diaeresis, of Antenor and Meriones twice each in the *Iliad*, and of Eurypylus, Polydamas, and Antilochus once each, again in the same position. In the *Odyssey* it is of course the most common epithet of Telemachus, occurring some forty-five times. Whallon has some comments on its distinctive use in the *Odyssey*. In *Iliad* 23 Antilochus is carried away by youthful desire for victory in the horse race, and forces Menelaus to hold back his horses, depriving him of his prize. In lines 570–85, Menelaus makes a furious speech of remonstrance to Antilochus, beginning:

'Αντίλοχε, πρόσθεν πεπνυμένε, ποῖον ἔρεξας

where the distinct statement of πρόσθεν πεπνυμένε cannot be in doubt. In reply, Antilochus apologizes for his youthful indiscretion, and yields the prize to Menelaus. That speech is introduced by—

τὸν δ' αὖτ' 'Αντίλοχος πεπνυμένος ἀντίον ηὔδα.

where it is again beyond doubt that, despite the entirely formulary position of the adjective, we are being told that Antilochus is showing himself to be a sensible man after all.

The original statement in *L'Épithète traditionnelle* of the theory of the fixed phrase made a rigid distinction between the fixed and the particularized epithet, the former having never a distinct meaning, but always fusing with its noun into a single idea, the latter always embodying a distinct predication. πολύτροπος occurs twice in the *Odyssey*, in the same position after the trochaic caesura, in the first line of the poem, and in 10. 330, where Circe says:

ἦ σύ γ' Ὀδυσσεύς ἐσσι πολύτροπος, ὅν τέ μοι αἰεὶ
φάσκεν ἐλεύσεσθαι χρυσόρραπις ἀργειφόντης,

recognizing him as the hero Hermes had said would come and vanquish her. This word, πολύτροπος, because it occurs rarely and has a clearly formulary equivalent in διίφιλος, is a particularized epithet, and its full | meaning must be felt. πτολιπόρθιον on the other hand in 9. 504, being a fairly common epithet of both Odysseus and Achilles, and occurring several times with the names of minor heroes, must be, according to Milman Parry, entirely ornamental. πτολιπόρθιον in fact is made an example of the impossibility of translating the fixed epithet adequately into a modern language. Odysseus heroically, if indiscreetly, identifying himself to the Cyclops as his ship pulls away, says: 'Cyclops, if any mortal man asks you about the sorry blinding of your eye, say that it was Odysseus πτολιπόρθιος who blinded you, Laertes' son, who has his home in Ithaca':

φάσθαι Ὀδυσσῆα πτολιπόρθιον ἐξαλαῶσαι,
υἱὸν Λαέρτεω, Ἰθάκῃ ἔνι οἰκί᾽ ἔχοντα.

The difficulty in translation, it is said in *L'Épithète traditionnelle*, is to render πτολιπόρθιον so as not to lead the reader to think of its meaning, and at the same time to render the two phrases of the following line so that their statements will be fully appreciated. But if we once abandon the rigid dichotomy between fixed and particularized epithets, the difficulty vanishes. It is hard to think of an epithet, in fact, which serves better to reveal the nonentity of the cave suddenly as one of the greatest heroes of the epic tradition. Moreover, the word is used in an entirely unusual position here, so much so that the poet has recourse to the unusual form πτολιπόρθιον instead of πτολιπόρθον. μεγαθυμὸν, before the bucolic diaeresis, or the unattested *μεγαθύμιον (no odder than καταθύμιος), might conceivably have been used. What is certain is that the poet actually used, in an unusual form and an unusual place in the line, an adjective which the most assiduous reader today, and surely the most practised member of Homer's audience, would need considerable special pleading *not* to recognize as a word which adds much to the scene and to the characterization of the hero. The two phrases in the next line are for that matter attested formulae, and in positions moré usual than that of πτολιπόρθιον. One might, if one wanted to run Milman Parry's argument into the ground, claim that they too mean no more than 'Odysseus'.

This is, of course, not to deny some distinction between the fixed and the particularized adjective. The distinction exists, and those adjectives which the evidence allows us to classify as fixed are often used in such a way that they add little to the meaning other than, as Milman Parry so well said, to remind us of the heroic nature of the world of epic poetry. But the distinction is not rigid, and there is no | absolute line of demarcation. πολύτροπος in *Odyssey* 1. 1 and 10. 330 certainly has appropriate meaning but, as its two uses in the same part of the line indicate, it conforms to a pattern, one of the myriad patterns that enable the poets of the oral tradition to improvise their wondrous song. If we dwell on its meaning too much, in the two places where it occurs, we may there too lose the sense of that eminent rapidity of Homeric poetry which Matthew Arnold defined and which was for Milman Parry the starting-point of his momentous researches into Homeric style.

Apostrophe

Sixteen times in the *Iliad* and fifteen times in the *Odyssey*, the poet addresses a character directly. An older criticism saw this as a deliberate poetic figure, designed to produce interest by variation and to focus attention on a particular person. The current explanation is that these direct addresses are not to be distinguished in meaning from third-person statements, since they are only ways of accommodating names of awkward metrical shape into standard formulary patterns.

Neither view is wholly satisfactory. Against the older and naïve view are two facts. First there is the occurrence of apostrophe with a character of as little importance as Melanippus, son of Hicetaon, in *Iliad* 15. 582:

> ὣς ἐπὶ σοί, Μελάνιππε, θόρ' Ἀντίλοχος μενεχάρμης

Secondly, all fifteen occurrences in the *Odyssey* are in the same formula for speech, used of Eumaeus: $- \cup \cup - \cup \cup -$ προσέφης Εὔμαιε συβῶτα, usually (13 times) in the line τὸν δ' ἀπαμειβόμενος προσέφης Εὔμαιε συβῶτα. This formula is used so often for Eumaeus that it comes to seem a reflex. Hence even Ameis in 1873 suggested that metrical as well as emotional reasons played a part in the choice of apostrophe.

The modern and technical explanation does not entirely satisfy either. The three characters of whom apostrophe is used with some frequency, Menelaus, Patroclus, and Eumaeus are all in other ways treated with particular concern by the poet; are represented as unusually sensitive and worthy of the audience's sympathy. Is this mere coincidence?

The notion of the imperious necessity of metre, as an explanation of an unusual mode of expression in Homer, is shakier than it has been | made to appear. Often, it is true, we are unable to find another noun–epithet or other formulary phrase to offer as an alternative to the one which appears in the poem. But this does not mean that the resources of the epic tradition made any alternative impossible. For example, the first of the six apostrophes to Patroclus in *Iliad* 16 reads:

> τὸν δὲ βαρὺ στενάχων προσέφης, Πατρόκλεες ἱππεῦ.

Could the poet have said, e.g.

> *Πάτροκλος δὲ βαρὺ στενάχων ἠμείβετο μύθῳ?

Here it may be pointed out that the position of βαρὺ στενάχων around the main caesura is not attested in Homer and is *in that sense* 'impossible'. Or could the poet have sung:

*τὸν δὲ βαρὺ στενάχων προσέφη Πάτροκλος ἀμύμων?

*Πάτροκλος ἀμύμων, suggested as a hypothetical nominative formula by J. B. Hainsworth, does not occur, though we know of no specific reason why not.

But the fact that we cannot in these or other cases exhibit an attested alternative does not mean that the tradition, or Homer, could not have provided one. It is time we stopped saying that the poet must have said a thing in this way because no other way existed for him to say it. We do not know what existed, either potentially or actually.

The case of apostrophe must be more complex than either the traditional or the modern view suggests. Let us consider the three characters with whom apostrophe is especially associated. Before Book 16, Patroclus plays a part in Books 1, 9, 11, and 15, being mentioned by name or patronym in those books twenty-one times. Apostrophe, however, is used for him only in Book 16, where it occurs six times, always at significant points of the action. The special place Patroclus occupies in the organization of the poem scarcely needs comment. This place, in our poem, depends on his character. He is the sweetest and most compassionate of the Homeric warriors. We see this most clearly in the moving lament for him spoken by Briseis in 19. 282–300, where she says of him: 'When Achilles slew my husband, you would not let me cry. You promised to make me Achilles' wedded wife, to bring me back to Phthia, and to give me a marriage feast there, among the Myrmidons. And so now with all my heart I weep for your death, for you were always sweet'—τῶ σ' ἄμοτον κλαίω τεθνηότα μείλιχον αἰεί. μείλιχον is in this sense a word reserved for Patroclus in the *Iliad*. | Elsewhere it is used almost entirely to mean 'gentle *words*' as opposed to 'harsh' (ten times), the other two occurrences being 'he was no gentle fighter in battle'. Only Patroclus as a man is μείλιχος, and the distinctive word defines that quality in him which ensures his death and with it the tragic plot of the poem.

Menelaus also signals this quality when in 17. 669 he calls for help in protecting the body of Patroclus: 'Ajax, you and your brother, and Meriones, let us remember the goodness of unhappy Patroclus; he was able to be sweet (μείλιχος) to everyone, while he was alive; now death and doom have found him.' The word I translate as 'goodness' is ἐνηής. Of its five occurrences in the *Iliad*, four refer to Patroclus. The single exception is in 23. 648 where the egotistical Nestor speaks of himself, and here the adjective occurs immediately after Nestor has himself spoken of Patroclus, so that some process of association appears likely.

To Patroclus is attributed in the poem a distinct character: kind, easily moved to pity, remarkably free from the sort of heroic self-assertion which many, and recently Professor Adkins, have sought to define for us. This character is manifested in the poem not only by his actions, but also by a distinct vocabulary.

The idle genetic speculations of earlier years did not fail to suggest that Homer might have invented the person of Patroclus altogether. That Patroclus first appears in the poem at 1. 307, described casually as 'the son of Menoetius', is enough to show that the audience was expected to know who he was. But the attempt to distinguish between Homer and the tradition will not in any case add much to our understanding of the poem. What is clear is that the character and function of Patroclus in the poem *as we have it* have been elaborated with great fineness and consistency.

Patroclus is unimportant in Book 1. The single brief mention in 1. 307, 'Achilles went off to his tent with the son of Menoetius and his comrades', merely looks forward to his later significance. His closeness to Achilles is made more evident in Book 9—he is listening to Achilles singing the glories of heroes when the Embassy arrives; but here too he plays no central part. It is in Book 11 that he becomes essential. Achilles sends him out to report on the condition of the Greeks—the one man who has too much of the milk of human kindness to tolerate the effects of Achilles' revenge on Agamemnon and the Greeks. When in Nestor's tent Patroclus ascertains that the wounded man of whom Achilles had caught a glimpse is indeed Machaon, his scouting mission is accomplished. But when Nestor makes him sit down and listen to the sad tale | of the Greeks, and incidentally to the longest of Nestor's tales of his own glorious past, Patroclus cannot resist, although he tries. And his encounter, as he finally leaves, with the wounded Eurypylus, neatly saved by the poet till the end of Book 11, completes the effect of Nestor's words on him. Now he *must*, as he tells Eurypylus in Book 15, persuade Achilles to help the Greeks.

He must do this, although he knows that he is disobeying Achilles, and that Achilles will be angry with him. He must because his love for Achilles is matched by his love for all his Greek comrades-in-arms. The tragic story, that is, is directly dependent on the precise character of Patroclus as Homer has developed it in the poem. The man who takes time out from the errand on which Achilles has sent him, first to listen to Nestor's garrulous and devious tale, and then to care for Eurypylus in the latter's tent, 'sitting by him and entertaining him with stories', is the same, and the only, man who in Book 16 could make so passionate an appeal to Achilles, blaming him precisely for the lack of that human concern which is so distinctly a mark of his own character—'I see it now: Peleus was not your father, Thetis not your mother: the grey sea gave you birth, and the sharp rocks, because your heart is empty of kindness': ... ὅτι τοι νόος ἐστὶν ἀπηνής. ἀπηνής is the contrary in meaning, and probably in etymology, of ἐνηής. Patroclus is in short the only kind of man who could bring about the plot of the *Iliad* as we have it.

ἦθος ἀνθρώπῳ δαίμων. The ἦθος of Patroclus is δαίμων to Achilles and Hector as well as to himself. That character is the sum of many precise touches, touches of words as well as of action, within the poem. And the person we extrapolate or infer from these touches is the one whom the poet six times addresses directly in the course of Book 16.

The statement of Milman Parry concerning apostrophe, which subsequent scholars have followed in finding for all examples of it only a metrical purpose, is in fact limited to the speech formulae:

$$-\cup\ \cup-\cup\ \cup-\ προσέφης\ Πατρόκλεες\ ἱππεῦ.$$

These occur three times, half of the apostrophes in Book 16.

Each is emotionally qualified, i.e. has a special, though still formulary, participle in the first half of the line:

16. 20   τὸν δὲ βαρὺ στενάχων προσέφης, Πατρόκλεες ἱππεῦ·
16. 744 τὸν δ' ἐπικερτομέων προσέφης, Πατρόκλεες ἱππεῦ·
16. 843 τὸν δ' ὀλιγοδρανέων προσέφης, Πατρόκλεες ἱππεῦ· |

The other three apostrophes, however, are more extended. The first, line 584, comes after a simile comparing Patroclus' rush upon the Trojans to the diving attack of a hawk:

ὡς ἰθὺς Λυκίων, Πατρόκλεες ἱπποκέλευθε,
ἔσσυο καὶ Τρώων, κεχόλωσο δὲ κῆρ ἑτάροιο,

The vocative is picked up by two second-person verbs in the following line. The syntax of the vocative line is not therefore a mere equivalent of a third-person statement.

The second extended apostrophe is preceded by a line with a second-person verb. It is the catalogue of Trojans slain by Patroclus just before his encounter with Apollo (16. 692–3):

Ἔνθα τίνα πρῶτον, τίνα δ' ὕστατον ἐξενάριξας,
Πατρόκλεις, ὅτε δή σε θεοὶ θάνατόνδε κάλεσσαν.

So that the poet was aware of the apostrophe before the line containing the vocative.

Milman Parry says of the three speech formulae ending in Πατρόκλεες ἱππεῦ that Πατρόκλεες is an artificial form chosen because no nominative form was available after the formula which runs from the beginning of the line to the hephthemi-meral caesura and ends in προσέφη. This was a trifle thought-less, because it goes against Parry's own correct principles to label one recurrent form in the *Kunstsprache* 'artificial' in relation to other forms. If, at the line ending, we assume that e.g. *Πάτροκλος ἀμύμων was for unknown reasons impossible, Milman Parry may have been right *within the limits of this particular formulary pattern*. But where the vocative of the third declension form begins the line, as in 693, the second of the three extended apostrophes, Πατρόκλεις, ὅτε δή σε θεοὶ θάνα-τόνδε κάλεσσαν, the choice of the vocative, as opposed to the nominative, was not forced, and we should not have daggered the text if, after a third-person ἐξενάριξας, we had found

*Πατροκλῆς nominative, although this form does not in fact occur elsewhere.

The third extended apostrophe, lines 787–9, contains the second-person pronoun twice:

> ἔνθ' ἄρα τοι, Πάτροκλε, φάνη βιότοιο τελευτή·
> ἤντετο γάρ τοι Φοῖβος ἐνὶ κρατερῇ ὑσμίνῃ

Only after the strong pause in the third line does the poet return to the third person:

> ὁ μὲν τὸν ἰόντα κατὰ κλόνον οὐκ ἐνόησεν· |

These lines occur at the end of an elaborate climactic passage in which two of the poet's crescendo devices are used at once. First the time-of-day crescendo: 'As long as Helios bestrode the middle heaven, / So long did fly spears on both sides, and the people fell, / But when Helios moved on to the unyoking of oxen / Then . . .' And then the three + one crescendo: 'Patroclus hurled himself at the Trojans with deadly purpose. / Thrice then he rushed at them, equal to rapid Ares / Shrieking horribly, and thrice nine men he killed; / But when at last the fourth time he flung himself like a god /—Then, for you, Patroclus, appeared the end of life . . .'

The crisis of the action of the whole book is marked by the sequence of these crescendos, and by the following apostrophe. The words of the apostrophe

> ἔνθ' ἄρα τοι, Πάτροκλε, φάνη βιότοιο τελευτή·

are perfectly formulary and within the tradition; as we can see from 7. 104, of Menelaus: ἔνθα κέ τοι, Μενέλαε, φάνη βιότοιο τελευτή. Its choice here, rather than some different way of resolving the crescendo, is vital to the effect of the passage.

The three extended apostrophes of Patroclus are elaborated, by second-person pronouns and verbs, beyond the vocative and beyond the line in which the vocative occurs. They appear to mark significant moments in the action, as we saw in examining the last of them—

> ἔνθ' ἄρα τοι, Πάτροκλε . . .

They are chosen to mark the particular pathos of Patroclus' death, a pathos which is part of his character in the poem.

What of the speech formulae? They are in themselves more apparently formulary, less significant, in that they are confined to a single line, and represent a minimal variation in the most common line-formula in the whole poem. But (1) they occur only in Book 16; (2) they occur in conjunction with the more striking apostrophes which we have been discussing: (3) they are emotional variants of the basic line-formula, e.g. τὸν δὲ βαρὺ στενάχων προσέφης, Πατρόκλεες ἱππεῦ as opposed to τὸν δ' ἀπαμειβόμενος προσέφης, Πατρόκλεες ἱππεῦ; and (4) two *at least* occur at moments of special significance: one as Patroclus answers Achilles in the conversation at the beginning of Book 16, the other to introduce Patroclus' last words to Hector at the end of the book. Only by entirely disregarding the context can these apostrophes be regarded as full equivalents, technically determined, of third-person formulae. They are different formulae, formulae of more emotional | content than the third-person formula which we find in 11. 837: τὸν δ' αὖτε προσέειπε Μενοιτίου ἄλκιμος υἱός, and the poet of the *Iliad* has used all six apostrophes to particular effect to help describe the glory and death of a hero whose precise character has largely motivated, and will continue to motivate, the action of the poem.

The case of Menelaus is a little less clear. Of the six apostrophes made to him, all contain the second-person pronoun as well as one or more second-person verbs; some appear more than others to mark significant moments in his career in the poem.

In Book 4, after Pandarus, persuaded by Athena, has shot his treacherous arrow, the poet turns to Menelaus (127–9):

οὐδὲ σέθεν, Μενέλαε, θεοὶ μάκαρες λελάθοντο
ἀθάνατοι, πρώτη δὲ Διὸς θυγάτηρ ἀγελείη,
ἥ τοι πρόσθε στᾶσα βέλος ἐχεπευκὲς ἄμυνεν.

If τοι is right—some manuscripts have οἱ—the apostrophe continues into the third line.

It is a crucial moment in the story of the poem. The shooting of Menelaus means the end of any possibility that the war between the Trojans and the Greeks can be settled by agreement. Morally, it puts the Trojans once and for all in the wrong. Representatives of both sides, Antenor and Diomedes, as Lesky points out, recognize this.

The moral basis of the plot is not the most important element in the poem that ends with Priam and Achilles lamenting together the bitterness of life, and Achilles saying, 'Nor has my father me to care for him in his old age; instead I sit here in Troy, making trouble for you and your children'. Nevertheless it is there, and Menelaus is for this element the central character. It is he who has been wronged, he who has most suffered. If things are to be made in any sense right, they must be made right for him. Alone of the Greeks, even including his brother, Menelaus has a cause for which he is fighting other than τιμή. He is asserting his rights, recovering his due. And the gods, especially Hera and Athena, who are clearly going to have their way, are on his side. It is ironic then that, here in Book 4, Athena must make him victim of Pandarus' arrow in order to achieve her purpose of not letting the war end easily for Troy. Menelaus is the last man to be in this way forgotten by the gods. They must after all protect him; he must live to regain his loss, since all this is being done for him. Hence the poet turns our full attention on him now at this crucial point, on him and the goddess who has not after all abandoned him—οὐδὲ σέθεν, Μενέλαε, θεοὶ μάκαρες λελάθοντο. |

The whole question of the outcome of the war arises again in Book 7 when Hector offers to stand in single combat against one of the Greeks. Achilles is not there, and no one of the Greeks feels like answering the challenge: 'They were ashamed to say no, but afraid to say yes', the poet tells us. Is Hector really superior to the Greeks? How can they, admitting to this, ever hope to take Troy? Such questions are in the air, when Menelaus in great bitterness rises to meet the challenge. The Greeks cannot, he sternly feels, accept this shame; indeed he himself will do battle with Hector.

Of such a battle the outcome could not be in doubt. And with Menelaus dead, the whole point of the war is lost. Again the larger dimensions of the story are evident when the poet uses for Menelaus the formula which he later uses so movingly for Patroclus:

ἔνθα κέ τοι, Μενέλαε, φάνη βιότοιο τελευτή.

One of the consistent features of Menelaus' character, evolving naturally from the story, is this particular concern of the gods, and specifically of Hera and Athena, for him. Hera, in 5. 715, seeing the Trojans get the upper hand, says to Athena, ἦ ῥ᾽ ἅλιον τὸν μῦθον ὑπέστημεν Μενελάῳ, as if their whole action were due to a promise to him alone. We might note, however, following some of the discussion in Reinhardt's excellent article on the Judgement of Paris, that the whole story of the mythical events leading to the Trojan War, while it is essential for the audience to have it somewhere in mind, is largely kept in the background of the *Iliad*. The poet, that is, is more concerned, in the actual narrative, with the character of Menelaus, and also of Helen and Paris, than he is with the folk-tale itself and with Menelaus as a counter in this folk-tale. The dearness of Menelaus to the gods becomes a part of this characterization in the poem. The use of apostrophe for him aids that characterization by making the audiences especially concerned, just as the use of apostrophe with Patroclus is vitally connected with Patroclus' character and with the special concern felt for him by his comrades-in-arms.

Of the four remaining instances of apostrophe to Menelaus, one occurs in Book 13, two in 17, and one in 23. Before we consider them, let us look at some of the other consistent features of Menelaus' character in the *Iliad*. Like Patroclus, Menelaus does not fit what may be thought the standard picture of the Homeric hero, because he shows no concern for τιμή or self-assertion of the usual hero type. His concern is moral, even in a slightly stuffy way. He wants to regain his rights, specifically Helen and the possessions stolen with her. He is neither an | outstanding warrior—μαλθακὸς αἰχμητής Apollo calls him in Book 17, and Agamemnon's terror in Book 7 when Menelaus offers to meet Hector's challenge, confirms this; nor a good counsellor: Antenor in Book 3 tactfully prefaces his admiration of Odysseus' speaking powers by saying that Menelaus of course spoke briefly but always to the point, and Agamemnon in Book 11 says to Nestor that if Menelaus seems at times unhelpful, it is not out of laziness, but because he waits for directions from his brother. Already in the Catalogue—here at least consistent with the rest of the poem—Menelaus is described as keeping his contingent next to that of Agamemnon, and moving among his men ἧσι προθυμίῃσι πεποιθώς. He has more zeal than true valour, and the poet continues to say that he was determined in his spirit to get revenge for the pain of Helen's theft:

$$\text{μάλιστα δὲ ἵετο θυμῷ}$$
$$\text{τείσασθαι Ἑλένης ὁρμήματά τε στονάχας τε.}$$

Two other characteristics: like Patroclus, Menelaus has a soft streak. In Book 6, in a scene meant to contrast with the following meeting of Glaucus and Diomedes, Menelaus overcomes an enemy who begs him for mercy, promising ransom if Menelaus will spare him. 'He persuaded Menelaus' heart within him'—Menelaus is always being persuaded—until brutal Agamemnon arrives and scolds his brother, saying 'come along, Menelaus, why do you care so much for these men'—τίη δὲ σὺ κήδεαι οὕτως / ἀνδρῶν—and urges the ruthless slaughter of all Trojans, even the unborn child in the mother's womb.

Menelaus is an object of κῆδος to the gods and even to the unfeeling Agamemnon—although here it is uncertain how much Agamemnon's worries are due to his fear that without Menelaus the expedition might collapse; and he is especially prone to feeling concern for others. ἀνὰ δὲ σχέο κηδόμενός περ Agamemnon says to him in Book 7 to dissuade him from presuming to fight Hector.

Unlike his brother, Menelaus feels a concern for those who are fighting on his behalf, a concern which he expresses in Book 3 and which motivates his offer to fight Hector in Book 7. His moment of glory comes appropriately in 17, the book which is the elaborate expansion of the theme of fighting for the body of a slain comrade. Menelaus stands over Patroclus' body like a mother-cow over a calf, risks his life, and wins heroic glory in an unwonted way in his desire to do the right thing by the fallen hero who was in so many ways the Greek leader most like himself. And it is Menelaus who along with Briseis gives the most important characterization of Patroclus. |

Finally—the more stuffy side of Menelaus—he is unusually conscious of himself as a middle-aged man. This feature of his character is connected with his great susceptibility to moral outrage. In Book 3 he demands that Priam come to witness the truce, because his sons are arrogant and untrustworthy—and if it is left up to them, the oaths may be violated. Then he generalizes: the minds of young men are capricious and unsound; an older man is needed to make sure things are done properly. In Book 15, Menelaus with a trace of condescension urges on Antilochus, saying that he is the youngest of the Achaeans and also the swiftest and bravest—where the sense is clearly: 'You are a good warrior, although you are so young'. In Book 23, this difference of generations becomes the keynote of the whole scene of the horse race. Nestor has persuaded Antilochus that although he is so young, he can win the race, if he plays it right. Menelaus feels confident of beating Antilochus because the latter's horses *lack youth*. When the racers approach the finish line, Idomeneus recognizes the front-runner first, and is abused by the lesser Ajax, who says, 'Look, how can you possibly see at such a distance: you are hardly the youngest of the Achaeans'. When Antilochus apologizes to Menelaus after having cheated him of his prize, he himself speaks of the thoughtlessness of the young:

κραιπνότερος μὲν γάρ τε νόος, λεπτὴ δέ τε μῆτις.

And in accepting his apology, and the prize, Menelaus says that he had never thought Antilochus foolish or hot-headed before; but on this occasion his youth got the better of his sense: νῦν αὖτε νόον νίκησε νεοίη.

Bearing in mind this fine and consistent characterization, let us return quickly to the four apostrophes to Menelaus that we have not yet discussed. I pass for the moment over the first of these, in Book 13. The next two appear in 17, the first after a simile comparing Menelaus to an eagle at the point when he is looking for Antilochus in order to tell him to bear the news of Patroclus' death to Achilles: 'So then Menelaus, ward of Zeus, did your sharp eyes / Rove every which way about the multitude of your comrades' (679–80):

ὡς τότε σοί, Μενέλαε διοτρεφές, ὄσσε φαεινὼ
πάντοσε δινείσθην πολέων κατὰ ἔθνος ἑταίρων.

And the second when Menelaus, having to choose between two loyalties, leaves off helping the Pylians and Antilochus and returns to protect the body of Patroclus: 'Nor did your spirit, Menelaus ward of Zeus, choose longer to fight with these embattled comrades-in-arms. . . .' (702–3):

οὐδ' ἄρα σοί, Μενέλαε διοτρεφές, ἤθελε θυμὸς
τειρομένοις ἑτάροισιν ἀμυνέμεν. . . .

These two apostrophes come in Book 17, after the book in which six apostrophes are used so powerfully for Patroclus. They both occur at moments where Menelaus' loyalty to and concern for his comrades are especially at issue; and each of them involves his relation to both Patroclus and to Achilles.

Menelaus' last apostrophe appears in the scene of the horse race with Antilochus which I discussed above, and specifically at the point when Menelaus accepts Antilochus' apology, and there is a neat resolution of his essential magnanimity and his hostility to the younger generation. Again there is a simile: his spirit is softened like dew upon the growing corn, when the fields spring up in their greenness—'so, Menelaus, was softened the spirit within you' (23. 600):

ὡς ἄρα σοί, Μενέλαε, μετὰ φρεσὶ θυμὸς ἰάνθη.

All these apostrophes appear in scenes which especially reveal aspects of Menelaus' character that the poet elsewhere is at pains to throw into relief; and they occur in connection with those persons, Patroclus and Achilles, whose relationship with Menelaus again throws light on his character. Did the poet carefully plan out these subtle details of the poem's architecture? Not exactly, we may guess. Rather he had developed, over the years in which he had sung countless versions of these stories, a precise conception of Menelaus' character and his relation to other characters which made these details, including the apostrophes, fall into place.

Finally, 13. 603. At a not very notable moment in the longest day's fighting, Menelaus meets an enemy who is a very obscure character indeed: 'Peisander came up to glorious

Menelaus. A fell doom led him to his death, to be vanquished
by you, Menelaus, in bitter combat'.

σοί, Μενέλαε, δαμῆναι ἐν αἰνῇ δηϊοτῆτι.

Here no special reason for the apostrophe irresistibly com-
mends itself to us. It may be that the poet wanted no more than
a variation from the κόρος of battle. Some odd features of the
scene, however, may lead us to a speculative conclusion.
Peisander is abruptly introduced, without patronym, without
description, without even being defined as a Trojan. The
narrative of the fight between the two men, however, is long
and elaborate. After it, at 628 ff., Menelaus makes a remarkable
| speech, especially remarkable in that we expect it so little. He
rages against all the Trojans, calling them filthy bitches,
blaming them for disregarding the wrath—the μῆνις—of
Zeus, god of hospitality. They took his wife and his goods
from him when they were her guests. Then he rises to a more
general pitch: the Trojans are men of ὕβρις; they never weary
of war. Why, men weary of all things, of sleep, of love, of
sweet singing, of the beautiful dance—

πάντων μὲν κόρος ἐστί, καὶ ὕπνου καὶ φιλότητος
μολπῆς τε γλυκερῆς καὶ ἀμύμονος ὀρχηθμοῖο

but the Trojans are never tired of battle.

Why does the poet develop a scene where Menelaus is made
so rhetorically and so beautifully to express his resentment
against the Trojans, when he is only killing a character of
singular obscurity? In 11. 122 ff. Agamemnon kills two sons of
Antimachus, and in this scene we learn that Antimachus had
accepted a bribe from Paris to oppose in council the return of
Helen before the war, and that when Menelaus and Odysseus
came on an embassy to Troy to seek for her return, this same
Antimachus had urged the Trojans not to allow their safe
return, but to kill them both. The sons of this reprehensible
character are named Hippolochus and Peisander. Now if the
scene in Book 13, where Menelaus receives an apostrophe, kills
an unidentified Peisander, and makes a striking speech of moral
outrage, had been earlier developed by the poet as a scene in
which Menelaus kills Peisander, the son of Antimachus, we

could better understand the elaboration. But in Book 13, when he has Menelaus fighting, the poet remembers that he has already had Peisander, son of Antimachus, and his brother killed by Agamemnon. He does not want to make the mistake he actually makes in the case of Πυλαιμένης the·Paphlagonian, of bringing a slain warrior back to life in a later part of the poem. He also wants the scene itself. So he puts in Peisander, but does not identify him, and then lets the scene follow as we have it.

This may be what happened, but it must remain speculative. In any case, the apostrophe in 13. 603 occurs in a scene especially revelatory of Menelaus' character, as do Menelaus' five other apostrophes, though these are not used in the closely controlled crescendo of pathos that we find in Book 16 in the case of Patroclus.

Of the character of Eumaeus, and its relation to the repeated second-person speech formulae used for him, there is less to say. The fact that the same second-person introductory phrase is used over and again for all sorts of speeches of Eumaeus makes the device, as I said earlier, | seem little more than a reflex; and perhaps here we have the point where too much examination of detail may distract us from the poetry rather than help us to understand it. None the less, it should be clearly noted that Eumaeus, the only character to be apostrophized in the *Odyssey*, is in many ways remarkably like Menelaus and Patroclus. He is altruistic, loyal, sensitive, vulnerable. And it is plausible that the length at which his character is developed, especially in Books 14 and 15, is due to the poet of the *Odyssey* rather than to the tradition, since this elaborate development is not strictly necessary to the fundamental plot of the poem. It is of course a valuable part of our *Odyssey*, in as much as Eumaeus becomes the type of the loyalty which Odysseus' good kingship in Ithaca has won from those worthy of appreciating it.

What is interesting in the treatment of apostrophe in our version is the way in which the poet settles on the second-person formula. It is used in 14. 55, the first time Eumaeus is mentioned by name. But the poem then employs a third-person alternative in 14. 121:

τὸν δ' ἠμείβετ' ἔπειτα συβώτης, ὄρχαμος ἀνδρῶν·

returns to the second person in 165, and adopts a different third-person alternative in 401: τὸν δ' ἀπαμειβόμενος προσεφώ- νεε δῖος ὑφορβός: The second half of this line, we note, is an exact replacement for the second-person line-ending προσέφης Εὔμαιε συβῶτα. But if we had not found it in the text, it would have been possible to argue, using the faulty reasoning to which I objected earlier, that no such third-person formula was possible.

Speeches of Eumaeus are then introduced fourteen times more in the poem, and every one by the second-person formula. Clearly the poet simplified his choice, and in this direction because the apostrophe, however mildly it is felt, was appropriate to the sense of Eumaeus' character, as distinct from that of the other loyal servants, developed less fully, like Philoetius, which he wished to impose on the audience. The individual occurrences of the formula do not, as in the cases of Menelaus and Patroclus in the *Iliad*, mark emotion special to the scene; but they derive from, and help to communicate, a special sense of the character in general, so that here too, however much the apostrophe becomes a reflex, it blinds us to the poetry to argue that we have in these expressions no more than words meaning: 'Eumaeus replied'. |

The degree of the audience's consciousness varies in the cases of all the characters of whom apostrophe is used. But whenever it is used, it bears some meaning, just as there is some meaning in the most fixed of fixed epithets. That these words and these devices should so neatly fill the exigencies of metre, that they should become sufficiently natural and even ritualistic not to slow down the rapidity of the epic story, and that they should have at the same time a meaning which adds to the complex characterization of the poems, is part of the genius of Homeric poetry. Such a convergence of values is analogous to what we find in all poetry, where an external form—metre or rhyme— pre-existing the individual poem and not varying with it, coexists with an internal form—the words and what they say—so that both are felt to be right.

It ought not to surprise us, though it has almost come to seem surprising, that this should be so in Homer. An older view of the poems, which saw them more or less entirely as the creation of a single man, was unable to account for the elaborate system of formulae, with their extension and their economy. The brilliant work of Milman Parry and that of many of his successors, which pointed this out, made the unreasonable assumption of a sort of monolithic tradition, a tradition allowing only those forms of expression that we actually find in the poems. The tradition must actually have been far more complex and far more flexible. It certainly was if the evidence of Serbocroatian poetry offers any useful analogy. The *Iliad* and *Odyssey* that we have, with their splendid coincidence of meaning and form, were the result of generations of selections from this fluid tradition, and of the long years over which Homer himself perfected his songs.

Index of Names

Index of Subjects

abstract language, 186–92, 194
accuracy of oral transmission, 110,
  111–15, 126, 127, 128, 130, 205
Achilles (*Iliad*), 5–7, 18, 34, 84, 102,
  118, 120, 121, 302–4, 306, 313,
  314, 318
Aeneas (*Aeneid*), 82, 85, 86, 87–96
Aeneid, 78–96, 283; destiny in, 85–6,
  88–91; Dido episode, 83, 87–8,
  92–3; identifications in, 84–8, 93;
  lamentation for Umbro, 78–82;
  themes of, 266–7; Underworld
  episode, 93–4
Agamemnon: in *Iliad*, 302–4, 305, 307,
  319–20; in Thucydides, 292–3
agricultural themes in Virgil, 265, 268,
  269–74, 276, 284
Alcestis, 53, 58
Alexandrian criticism, 204
ambiguities, 186
analogy, 225, 227
Analytic criticism, 203–8, 248
Anchises (*Aeneid*), 89
Antilochus (*Iliad*), 308–9, 321
Antimachus (*Iliad*), 323
antithetical language, 180–1, 190–1
Apology (Plato), 102
apostrophe, 310–26
'Archaeology', (Thucydides), 192–3,
  287–9, 290–5
archaism, 40–1, 42, 148
Archidamian War, 176
Aristaeus episode (*Georgics*), 277–85
aristocratic values, 101, 102
art, idea of, 266, 267, 277, 281, 284–5
Athena (*Iliad*), 317, 318, 319
Athens/Athenians, 144–5, 146–7, 162,
  176, 182, 183, 287, 288, 293, 294,
  295, 297, 298, 299–300
attitudes, conflicts of, 117–18
Augustus, Emperor, identifications in
  Aeneid, 83, 86–7
authorship of Homer's works, 106–7,
  111–12, 119, 125–6, 200–1,
  219–20, 249–50
Axel's Castle (Wilson), 265

bees, 274–6, 278, 281–2
Bible, 52, 55–6, 199; as source of poetic

language, 40; medical influences
  on, 170
'Boeotian School' theory, 132

catalogue technique, 78, 80, 132
cattle, 270–2
central problems of history,
  Thucydides' conception of, 192–4
Chanson de Roland, 242
characterization in Homer, 303–4, 305,
  308–9, 312–14, 319–21, 322, 324,
  325
choral poetry, 62–5
civilization, concept of, 193–4, 292–3,
  294–5, 297, 298, 299
clarity, intellectual, 179–80
Classics, 141–7, 177
Cleopatra, identifications in *Aeneid*, 88,
  93
Close Reading, 141, 142
comparative studies deriving from
  Parry, 243–4
consciousness of Homer's audience,
  302, 303–4, 325
Corcyrean revolution, 160, 161, 193,
  286, 295
Corinthians, 152–3
creative phase of oral tradition, 128–9
criticism, literary, 141, 143
Cyclic Epics, 128
Cyclops episode (*Odyssey*), 32–4, 309
Cyrene (*Georgics*), 278, 279, 282, 285

Daphnis (Theocritus' *Idyll 1*), 17–20
dating of Homer, 107
De Rerum Natura (Lucretius), 267
Defence of Helen (Gorgias), 190–2
degenerate phase of oral tradition,
  129–30, 131
destiny, 85–6, 88–91
destruction, 193, 290, 294–5, 299–300
detachment, 56–7
development of Greek language,
  178–9, 180–1
dialect-mixture of Homeric poetry,
  210–11
dictation theory of Homer, 109, 110,
  111, 139
Dido (*Aeneid*), 83, 87–8, 92–3

Indexes compiled by Peva Keane